Sustainability through
Energy-Efficient Buildings

Sustainability through Energy-Efficient Buildings

Edited by
Amritanshu Shukla
Atul Sharma

CRC Press
Taylor & Francis Group
Boca Raton London New York

CRC Press is an imprint of the
Taylor & Francis Group, an **informa** business

CRC Press
Taylor & Francis Group
6000 Broken Sound Parkway NW, Suite 300
Boca Raton, FL 33487-2742

First issued in paperback 2020

© 2018 by Taylor & Francis Group, LLC
CRC Press is an imprint of Taylor & Francis Group, an Informa business

No claim to original U.S. Government works

ISBN-13: 978-0-367-57186-3 (pbk)
ISBN-13: 978-1-138-06675-5 (hbk)

Library of Congress Cataloging-in-Publication Data

Names: Shukla, Amritanshu, editor. | Sharma, Atul (Professor of environmental studies), editor.
Title: Sustainability through energy-efficient buildings / edited by Amritanshu Shukla, Atul Sharma.
Description: Boca Raton : Taylor & Francis, CRC Press, 2018. | Includes bibliographical references and index.
Identifiers: LCCN 2017052525| ISBN 9781138066755 (hardback : alk. paper) | ISBN 9781315159065 (ebook)
Subjects: LCSH: Buildings--Energy conservation.
Classification: LCC TJ163.5.B84 S864 2018 | DDC 696--dc23
LC record available at https://lccn.loc.gov/2017052525

Visit the Taylor & Francis Web site at
http://www.taylorandfrancis.com

and the CRC Press Web site at
http://www.crcpress.com

Contents

Foreword I.. vii
Foreword II.. ix
Editors.. xi
Contributors... xiii

1. Introduction to Energy-Efficient Building Development and Sustainability 1
   R. Singh and V.V.N. Kishore

2. An Introduction of Zero-Energy Lab as a Testbed: Concept, Features, and
   Application.. 33
   Tingzhen Ming and Yong Tao

3. Building Envelopes: A Passive Way to Achieve Energy Sustainability
   through Energy-Efficient Buildings 59
   Manoj Kumar Srivastava

4. Passive and Low Energy Buildings.. 73
   Lu Aye and Amitha Jayalath

5. Energy-Efficient Building Construction and Embodied Energy.................. 89
   R. Singh and Ian J. Lazarus

6. Building Integrated Photovoltaic: Building Envelope Material and Power
   Generator for Energy-Efficient Buildings 109
   Karunesh Kant, Amritanshu Shukla, and Atul Sharma

7. Energy Conservation Potential through Thermal Energy Storage Medium
   in Buildings 131
   C. Veerakumar and A. Sreekumar

8. Silica Aerogel Blankets as Superinsulating Material for Developing Energy
   Efficient Buildings... 151
   Kevin Nocentini, Pascal Biwole, and Patrick Achard

9. Heating Ventilation and Air-Conditioning Systems for Energy-Efficient
   Buildings 165
   Karunesh Kant, Amritanshu Shukla, and Atul Sharma

10. Improving Energy Efficiency in Buildings: Challenges and Opportunities
    in the European Context.. 181
    Delia D'Agostino

11. Advances in Simulation Studies for Developing Energy-Efficient Buildings 209
Karunesh Kant, Amritanshu Shukla, and Atul Sharma

12. Advances in Energy-Efficient Buildings for New and Old Buildings 235
A.K. Chaturvedi, Siddartha Jain, Deep Gupta, and Mridula Singh

13. Role and the Impact of Policy on Growth of Green Buildings in India 259
Manish Vaid and Sanjay Kumar Kar

14. Energy-Efficient Buildings: Technology to Policy and Awareness 273
Saurabh Mishra

Index 285

Foreword I

The rapidly progressing world is demanding more and more energy to be utilized for different sectors, raising serious concerns over supply difficulties and exhaustion of energy resources with heavy environmental issues. The buildings sector is responsible for approximately 31% of global final energy demand, approximately one-third of energy-related CO_2 emissions, approximately two-thirds of halocarbon, and approximately 25%–33% of black carbon emissions. Energy consumption in the residential sector includes all energy consumed by households, namely, energy used for heating, cooling, lighting, water heating, and other consumer needs. This consumption is duly affected by income levels, energy prices, location, building and household characteristics, weather, efficiency and type of equipment, energy access, availability of energy sources, and energy-related policies, among other factors. As a result, the type and amount of energy consumed by households can vary significantly within and across regions and countries. As per data available from the U.S. Energy Information Administration, energy consumed in the buildings sector worldwide is increasing by an average of 1.5% per year from 2012 to 2040. This is in sync with the growth in population, global and local climate change, increasing demand for building services and indoor comfort levels, together with the rise in man-hours spent inside buildings, which is expected to continue into the near future. This is quite challenging while measuring against the aggregate energy production from different resources, making energy efficiency in buildings a prime concern of today for energy policy/technology researchers.

Among other energy sources used for the building sector, reliance on electricity is the major provider, and it is supposed to remain same in the years to come. The electricity share of the world's residential energy consumption is predicted to grow from 39% in 2012 to 43% in 2040, and by 2025 electricity is expected to surpass natural gas as the leading source of residential delivered energy. Seeing from other side, almost 60% of the world's electricity is consumed only for residential and commercial buildings purposes. The rapid penetration of the air conditioning in the developing countries further accelerates the electricity demand. China and India continue to lead world residential energy demand growth, mainly as a result of their relatively fast-paced economic growth. In 2040 their combined residential energy consumption, trending to more than double their 2012 total will account for nearly 27% of total world residential energy consumption. With uncertainty over the availability of fossil fuels into the future, rising demand for fossil fuels, rising concerns over energy security (both for general supply and specific needs of facilities), and the potential that greenhouse gases may be negatively affecting the world's climate, it is essential to find ways to reduce electricity load, increase efficiency, and utilize renewable energy resources in facilities of all types. More efficient energy and material use, as well as renewable energy supply in buildings, may be critical to tackling the sustainability-related challenges. Recent major advances in building design, know-how, technology, and policy have made it possible for global building energy use to decline significantly. A number of low-energy and passive buildings, both retrofitted and newly constructed, already exist, demonstrating that low level of building energy performance is achievable.

Many of these aspects are well researched and documented, such as the engineering aspects, policy aspects, economical aspects, but in pieces. Surprisingly, there is lack of understanding about the energy issues for buildings when put up in a single frame.

Also, limited data exist on how energy technologies have progressed rapidly and how well they have served end-uses. Another major knowledge gap lies in region-specific costs of new buildings in relation to their energy performance and region-specific costs of retrofits of existing buildings in relation to the savings in energy use achieved. Sufficient knowledge is also required about the best practices, that is, the most sustainable means for providing energy services in each developmental, cultural, geographical building sector. Without a doubt "Sustainability" and "Energy-efficient buildings" are two very important challenges to be achieved for mankind and make it a quite cumbersome but must-do task to be achieved for today's world, especially when put up together. Editors of this book, *Sustainability through Energy-Efficient Buildings,* Dr Amritanshu Shukla and Dr Atul Sharma have aptly taken up this challenge and have come out with a book that discusses the significant advances being made in the direction of developing energy-efficient buildings and developing a green, clean, and sustainable society. Such effort is the need of a time where various research works in this particularly important area are to be viewed in a most composite and comprehensive manner so that the knowledge acquired so far can be much more useful for students, academicians, and professionals/practioners working in this field. I congratulate both Drs Shukla and Sharma for bringing out a book which will be of immense help in achieving the goal of managing growth in energy consumption through improvements in the commercial and residential building sector.

Prof. Manthos Santamouris
Anita Lawrence Chair in High Performance Architecture, School of Built Environment, University of New South Wales, Sydney, NSW, Australia Visiting Professor, Metropolitan University of London, U.K. Visiting Professor, Univ Boltzano, Italy, Visiting Professor, Tokyo Polytechnic University, Japan Visiting Professor, Cyprus Research Institute, Cyprus Co-Editor in Chief-Energy and Buildings Journal

Foreword II

Over the last years climate change and the economic crisis have been affecting the world. Rather than being only negative, the increase of global environmental awareness has paved the way toward a low-carbon economy, while the economic recession has called for the adoption of a new growth model. A critical issue of the global debate is whether double dividend can be achieved, namely economic growth and environmental objectives.

Accelerating the transition to a low-carbon competitive economy is both a tremendous opportunity and an urgent necessity for Europe. The European Union is well placed to lead this transition through specific policies and actions aiming to underpin our European targets. The new momentum and clear direction stemming from the Paris Agreement is further sustained by the European Commission's adoption of the "Clean Energy for all Europeans" package on 30 November 2016.

The "Clean Energy for all Europeans" package puts energy efficiency first. Saving energy and improving energy efficiency is a prerequisite for the implementation of any other policy, as energy savings is the biggest reservoir of energy sources. Energy efficiency brings exceptional multiplier benefits to the economy, strengthening Europe's competitiveness, helping to reduce energy dependence, contributing to GDP growth and job creation, increasing health conditions in households, and reducing the level of greenhouse gas emissions.

Buildings can significantly contribute in this direction. The building sector has the biggest untapped energy-saving potential. Improving the energy performance of the European building stock remains high in the political agenda, not only because buildings are responsible for 40% of energy consumption and 36% of greenhouse gas emissions in the European Union, but because almost 75% of the buildings in Europe are energy inefficient.

The Energy Performance of Buildings Directive, together with the relevant measures under the Energy Efficiency Directive and the Ecodesign and Energy Labelling Directives, provides a comprehensive regulatory framework for energy efficiency in buildings in the European Union. Clearly, no regulation can succeed alone without the stimulation of a dialog among market stakeholders, the critical thinking and creativity of the scientific and academic society, the support of the institutional representatives of citizens, and, eventually, the actual involvement of every citizen individually.

The goal we have set is ambitious, and so is the effort needed to create more sustainable cities and nearly zero-energy buildings, but the benefits to gain are more important and will bring advantages in terms of environmental protection, economic growth, and quality of life. This book, *Sustainability through Energy-Efficient Buildings*, constitutes an excellent initiative, sharing recent advances and knowledge on energy-efficient concepts and addressing the latest technological challenges and policy issues.

I strongly recommend this book to all stakeholders involved in the construction supply chain and interested in energy efficiency and to all those who want to contribute to the decarbonization of our economy by 2050.

Paul Hodson
DIRECTORATE-GENERAL FOR ENERGY, EUROPEAN COMMISSION
Directorate C - Renewables, Research and Innovation, Energy Efficiency
C.3 - Energy efficiency, Head of Unit, Brussels

Editors

Dr. Amritanshu Shukla completed his masters in physics from the University of Lucknow, India and earned his PhD from IIT Kharagpur, India in January 2005. He did his postdoctoral research work at some of the premier international institutes, namely the Institute of Physics, Bhubaneswar (Department of Atomic Energy, Government of India); University of North Carolina, Chapel Hill; University of Rome/Gran Sasso National Laboratory, Italy; and Physical Research Laboratory, Ahmedabad (Department of Space, Government of India). He is currently an associate professor in physics at Rajiv Gandhi Institute of Petroleum Technology (RGIPT) (set up through an Act of Parliament by the Ministry of Petroleum & Natural Gas, as an "Institute of National Importance" on the lines of IITs).

Dr. Shukla's research interests include nuclear physics and physics of renewable energy resources. He has published more than 100 research papers in many international journals and international and national conference proceedings. He has delivered invited talks at various national and international institutes. Currently he is working on a number of national and international projects as well as collaborating with peers from India and abroad on the topics of his research interests.

Dr. Atul Sharma earned a MPhil in energy and environment (August 1998) and his PhD from the School of Energy and Environmental Studies, Devi Ahilya University, Indore (M.P.), India. Afterward, he worked as a scientific officer at Regional Testing Centre Cum Technical Backup Unit for Solar Thermal Devices at the School of Energy & Environmental Studies, Devi Ahilya University, Indore funded by Ministry of Nonconventional Energy Sources, New Delhi, Government of India and later as a research assistant at Solar Thermal Research Center, New & Renewable Energy Research Department at Korea Institute of Energy Research, Daejon, South Korea (April 1, 2004–May 31, 2005). Dr. Sharma was a visiting professor at the Department of Mechanical Engineering, Kun Shan University, Tainan, Taiwan, R.O.C (August 1, 2005–June 30, 2009).

Dr. Sharma is currently an associate professor at Rajiv Gandhi Institute of Petroleum Technology (RGIPT), which has been set up by the Ministry of Petroleum & Natural Gas, Government of India. He has published 44 research papers in various international journals and 67 in various international and national conference proceedings. He also published several patents related to the PCMs technology in Taiwan region only. He is currently working on the development and applications of phase change materials, green building, solar water heating systems, solar air heating systems, solar drying systems, and so on. Dr. Sharma is conducting research at the Nonconventional Energy Laboratory (NCEL), RGIPT and currently engaged with three Department of Science & Technology (DST), New Delhi-sponsored projects at his lab. Further, he served as an editorial board member and reviewer for many national and international journals, project reports, and book chapters.

Contributors

Patrick Achard
MINES ParisTech
PSL Research University
PERSEE—Centre procédés, énergies
   renouvelables et systèmes énergétiques
Sophia Antipolis Cedex, France

Lu Aye
Renewable Energy and Energy Efficiency
   Group
The University of Melbourne
Melbourne, Victoria, Australia

Pascal Biwole
MINES ParisTech
PSL Research University
PERSEE—Centre procédés, énergies
   renouvelables et systèmes énergétiques
Sophia Antipolis Cedex, France
CNRS, Jean Alexandre Dieudonné
   Laboratory
Université Côte d'Azur
Nice, France

A.K. Chaturvedi
Roorkee College of Engineering
Roorkee Institute of Technology
Roorkee, India

Delia D'Agostino
Directorate C—Energy Efficiency and
   Renewables
Joint Research Centre
European Commission
Ispra, Italy

Deep Gupta
Roorkee College of Engineering
Roorkee Institute of Technology
Roorkee, India

Siddartha Jain
Roorkee College of Engineering
Roorkee Institute of Technology
Roorkee, India

Amitha Jayalath
Renewable Energy and Energy Efficiency
   Group
The University of Melbourne
Melbourne, Victoria, Australia

Karunesh Kant
Nonconventional Energy Laboratory
Rajiv Gandhi Institute of Petroleum
   Technology
Jais, India

Sanjay Kumar Kar
Department Management Studies
Rajiv Gandhi Institute of Petroleum
   Technology
Jais, India

V.V.N. Kishore
Department of Energy and Environment
TERI University
New Delhi, India

Ian J. Lazarus
Department of Physics
Durban University of Technology
Durban, South Africa

Tingzhen Ming
School of Civil Engineering and
   Architecture
Wuhan University of Technology
Wuhan, China

Saurabh Mishra
Department of Sciences and Humanities
Nonconventional Energy Laboratory
Rajiv Gandhi Institute of Petroleum
 Technology
Jais, India

Mridula Singh
Roorkee College of Engineering
Roorkee Institute of Technology
Roorkee, India

Kevin Nocentini
MINES ParisTech
PSL Research University
PERSEE–Centre procédés, énergies
 renouvelables et systèmes
 énergétiques
Sophia Antipolis Cedex, France

Atul Sharma
Nonconventional Energy Laboratory
Rajiv Gandhi Institute of Petroleum
 Technology
Jais, India

Amritanshu Shukla
Nonconventional Energy Laboratory
Rajiv Gandhi Institute of Petroleum
 Technology
Jais, India

R. Singh
Department of Physics
Durban University of Technology
Durban, South Africa

A. Sreekumar
Department of Green Energy Technology
Pondicherry University (A Central
 University)
Puducherry, India

Manoj Kumar Srivastava
Department of Physics
Army Cadet College
Indian Military Academy
Dehradun, India

Yong Tao
College of Engineering and Computing
Nova Southeastern University
Fort Lauderdale, Florida

Manish Vaid
Observer Research Foundation
New Delhi, India

C. Veerakumar
Department of Green Energy Technology
Pondicherry University (A Central
 University)
Puducherry, India

1

Introduction to Energy-Efficient Building Development and Sustainability

R. Singh and V.V.N. Kishore

CONTENTS

1.1 Introduction 1
1.2 Growth in Building Sector 3
1.3 Building's Energy Consumption and Greenhouse Gas Emissions 4
1.4 Building Sustainability 6
1.5 The Energy Conservation Building Codes and Green Building Rating Tools 8
1.6 Cost Effective Energy-Efficiency Concepts for Energy-Efficient/Green Buildings 18
    1.6.1 Improved Ventilation 21
    1.6.2 Shading of Glazed Components 21
    1.6.3 Insulation of Opaque Components 22
    1.6.4 Green Roof Concept 22
1.7 Benefits and Impact on the Society 25
1.8 Conclusion 26
References 27

1.1 Introduction

The fifth assessment report by the Intergovernmental Panel on Climate Change (IPCC) confirmed with 95% certainty that global warming is mainly a result of anthropogenic greenhouse gas (GHG) emissions. It has also been clearly mentioned in the report that the human activities disrupted the climate significantly and led to severe risks that are pervasive and irreversible impacts for the people and ecosystems [1]. Therefore, appropriate actions were suggested to be taken urgently to stabilize the rise in the global temperature below 2°C relative to preindustrial level. The modern life style, increased population, and economic status have increased the energy demand immensely around the globe. As in the present scenario, almost every activity performed by the human consumes energy to complete the task. These energy consuming activities have increased the energy demand tremendously and created unsustainability in the energy sector globally. The energy demand will continue grow in future too if proper attention is not be given to the sector to minimize the demand as well as to shift from conventional to renewable energy sources. Figure 1.1 shows total energy consumption between years 2000 and 2015 for different regions [2]. The energy consumption for Latin America, Africa, Pacific, Middle East, and CIS remains between 0 and 1000 Mtoe with substantial increase year after year. Major change can be observed for Asia, where total energy consumption increased from 3000 to approximately

1

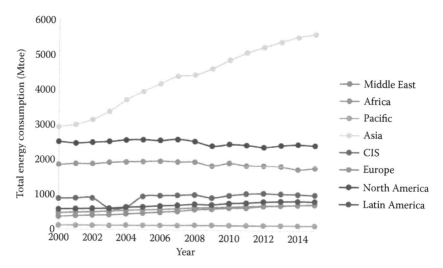

FIGURE 1.1
Regional total energy consumption trend. (From www.enerdata.net.)

6000 Mtoe. The energy consumption for Europe and CIS does not show much variation. The electricity consumption for different regions, between years 2000 and 2015, is shown in Figure 1.2 [2]. A similar trend can be observed with highest increase (6.5%) in the Asian electricity consumption followed by Middle East, Africa, and Latin America with 5.8%, 3.4%, and 3.2% increase respectively. This increased energy consumption resulted in higher GHG emissions and issue of climate change, which eventually created severe issues like catastrophic drought, heat waves, and stronger hurricanes that could be life-threatening.

Recent EIA projection for a period of 28 years, between the years 2012 and 2040, clearly indicates a 48% increase in the worldwide energy demand. The energy demand, which was 549 quadrillion British thermal units (Btu) in the year 2012, is estimated to be 815 quadrillion Btu in 2040 (Figure 1.3). The demand for conventional fuels will grow continuously despite significant increase in the use of renewable energy too. Almost stable consumption

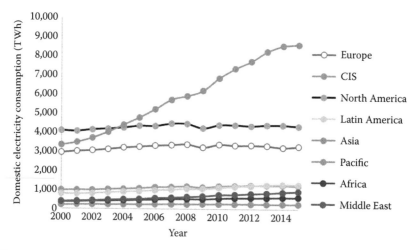

FIGURE 1.2
Domestic electricity consumption. (From www.enerdata.net.)

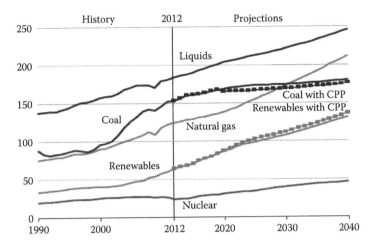

FIGURE 1.3
World energy consumption by energy source, 1990–2040 (quadrillion Btu). (Courtesy of U.S. Energy Information Administration, Washington, DC.)

of coal (between 150 and 200 btu) is one of positive and persuasive outcomes of the projection indicating shift in the energy consumption behavior and efficient technology adaptation [3]. However, increased global energy consumption would still be a significant concern for the sustainability of the society and the energy sector. Continuous advancement in the technologies regarding their efficiency, adaptation of energy efficiency and conservation measures, and increased use of renewable energy sources in passive as well as active mode would possibly ensure the sustainability. Recently, Paris Climate Conference (COP21) also reached an agreement to set out an action plan to minimize the effects of climate change by keeping the increase in global average temperature well below 2°C. To achieve the above goal of keeping the temperature within the desirable limit, the energy consumption and emission must be reduced in all economic sectors including buildings [4].

The building sector consumes approximately 30%–40% of total global energy production and has enormous energy conservation opportunities to lower the energy demand like in other major energy consuming sectors. This chapter aims to explore and discuss the growth in the building sector and current and future energy consumption trends as a result of growth in the building sector. Moreover, energy conservation strategies, standards, and procedures that are being implemented in the building sector to reduce the overall energy consumption, negative impact on the environment, and future sustainability of the sector are discussed in this study.

1.2 Growth in Building Sector

The continuous population growth has resulted in higher demand of residential dwellings as well as the commercial floor space around the world. The demand for building infrastructure has been observed to be prominently and substantially high particularly in Asia and so is the energy consumption. The trend is expected to be the same in future too. Recently, the Global Construction 2030 forecast report indicated a worldwide construction growth close to 85%, which is equivalent to $15.5 trillion, by 2030. Particularly, three countries: China, US, and India leading the way and would account for 57% of total global growth. Over the next 15 years, the growth in the Chinese construction market is expected to increase marginally while US construction market will grow little faster than China. The Indian construction market probably will overtake Japan to become the world's third largest construction market by 2021. The growth in Indian construction market is expected to be almost twice as fast as in China by 2030. Moreover, India's urban population is expected to grow by a staggering 165 million by 2030, swelling Delhi by 10.4 million people to become the world's second largest city [5]. A similar growth in the sector can be expected in other cities. In Canada, about 12.5 million residential units and 430,000 commercial and institutional buildings account for approximately 33% of total energy use, 50% of natural resource consumption and responsible for 30% of Canadian GHG emissions and produce about 25% of the nation's landfill waste [6]. The above data indicate that the energy consumption will continue to grow because of faster growth in the construction sector demanded by increased population.

1.3 Building's Energy Consumption and Greenhouse Gas Emissions

Design, fabrication, construction, and operation of buildings, in which we live and work, are responsible for consumption of energy and natural resources. As explained previously, the buildings consume a significant portion of total energy globally, over one-third of final energy, and are equally responsible for carbon dioxide (CO_2) emissions [7]. The energy consumed by the buildings sector consists of residential and commercial end users and accounts for 20.1% of the total delivered energy consumed worldwide. In 2010, the building sector accounted for approximately 117 exajoules (EJ) or 32% of global final energy consumption, 19% of energy-related CO_2 emissions and 51% of global electricity consumption [8]. Moreover, final energy use of the buildings sector is expected to grow from approximately 117 exajoules per year (EJ/yr) in 2010 to 270 EJ/yr in 2050. If only currently planned policies are implemented, the final energy use in buildings that could be locked-in by 2050 is equivalent to approximately 80% of the final energy use of the buildings sector in 2005 [9]. Recently, Güneralp et al. [10] projected the energy use for heating and cooling by the middle of the century between 45 and 59 EJ/yr (with an increase of 7%–40% since 2010). In this study, reason of the variability could be uncertainty in the future urban densities of rapidly growing cities in Asia and particularly in China. The electricity share of world residential energy consumption will grow from 39% in 2012 to 43% in 2040. China and India continue to lead world residential energy demand growth mainly as a result of their relatively fast-paced economic and population growth. In 2040 their combined residential energy consumption will be more than double 2012 total consumption and will account nearly 27% of total world residential energy consumption [3]. Therefore, this sector needs urgent attention to improve the energy efficiency and integration of renewable energy technologies. The electricity savings in buildings would have far-reaching benefits for the power sector and will translate into avoided electrical capacity additions, as well as reduced distribution and transmission network expansion, with potentially huge savings for utilities [7].

Commercial energy consumption occurs in enterprises engaged in commercial activities (the service sector). The energy is consumed by heating and cooling systems, lights, refrigerators, computers, and other equipment in the buildings where businesses, institutions, and other organizations are located. The commercial sector/service sector buildings include retail stores, office buildings, government buildings, restaurants, hotels, schools, hospitals, and leisure and recreational facilities. Total world delivered commercial sector (also included some nonbuilding energy use such as traffic lights and water and sewer systems) energy consumption is expected to grow by an average of 1.6%/year from 2012 to 2040 and the sector is the fastest-growing energy demand sector (Table 1.1). The commercial sector energy demand in the Organization for Economic and Cooperative Development (OECD) economies is higher than the non-OECD economies. In 2012, OECD commercial energy use was 116% higher than non-OECD use and in 2040 it is expected to remain 50% despite considerably larger and more rapid growth in population in the non-OECD countries. The rate of energy use in non-OECD commercial sector is expected to be double the rate for the more mature energy-consuming economies of the OECD.

China's commercial energy use was projected highest in the non-OECD region while India's commercial energy consumption would grow fastest. The slowest commercial energy use has been estimated for Russia in the non-OECD region from 2012 to 2040, mostly because of slowest economic growth in the region and a declining population growth. On average, electricity use accounted for 53% of non-OECD commercial sector fuel use in 2012 and its share is expected to continue growing close to 67% by 2040. The average annual change in the Indian commercial sector has been projected 3.7%, which is highest in Asia and much higher than total non-OECD average (2.4%) between 2012 and 2040. Overall, the building sector has a profound impact on the natural environment, public health, economy, and productivity and the impact will continue to grow in future. Globally, the construction industry consumes 40% of total energy production, 12%–16% of all water available, 32% of nonrenewable and renewable resources, 25% of all timber, 40% of all raw materials, produces 30%–40% of all solid wastes, and emits 35%–40% of CO_2 [11,12], which has resulted in a rising global awareness of the importance of sustainability in the construction industry [13,14]. The energy consumption in buildings worldwide is expected to increase by an average of 1.5% per year from 2012 to 2040 [3]. The building sector in the United States accounts for 39% of total energy use, 68% of total electricity consumption, 30% of landfill waste, 38% of carbon dioxide emissions and 12% of total water consumption [15].

Being one of the major energy consuming sectors, the building sector is also one of the major emitters of total global GHG emission (Figure 1.4). In this sector, an increase in the total annual anthropogenic GHG emissions by about 10 $GtCO_2$-eq between 2000 and

TABLE 1.1

Annual Average Percentage Change in Commercial Sector Delivered Energy Consumption by Region per Year

Region	Average Annual Percent Change			
	2012–2020	2020–2030	2030–2040	2012–2040
OECD	1.6	0.9	0.8	1.1
Americas	1.1	0.8	1.0	0.9
Europe	1.9	1.2	0.8	1.3
Asia	2.4	0.9	0.6	1.2
Non-OECD	3.0	2.5	1.9	2.4
Europe and Eurasia	2.1	1.5	0.9	1.4
Asia	3.7	2.9	2.1	2.9
Middle East	2.4	2.5	2.0	2.3
Africa	3.3	3.1	3.1	3.2
Americas	2.1	1.9	1.7	1.9
Total world	2.1	1.5	1.2	1.6

Source: EIA, *International Energy Outlook 2016*, U.S. Energy Information Administration, 2016.

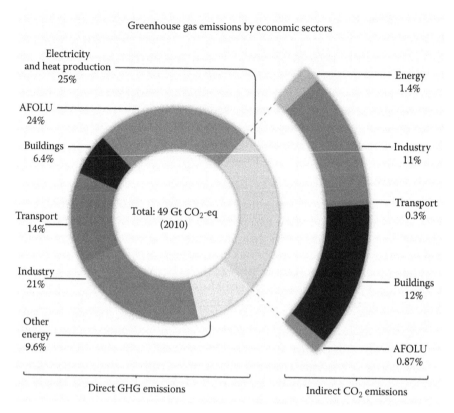

FIGURE 1.4

Total anthropogenic greenhouse gas (GHG) emissions (gigatonne of CO_2-equivalent per year, $GtCO_2$-eq/yr) from economic sectors in 2010. (Figure 1.7 from IPCC, Climate change 2014: Synthesis report, Contribution of working groups I, II and III to the fifth assessment report of the Intergovernmental Panel on Climate Change, in *Core Writing Team* R.K. Pachauri and L.A. Meyer (Ed.), IPCC, Geneva, Switzerland, 2014.)

2010 was reported [9]. The buildings sector also accounted for approximately one-third of black carbon emissions, and an eighth to a third of F-gases [9]. The direct and indirect CO_2 emissions from buildings have been projected to increase from 8.8 $GtCO_2$/yr in 2010 to 13–17 $GtCO_2$/yr in 2050 in baseline scenarios.

Significant reduction in energy consumption and CO_2 emissions is a challenging goal that can possibly be achieved through integral implementation of best available technology and intelligent public policy. It should be ensured that all available options are tapped effectively and may require unprecedented efforts and coordination among a diverse set of stakeholders including policymakers, builders, technology developers, manufacturers, equipment installers, financial institutions, businesses, and household consumers [7].

1.4 Building Sustainability

To make the building sector sustainable, the development of energy-efficient, green, and net zero energy buildings should well be considered as one of the viable choices. The energy-efficient buildings (EEBs) and green buildings are about resources, efficiency, lifecycle effects, and building performance with the least environmental impact. Here it should be made clear that green building development (unlike EEB) is not only implementing energy-efficient measures in designs and technologies but also includes the efficiency and conservation measures at every step in the entire development process, including planning and well-being of workers and users. One can simply understand that the green building movement aims to reduce resource and energy consumption, increase the use of renewable energy, minimize environmental degradation and the production of waste, and maximize occupant health and comfort. These goals can be achieved using sustainable building products and practices (e.g., modern building site environmental management techniques, utilization of recycled or locally sourced building materials, efficient building designs to maximize daylighting, centrally controlled smart HVAC systems, etc.). Also, significant weightage is given to the use of non-toxic interior paints and finishes; water-conserving toilets, faucets and showerheads; improved insulation materials and techniques; maximization of open space and promotion of air circulation; and numerous other techniques. The green building may not be independent of grid unlike net zero energy buildings. In net zero energy building, all energy needs of the building are fulfilled by generating electricity on-site. However, all possible and desirable low-cost energy efficiency measures are implemented to reduce energy demand to its minimum level before integrating renewable energy generating technologies to meet the remaining electricity need. Also, it is not essential to follow the green building development procedure to develop net zero energy buildings. However, net zero energy building may or may not be able to get the best possible green building rating. Overall, the sustainability level of three different building development concepts may possibly follow the route presented in the following Figure 1.5.

Moreover, recent advances in technology, design practices and know-how, coupled with behavioral changes, can achieve a two- to ten-fold reduction in energy requirements of individual new buildings and a two- to four-fold reduction for individual existing

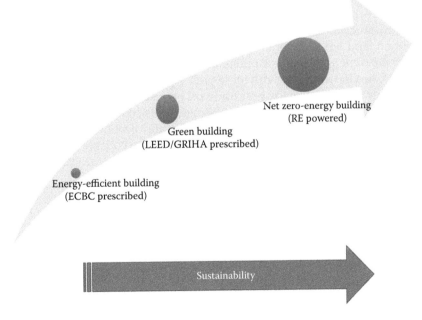

FIGURE 1.5
Possible sustainability of three different building development concepts.

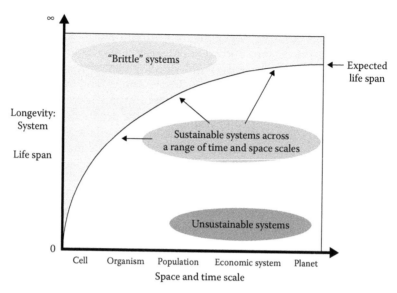

FIGURE 1.6
Sustainability as scale (time and space) dependent concepts. (From Costanza, R., and Patten, B.C., *Ecol. Econ.*, 15, 193–196, 1995.)

buildings largely cost-effectively or sometimes even at net negative costs [1]. Bearing the energy consumption pattern and energy saving potential in mind, substantial attention is being given, around the world, to ensure the sustainability in the building sector. To fast-track the progress in the sector, different projects and programs have been initiated at voluntary and mandatory levels. For instance, the energy conservation building codes and green building concepts and rating systems were developed and implemented aiming to reduce the overall impact of the built environment on human health and the natural environment by efficient use of energy, water, and other resources as well as minimizing waste, pollution, and environmental degradation [16]. As the environment, society, and economy are the three pillars of sustainability, the majority of sustainability rating systems have been developed in line with these pillars [17]. Mateus and Bragança [18] defined the sustainable development as the best trade-off between the three pillars that strive for greater compatibility. The sustainability assessment tools contribute to balancing between these dimensions or pillars (environmental, social, and economical) and to enhancing practicality and resiliency. Therefore, they should be able to consider constant technological development and multi-level applications [19]. The concept of sustainability as a scale dependent on time and space is illustrated in Figure 1.6.

1.5 The Energy Conservation Building Codes and Green Building Rating Tools

In recent years, energy crises and climate change issues have occurred around the world. Most governments, researchers, and policy-makers have started giving considerable attention to this area by addressing the problems through implementing effective energy conservations measures and including alternative renewable energy solutions. The energy conservation building codes designed and stipulated in most countries are meant to make the building sector energy efficient. The codes prescribe minimum standards for building design, equipment, mechanical, and lighting and power systems for new and existing building constructions to improve the energy efficiency. The climatic conditions and local user's acceptability of different parameters were taken into consideration to maintain the indoor thermal, visual, and acoustic comforts. Essential measures stated in the codes for different types of buildings were deduced through detailed analysis of the whole building performance For example, the U-value, SHGC, and Tv for opaque and transparent components in different Indian climatic zones, prescribed in energy conservation building code (ECBC) are given in Table 1.2. The detail of ECBC for other parameters can be seen from Reference 21. For other regions, the prescribed values of different components for code compliance buildings can be obtained from References 21–26.

The ECBC is being implemented in different countries on voluntary, mandatory, and mixed bases as can be seen in the following Figure 1.7 for existing and new residential as well as for commercial buildings. The mixed basis means that in some parts of the country the code is mandatory while in others, owners and developers choose to implement or not in their building. The maps clearly indicate that the codes have not been implemented in the existing buildings in major parts of the world. The code is mandatory mostly in Europe and Asia, and some regions in America, for both residential and non-residential buildings. In India, Middle East, and South Africa, only new buildings, including residential and non-residential, are being constructed code compliant on a voluntary basis. In South America, the code is being used on a voluntary basis in residential new buildings only. The mixed practices can be observed in US and Canada. Clearly, significant energy can be saved if the code is implemented in rest of the world. The status of building energy code stringency, technical requirements, enforcement, compliance, and energy intensity in each of the countries and national best practices for each element have been reported in Reference 25.

Approximately 10–15 EJ savings (230–350 Mtoe) in 2030 has been projected if energy consumption in new buildings is halved compared to the base scenario in WEO [28].

TABLE 1.2

U Values for Opaque and Glazed Components along with SHGC and VLT Prescribed for ECBC Compliance Buildings in Different Indian Climates

	Composite	Hot and Dry	Warm and Humid	Temperate	Cold
Wall (U value W/m² K)					
ECBC 2007 8 h	0.44	0.44	0.44	0.44	0.35
ECBC 2007 24 h	0.44	0.44	0.44	0.44	0.37
Roof (U value W/m² K)					
ECBC 2007 8 h	0.4	0.4	0.4	0.4	0.4
ECBC 2007 24 h	0.26	0.26	0.26	0.26	0.26
Glazing (40%)					
ECBC 2007 Standard	U-3.3 SHGC-0.25 VLT-0.2	U-3.3 SHGC-0.25 VLT-0.2	U-3.3 SHGC-0.25 VLT-0.2	U-6.8 SHGC-0.40 VLT-0.2	U-3.3 SHGC-0.51 VLT-0.2

Source: ECBC, Energy conservation building code, https://beeindia.gov.in/content/ecbc, 2007.

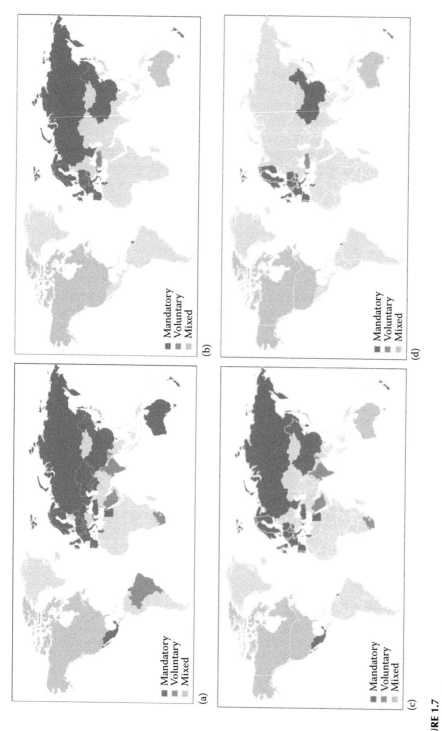

FIGURE 1.7

Status of building energy codes implementation (a) new residential buildings, (b) existing residential buildings, (c) new non-residential buildings, and (d) existing nonresidential buildings. The maps are without prejudice to the status of or sovereignty over any territory, to the delimitation of international frontiers and boundaries and to the name of any territory, city or area. (From UNDP IEA, Modernising building energy codes: To secure our global energy future, IEA, 22–23, 2013.)

These savings would be beneficial both for the owners and for society. Based on the WEO 2006 and the values for 2004, the savings would be between 15 and 40 EJ (nearly 950 Mtoe) in 2030, if strong measures are taken for improvement of energy efficiency by refurbishment and major renovation [28]. The code implementation could even be highly cost effective in both types of existing buildings. It is not necessary that the existing building must be completely code compliant, but a few components can be modified to the code compliance level with minimal investment. Moreover, a rebate policy could expedite the implementation of the code in all type of buildings. Recently, dynamic energy efficiency policies for new buildings (residential and commercial) have been discussed and compared by reviewing 25 best practice building energy efficiency codes using 15 criteria developed with some of the world's leading experts in the field [24]. The comparative chart with the criteria is shown in the Figure 1.8. The rating is completely dependent on the selected criteria. If number of criteria is changed, the rating for each country would change. One can draw different rating graphs (similar to Figure 1.8) online by selecting criteria as per own choices and preferences.

The compliance cycle to ensure the effective implementation of codes is explained in the following Figure 1.9.

For improvement in the energy performance of buildings and conservation of natural resources during the entire building development process, a new concept of green building design and rating system has been developed and suggested to implement. The concept has not only emphasized the energy-efficient designs but also suggests appropriate measures to take into account for efficient construction of the building, use of local resources efficiently, minimization of wastes during construction as well as while using the building, site planning, and efficient use of the occupied space. The constructed building is rated and certified based on acquired points to ensure the efficiency in the design and development process. The points have been prescribed for incorporating and following each suggested measure at each step of building development. The rating tools are used by the building industry to evaluate, enhance, and/or promote developments' sustainability. These tools provide guidance and/or better insights into sustainability through information analysis, valuations, and comparisons [29].

Quite a few green building rating tools and programs have been developed around the globe considering regional priorities of comfort, bylaws, and climatic conditions. Leadership in Energy and Environment Development (LEED) is one of them and is accepted globally. The main aim of such programs is to develop efficient and cost-effective building infrastructure to improve conventional design and construction practices and standards. In addition, the programs exist to increase productivity and contribute to healthier living and working environments for building occupants [30]. Other well-known tools such as BREEAM (United Kingdom), DGNB (Germany), Green Star (Australia), Green Mark (Singapore), CASBEE (Japan), and GRIHA (India) have helped the professionals quantify environmental performance in an explicit way. The measurement and benchmark "green building," which lays the foundation for the common attributes of different green building tools is based on reducing negative impacts as shown in Figure 1.10 [31].

The tools mainly rate the buildings based on the acquired credits and score prescribed for design features that can minimize a building's environmental impact such as location and site; conservation of water, energy, and materials; and occupant comfort and health. For example, under the LEED-NC 2009 (also known as LEED v3.0), buildings are judged

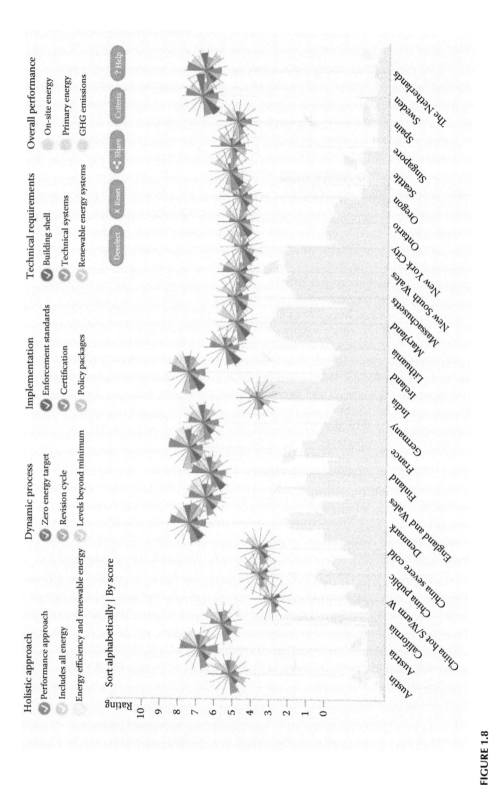

FIGURE 1.8
Components needed to develop ambitious and dynamic energy efficiency policies for new buildings. (From GBPN, http://www.gbpn.org/databases-tools/purpose-policy-tool-new-buildings, accessed July 10, 2017.)

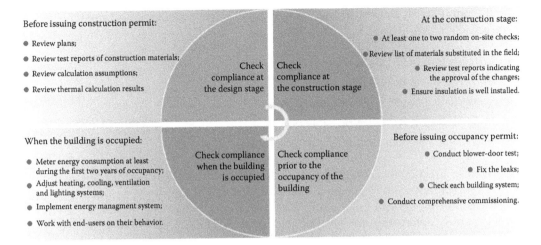

Before issuing construction permit:

● Review plans;
● Review test reports of construction materials;
● Review calculation assumptions;
● Review thermal calculation results

Check compliance at the design stage

Check compliance at the construction stage

At the construction stage:

● At least one to two random on-site checks;
● Review list of materials substituted in the field;
● Review test reports indicating the approval of the changes;
● Ensure insulation is well installed.

When the building is occupied:

● Meter energy consumption at least during the first two years of occupancy;
● Adjust heating, cooling, ventilation and lighting systems;
● Implement energy managment system;
● Work with end-users on their behavior.

Check compliance when the building is occupied

Check compliance prior to the occupancy of the building

Before issuing occupancy permit:

● Conduct blower-door test;
● Fix the leaks;
● Check each building system;
● Conduct comprehensive commissioning.

FIGURE 1.9
Compliance cycle to ensure effective implementation of building energy code. (From UNDP IEA, Modernising building energy codes: To secure our global energy future, IEA, 2013.)

FIGURE 1.10
The benchmark of green building based on reducing negative impacts. (From Gou, Z.H., and Xie, X.H., *J. Clean Prod.*, 153, 600–607, 2017.)

via a 100-point credit system in five categories. The five categories and their respective points are as follows [32]:

- *Sustainable sites*: 26 possible points
- *Water efficiency*: 10 possible points
- *Energy and atmosphere*: 35 possible points
- *Materials and resources*: 14 possible points
- *Indoor environmental quality*: 15 possible points

There are also 10 possible incentive points included in two additional categories for innovative strategies: innovation in design (6 points) and regional priority (4 points). Application documents and proofs can be submitted at the design phase. The number of credits or scores generally determines the level of achievement. These assessment or measurement tools have individually and collectively made significant contributions to understanding of building-related environmental impacts. The continuous improvement in the content and relevant applications in the tools/methods is being done according to the market needs and due to suggestions and critiques raised by researchers.

The green buildings generally have higher levels of indoor air quality and measures of lifecycle impact of choices of building materials, furnishings, and furniture are taken into account [33]. Therefore, the demand for more sustainable green buildings has increased in the across the globe [34] as implementation of the green building concepts in design offers several environmental, economic, and social benefits to the construction industry [35]. Moreover, green building helps to reduce contributing to global warming and climate change by minimizing CO_2 emissions and other pollutants, protecting the ecosystem, using renewable natural resources, improving health, comfort, and well-being, alleviating poverty, improving economic growth, raising rental income, decreasing healthcare costs, and so on [36]. Recently, Amos Darko et al. [37] identified generic drivers for stakeholders to pursue green buildings. They have classified these drivers into five main categories (Figure 1.11), that is, external drivers, corporate-level drivers, property-level drivers, project-level drivers, and individual-level drivers. This can encourage policymakers to advocate benefits and understanding of green building impact and help to further promotion of the green building concept. The top five drivers for green building concept are energy-efficiency; reduced environmental impact; water-efficiency; occupants' health, comfort, and satisfaction; and company image/reputation [38].

The most recent World Green Building Trends report by Dodge Data & Analytics shows that developing markets will see the largest potential for green building growth. Absorption, adsorption and various advanced thermodynamic cycles could be considered as there is renewed interest in thermally activated cooling and heating systems. Microchannel systems could help improve heat and mass transfer, leading to reduction of fluid inventories, material utilization, and environmental impact.

Despite global recognition and acceptance of current green building concepts and practices, some researches also observed weaknesses in these concepts and aimed to develop alternative frameworks for defining and benchmarking sustainable buildings. Four of the most representative frameworks, developed by architectural companies and research institutes, are shown in Figure 1.12: Arup's Sustainable Project Appraisal Routine (SPeAR) [39], REGEN [40], LENSES (Living Environments in Natural, Social and Economic Systems) [41], and the Perkins+Will framework [42].

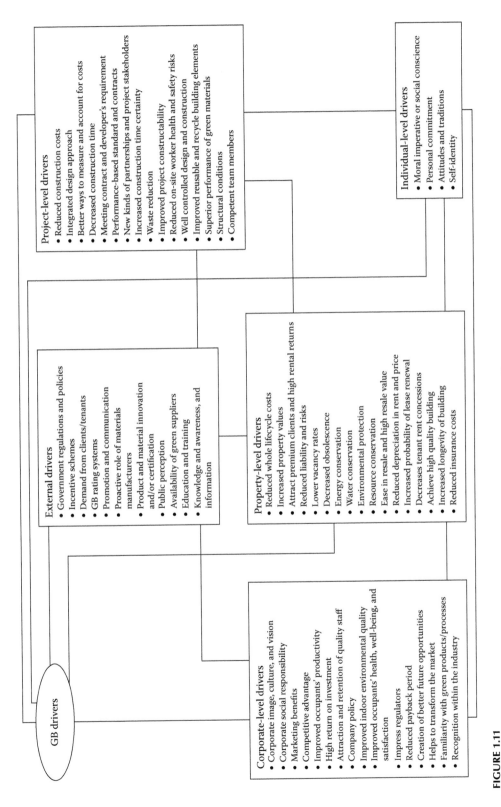

FIGURE 1.11

Conceptual framework for GB drivers. (From Darko, A. et al., *Habitat Int.*, 60, 34–49, 2017.)

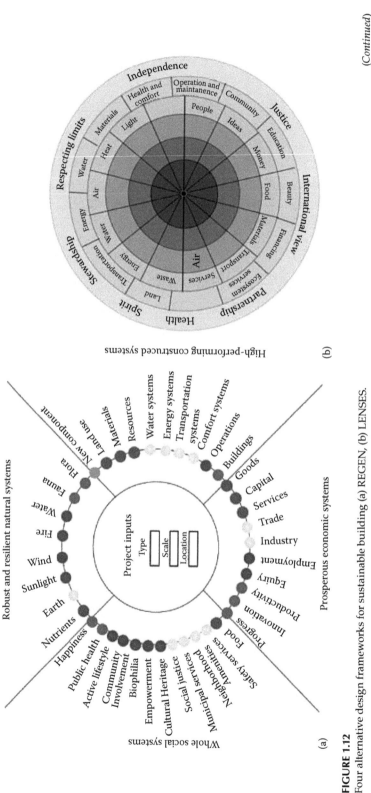

(Continued)

FIGURE 1.12

Four alternative design frameworks for sustainable building (a) REGEN, (b) LENSES.

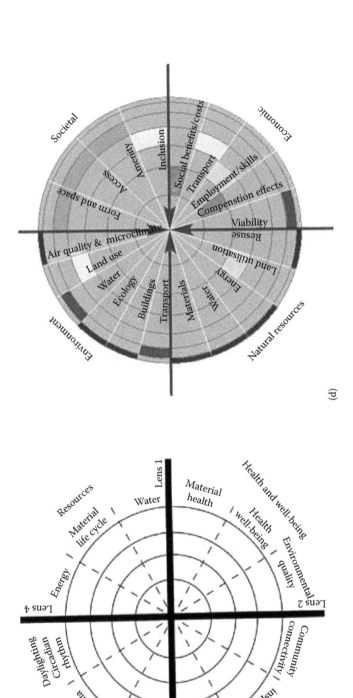

FIGURE 1.12 (Continued)
Four alternative design frameworks for sustainable building (c) Perkins and Will, and (d) SPeAR. (From Gou, Z.H., and Xie, X.H., *J. Clean Prod.*, 153, 600–607, 2017.)

The most explicit one is the Perkins+Will framework that sets resource-related design strategies within cycles from nature and back to nature [42]. It sets human needs, interactions, and resource flows within and interdependent of the constraints and opportunities afforded by natural systems.

1.6 Cost Effective Energy-Efficiency Concepts for Energy-Efficient/Green Buildings

There are several active and passive concepts that have been explored to improve the energy performance of buildings around the world. It has been observed from the literature and stated previously in this study that ECBC implementation has been ignored in some of the countries even in new construction, while existing buildings were not given much attention regarding improving energy efficiency through ECBC implementation. Moreover, the implementation of codes is voluntary in most parts of the world and do not attract owner or developer attention as desirable to get the maximum benefits for the society. One of the most important reasons could be significantly high initial cost associated with most of the options. Therefore, it is highly desirable to discuss the low-cost and effective energy conservation concepts that can be equally applicable in both the existing and newly planned buildings. Major energy consuming processes in buildings are heating, cooling, ventilation, and lighting. Therefore, in this section, we will highlight the retrofitting options to reduce the heating, cooling, ventilation, and lighting energy use.

High-performance retrofits are key mitigation strategies in countries with existing building stocks, as buildings are very long-lived and a large fraction of 2050 developed country buildings already exist (robust evidence, high agreement). Reductions of heating/cooling energy use by 50%–90% could be achieved using best practices. It has been observed that very low-energy construction and retrofits can be economically attractive. With ambitious policies, it is possible to keep global building energy use constant or significantly lower by mid-century compared to baseline scenarios which anticipate an increase of more than two-fold. In general, deeper reductions are possible in thermal energy uses than in other energy services mainly relying on electricity. Several factors beyond the building's constitutive elements affect a building's thermal performance, including climatological considerations, building shape, and surrounding urban morphology. These factors dictate the amount of precipitation, thermal energy (direct and indirect), and wind (intensity and prevailing direction) that a building is exposed to in the external environment. In fact, deliberate energetic urban planning represents the best starting point for an energetically optimized urban environment and building composition, as shown by Taleb et al. [43]. Aboulnaga et al. [44] also discusses the effect of urban patterns on building energy consumption by comparing four similar but differently oriented two-story residential buildings in Al Ain. Based on their results, the authors recommend inclusion of energetic considerations in the development of urban patterns. This recommendation is justified by reported energy savings (ES) of up to 55% accomplished by limiting fenestration to only two elevations, and restricting the window to surface area ratio to about 1:6. Giuliano Dall'O' et al. [45] have estimated real energy saving potential of retrofitting the housing stock. The saving potentials shown in Figure 1.13 were estimated for the Business as usual (BAU) and maximum savings scenarios for retrofitting the windows, external walls and roofs. The gap between the BAU and maximum savings scenarios was estimated quite large. The BAU scenario

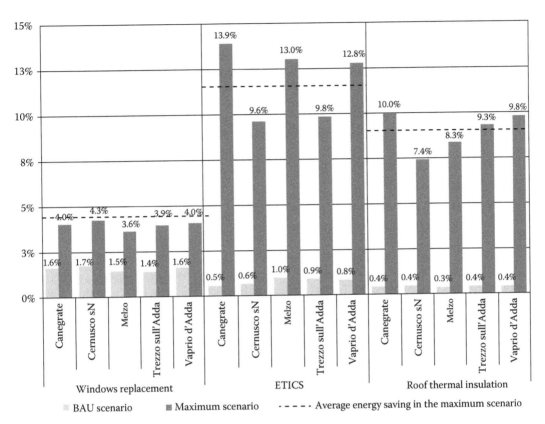

FIGURE 1.13
Potential energy savings in the BAU and maximum savings scenarios. (From Dall'O', G. et al., *Sustain. Cities Soc.*, 4, 12–21, 2012.)

could only save up to 2.7%, while in the maximum scenario possible energy consumption can be reduction up to 24.8% in the residential sector.

The geometric optimization at the individual building scale, which consists of configuring the building's shape and volume distribution, should be optimized for minimum energy use. However, these considerations, often constrained by the specific characteristics of the planned building and the size, shape, and orientation of the building plot, can be quite restrictive in the case of high-rise office and commercial buildings and in the light of local bylaws. In these buildings, space needs to be maximized for a given footprint that leave only limited room for layout optimization [46]. In accordance, low-rise residential buildings and villas offer the highest shape flexibility and thus best potential for energetic shape optimization. The directionality of solar radiation due to the sun's path implies that there are favored directions in which the sun will penetrate deeper into a building through the glazing, thereby increasing the solar heat gain. Restricting fenestration on the exposed sides and/or applying solar control measures to the glazing will improve thermal performance of the envelope. The new green building legislations that are currently being implemented incorporate this insight by prescribing maximum window areas for the different orientations and appropriate solar control glazing. While specific improvements in glazing and shading are presented in a later section, the building layout can also foster solar control and improved ventilation.

Considering climate change, worldwide heating energy demand is projected to decrease by 34% by the year 2100, while cooling demand is estimated to increase by 72% over the same time period [47]. Thus the severe cooling need of buildings in arid regions will increase over the next century, further impacting the often poor energetic and sustainability balance sheet of these regions and underscoring the necessity of energy efficiency measures. The retrofitting not only reduces the energy used, lowering the energy bills, but also improves indoor air quality and reduces external noise, increasing the market value. In order to improve energy efficiency of existing buildings, the most significant identified options are [48]:

- Replacement of windows
- Additional façade insulation
- Additional roof insulation
- New sealing to reduce ventilation losses

These retrofit measures are the best available techniques currently available.

Graziano Salvalai et al. [49] defined the most promising renovation strategies applicable to the different school building types (Table 1.3) in Lecco municipality from both the economic and the energy point of view. The presented research was based on the classification and analysis of the study samples, consisting of 38 school buildings that differ in educational level, age of construction, and typological design. Figure 1.14 shows the energy-saving potential in different cluster types through three different retrofitting strategies. It is clear from these results that plant replacement is one that can influence the energy consumption most followed by insulation and window replacement in Lecco province in Italy. A detail of energy saving using insulation in buildings can be found in Reference 50.

TABLE 1.3

Studied Building Designs and Respective Cluster Type

Cluster	C1	C2	C3	C4	C5	C6	C7	C8	C9
Design type									

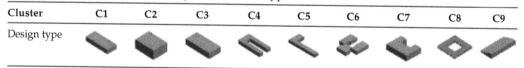

Source: Salvalai, G. et al., *Energ. Build.,* 145, 92–106, 2017.

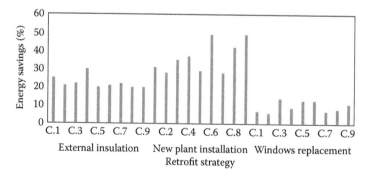

FIGURE 1.14

Energy savings in school buildings through different retrofit strategies. (From Salvalai, G. et al., *Energ. Build.,* 145, 92–106, 2017.)

Touraj Ashrafian et al. [51] introduced a new methodology to define proper retrofit measures for building owners and encouraging them to act as financiers. The three case study buildings with the same geometry but different envelope properties were analyzed for three different climatic conditions of Turkey. Results revealed that, except for cold climates, the actions that provide most ES are not beneficial when considering payback period and global cost issues. Moreover, even without any improvement in the active systems of a building, up to 30% savings can be achieved in primary energy consumption. However, these measures may represent suboptimal measures when compared with other important items such as investment cost, global cost, and payback period. Some of the low-cost options equally effective in existing and new buildings are discussed below:

1.6.1 Improved Ventilation

Night cooling, the combined effect of both natural or mechanical night ventilation, and building thermal inertia was proven to be an effective measure to reduce cooling loads [52–55]. The heat absorbed by the building exposed thermal mass during the day is released to the indoor air at night, after which it is purged by night ventilation. Meanwhile, external fresh air cools down the thermal mass which then acts as a heat sink in the following day [52]. The efficiency of night cooling depends on the thermal properties of the building and on the local climate conditions, that is, nighttime wind speed and temperature swing of the ambient air [54–57]. The effect of the increased night ventilation rate (ACH_N), varying from 0.5 to 20 h^{-1}, on the ES for different European climates has been estimated as shown in Figure 1.15 [58].

1.6.2 Shading of Glazed Components

Radiation control strategies are required in order to reduce the cooling need in summer as well as for improving indoor visual comfort by avoiding glare. Bellia et al. [59] have classified solar shading systems for buildings as the following: (a) fixed, (b) movable, and (c) other types. Any one of these categories included technologies placed at the external side of the transparent component, in an intermediate position or at the inside. Both thermal and visual comforts are impacted by these shading options when applied to glazed components. Further, Bellia et al.

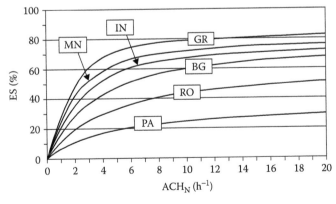

FIGURE 1.15
Effect of night ventilation rate (ACH_N) on the energy savings (ES) in different European climates (e.g., BG-Bergamo; GR-Groningen; MN-Munich; IN-Innsbruck; RO-Rome; PA-Palermo). (From Ramponi, R. et al., *Appl. Energ.*, 123, 185–195, 2014.)

[60] described a theoretical approach of performing transient energy simulations, which effects of the control strategy on both the cooling demands of a building and the need of electricity for artificial lighting. The correlation between the strategy of shading control, indoor temperature, and artificial lighting needs (that causes, beyond a demand of electricity, also a thermal gain inside the building) has been discussed, with reference to a real German case study, by experiments and simulations also by Krone et al. [61]. A very innovative system is the one presented, recently, by Shen and Xianting Li [62], which investigated the performances of cooling pipes embedded in venetian blinds, in order to cool the windows' shading devices.

1.6.3 Insulation of Opaque Components

Conduction losses through the wall and windows are mitigated in both cases by lowering the U value of the envelope. However, a different treatment is required for radiative losses or gains, as in cold climates they are desirable to reduce heating load, whereas in hot climates they increase the cooling load. In both cases, it is important to first optimize the envelope to minimize energy consumption and then address the active heating or cooling systems, as these can consequently be sized for a smaller load. There are two primary mechanisms that contribute to heat transmission through opaque building enclosures: first, the conduction through the walls due to the temperature difference from inside to outside ambient temperature; second, additional conduction through the walls due to radiative solar heating of the outside surface of the walls. The mitigation approaches for both mechanisms differ, as the first requires increased insulation, and the second implies decreasing the amount of solar energy absorbed by the outer surface of the wall. The insulation levels become an important parameter in energy conservation in residential as well as commercial buildings. The insulation level significantly controls the thermal energy exchange through opaque components of the buildings and eventually influences the cooling and heating energy consumptions. The heat loss in buildings can represent up to 30% of the total building's cooling load [63,64]. Therefore, improving the insulation becomes a critical driver to attain higher energy efficiency. Radhi [64] conducted a study to assess the impact of using AAC blocks on the energy performance of UAE residential buildings. In this study, the author compares 5 different insulation systems to ascertain the effectiveness of AAC blocks as a building/wall insulation material for residential applications. Energetically, 200 mm AAC block is superior to similar wall structures composed of concrete block and red clay brick and only slightly worse (0.05%) than 200 mm sand cement block with 25–50 mm EPS insulation. However, financially it only becomes viable in Sharjah, where the electricity costs are 10 times those of Abu Dhabi (0.013$/kWh in Abu Dhabi vs. 0.13$/kWh in Sharjah, as reported by Radhi [64]). For example, decrease in U-Value of the walls from 2.32 to 0.3 $W/m^2\,K$ by adding 35 mm of polystyrene, and similarly for the roof from 0.6 to 0.2 $W/m^2\,K$ yielded in a 19.3% reduction of cooling energy in Al Ain by Radhi [65].

1.6.4 Green Roof Concept

A green roof is a roof that is partially or completely covered with vegetation and growing media planted over a waterproof membrane. These roofs can filter pollutants out of rainwater, reduce heating and cooling costs, and increase roof lifespan; however, these benefits do come at an increased cost compared to more conventional roofing systems, such as asphalt shingles. Green roofs can greatly reduce stormwater runoff from the roof [66], and adding a cistern can further reduce the runoff discharged from a property into the environment. Green roofs have been studied for their positive effect on saving building energy costs. A green roof offers a building and its surrounding environment many benefits. These

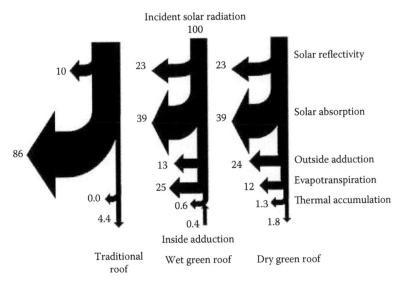

FIGURE 1.16
Comparison of the energy exchanges of the dry or wet green roof with a traditional roof, summer season. (From Lazzarin, R.A. et al., *Energ. Build.*, 37, 1260–1267, 2005.)

include stormwater management [67,68], improved water runoff quality [69], improved urban air quality [70], extension of roof life [71], and a reduction of the urban heat island effect [66]. Other benefits also include enhanced architectural interest and biodiversity [72]. Figure 1.16 enumerates the various phenomena involved in the energy balance of the solar radiation received by a dry green roof, a wet green roof, and a traditional roof. Although wet soil green roofs are disadvantageous as they are poor thermal insulators, they are advantageous in hot and dry climates where evapotranspiration is high. The wet green roofs have almost double the amount of evapotranspiration compared to dry green roofs, making them actually remove heat from the building and acting as a passive cooler [73].

The green roof can help cool a building during the summer due to the evapotranspiration effect from plants and the evaporation of moisture in the soil. During the winter months, the greater insulation property of the green roof prevents the heat from escaping. The insulation properties of the green roof have been extensively documented for commercial buildings [74–76]. There are two main classifications of green roofs: extensive and intensive. Extensive green roofs have a thin substrate layer with low level planting, typically sedum or lawn and can be very lightweight in structure. Intensive green roofs have a deeper substrate layer to allow deeper rooting plants such as shrubs and trees to survive.

Sedum is a common and very suitable plant for using on an extensive green roof. Sedums are succulents; they store water in their leaves, leaving them highly drought resistant. They are small plants that grow across the ground rather than upwards, offering good coverage and roof membrane protection. Sedum is installed in mats, which can be simply rolled onto a roof after the waterproofing and drainage layers have been installed. This means they require minimum maintenance and are easy to install as part of a roof system. Herman (2003) [77] reports that 13.5 km² of green roofs exist in Germany, which equates to 14% of all flat roofs. Of these green roofs, 80% are extensive systems, offering the most cost effective solution over intensive types [78]. Extensive roofs are the preferred option for retrofitting onto existing buildings as the structural capacity of the roof will often not have to be increased [67,79]. Extensive green roofs are relatively maintenance free and will readily survive in European climates. Gaffin (2005) [80] suggested that green roofs cool as effectively as the brightest possible white roofs, with an equivalent albedo of 0.7–0.85, compared with the typical 0.1–0.2 of a bitumen/tar/gravel roof [81]. A reference roof of the same type (steel deck with thermal insulation above) without greening was used for a comparison. By measurement they found that the heat gain through the green roof was reduced by an average of 70%–90% in the summer and heat loss by 10%–30% in the winter (Lui and Minor (2005) [82]). Santamouris et al (2007) [83], for example, investigated the energy saving potential of green roofs on a nursery school in Greece. They found that the building cooling load was reduced by between 6% and 49%, and the building with the insulated roof had a higher percentage of comfortable internal temperatures. The work of Nichaou et al. (2001) [84] is the most relevant here (Table 1.4). They determined how a green roof could save energy in buildings with different degrees of existing insulation.

Wong et al (2003) [85] conducted field experiments on a rooftop in Singapore to record the temperature at various depths of green roofs with differing plants. They modeled an insulated and noninsulated (exposed) roof and estimated the effects of each planting type on each roof. The results are shown in Figure 1.17. It is interesting to see the improved savings that result from more intensive planting for the noninsulated roof.

TABLE 1.4

Energy Saving Potential of Green Roof on Low, Moderately and Heavily Insulated Buildings in Athens, Greece

Roof Construction	Annual Energy Saving (%)		
	Heating	Cooling	Total
Well insulated	8%–9%	0%	2%
Moderately insulated	13%	0%–4%	3%–7%
Non-insulated	45%–46%	22%–45%	31%–44%

Source: Niachou, A. et al., *Energ. Build.*, 33, 719–729, 2001.

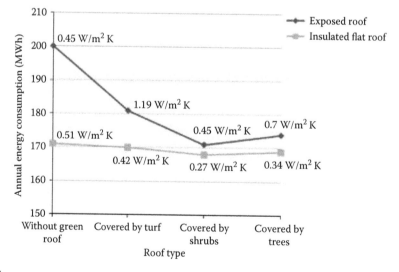

FIGURE 1.17

Comparison of annual energy consumption for different types of roofs on a commercial building. *U*-Values included as data labels. (Data from Wong, N.H. et al., *Energ. Build.*, 35, 353–364, 2003.)

The current rate of demolition and replacement of the UK's building stock is too low to combat climate change by improving the quality of new developments alone. To make an impact, the wide scale installation of green roofs will therefore have to take place in a way that has a significant impact; hence, the majority of green roofs will need to be retrofitted [86]. In a report compiled for Manchester City Council, Drivers Jonas Deloitte [87] state that although retrofitting of green roofs is technically feasible for most commercial or institutional buildings. For most commercial buildings, the additional loads associated with an extensive green roof (typically about 120–150 kg/m^2) do not require any additional strengthening. Wilkinson and Reed (2009) [88] analyzed the potential of retrofitting green roofs to existing buildings in the Central Business District of Melbourne, Australia. The concrete slab had an estimated capacity of 8–10 kN/m^2, enough to support a substrate depth up to 800 mm (Stovin et al. (2007) [68]). The green roof standards and valuation methods were independently investigated from engineering perspectives and aesthetic viewpoints [89–91]. The optimal area of green roof to balance cost, benefit, and risk remains unknown. To achieve such a complex systems analysis, a grey stochastic programming model was prepared to address the optimal design strategies under mixed uncertainties. The study identifies the optimal green roof area that keeps within the cost of a conventional home over a specific lifetime, such as 50 years [92]. Danielle Maia de Souza et al. [93] compared the life cycle impacts of ceramic and concrete roof coverage over 1 m^2, with an assumed lifetime of 20 years in Brazil. The results show that ceramic tiles appear to have less impact than concrete tiles on Climate Change, Resource Depletion and Water Withdrawal, while for the remaining damage categories, Human Health and Ecosystem Quality, the difference between the two alternatives was too low to be considered significant. The use of wood chips led to significant impacts, mainly related to respiratory inorganics. There are many other options, which could be applied to improve the energy performance of the building and bring them under EBCB compliance and green building certification. However, our study is limited to the above explained options as of their economic and technical viability in both new and existing building stocks.

1.7 Benefits and Impact on the Society

It is a well-known fact that energy is the backbone of the modern society and efficient use of different forms of energy, particularly electricity, is essential for sustainability. Therefore, despite several barriers and risks in development of EEB and green buildings, these concepts play a significant role in the overall sustainability of the society [94,95]. There are many research reports that affirm EEB and green buildings can outperform conventional buildings in many aspects [96–99]. Nevertheless, the high upfront cost of the building has been frequently cited as a hurdle to a widespread adoption of green buildings. Yet, there is no consensus about the cost of the green buildings and a significant gap has been observed in the quantified cost premium range. In more than 90% of reported cases cost premiums through empirical investigations was estimated between −0.4% and 21%. In two studies, the cost of the green buildings was estimated less than their conventional counterparts [100]. The benefits to building owners and investors have been widely cited. Lau et al. [101], stated that a low energy office building with green features such as solar design and utility-interactive Photovoltaic (PV) system can save up to 50% in energy cost compared to conventional buildings. Kats et al. [102] concluded that the operational cost of green buildings is lower and save energy on an average of 30%. They also indicated other benefits: reduced water consumption, maintenance cost, improved health and productivity, and financial benefits that are 10 times as high as the average cost premium which equals 1.84%. Madew [103] identified benefits of green building as 60% reduction in water and energy consumption, 1%–25% productivity increase, minimum 14% higher rate of return, 10% higher market value for asset, and 5%–10% higher rental rate. In comparison to the US national buildings average performance, GSA Public Buildings Service [104] estimated 26% less energy use, 13% lower aggregate maintenance cost, 27% higher occupants' satisfaction, and 33% fewer CO_2 emissions in commercial green buildings. Ries et al. [105] observed an increase of about 25% in productivity and about 30% lower energy consumption in a green precast concrete manufacturing facility certified by LEED rating system. Yudelson [106] listed a few benefits of green buildings including 30%–50% typical energy and water saving, reduced maintenance cost, increased property value, a typical 3%–5% improvement in productivity, 5% reduced absenteeism. It has also been further argued that a green building can annually reduce the sick building syndrome of its occupants by 41.5%.

1.8 Conclusion

A 85% growth has been forecasted in the building construction sector worldwide by 2030. The value of the growth would be equivalent to $15.5 trillion. Three countries (China, US, and India) would account for 57% of total global growth and growth in US construction market is expected little faster than China over the next 15 years. Moreover, Indian construction market probably will overtake Japan to become world's third largest construction market by 2021, and growth rate is expected almost twice of China by 2030. As a result final energy use of the global buildings sector may grow from approximately 120 EJ/yr in 2010 to 270 EJ/yr in 2050. In 2012, commercial energy use in OECD countries was 116% higher than non-OECD use and the value is expected to remain 50% greater in 2040 despite considerably larger and more rapidly growing populations in the non-OECD countries. The average annual change in the Indian commercial sector was projected 3.7%, highest in Asia and much higher than total non-OECD average (2.4%) between 2012 and 2040. Also, the direct and indirect CO_2 emissions from buildings have been projected to increase from 8.8 GtCO_2/yr in 2010 to 13–17 GtCO_2/yr by 2050 in baseline scenarios.

To reduce the energy consumption and GHG emission in the building sector, various policies, standards, and processes (including ECBCs and green building rating systems) have been developed and being implemented around the globe. If only currently planned policies are implemented, the final energy use in buildings that could be locked-in by 2050 is equivalent to approximately 80% of the final energy use of the buildings sector in 2005. Moreover, recent advances in technology, design practices, and know-how, coupled with behavioral changes, can help to achieve a two- to ten-fold reduction in energy requirements of individual new buildings and a two- to four-fold reduction for individual existing buildings cost-effectively or sometimes even at net negative costs. The ECBC prescribed the design standards for effective and significant reduction of energy usages in both new and existing buildings; however, these are implemented on a voluntary basis in most of the countries. Moreover, green building rating tools such as LEED and GRIHA not only prescribe technical detail to lower the energy demand, ensure indoor air quality and the well-being of occupants but also suggest integration of renewable energy technologies and implementation of efficient resource utilization during the entire building construction process and many more sustainable parameters. The effective implementation of the prescribed processes and techniques in the above standards, codes, and tools would certainly help ensure the sustainability in the building sector and society.

References

1. IPCC. *Climate Change 2014: Synthesis Report*. In: Meyer RKPaL (Ed.). IPCC, 2016.
2. Consulting E. *Global Energy Statistical Yearbook 2016*. In: Enerdata. France, 2016.
3. EIA. *International Energy Outlook 2016*. U.S. Department of Energy, Washington DC, 2016.
4. Pablo-Romero MD, Pozo-Barajas R, Yñiguez R. Global changes in residential energy consumption. *Energ. Policy* 2017;101:342–352.
5. Robinson G. Global construction 2030: A global forecast for the construction industry to 2030. https://www.ice.org.uk/ICEDevelopmentWebPortal/media/Documents/News/ICE%20News/ Global-Construction-press-release.pdf (Accessed: April 20, 2017).
6. Sangster W. Benchmark study on green buildings: Current policies and practices in leading green building nations. http://www3.cec.org/islandora-gb/en/islandora/object/greenbuilding% 3A143/datastream/OBJ-EN/view, (Accessed: April 20, 2017).
7. IEA. *Transition to Sustainable Buildings: Strategies and Opportunities to 2050*. Paris, France, 2013.
8. Lucon O, Ürge-Vorsatz D, Zain Ahmed A, Akbari H, Bertoldi P, Cabeza LF, Eyre N et al. Buildings. *Climate Change 2014: Mitigation of Climate Change Contribution of Working Group III to the Fifth Assessment Report of the Intergovernmental Panel on Climate Change*. R.K. Pachauri and L.A. Meyer (Ed.), New York: Cambridge University Press, 2014.
9. IPCC. Climate change 2014: Synthesis report. *Contribution of Working Groups I, II and III to the Fifth Assessment Report of the Intergovernmental Panel on Climate Change*. In: Core Writing Team RKPaLAMe (Ed.). Geneva, Switzerland, IPCC, 2014.
10. Güneralp B, Zhou Y, Ürge-Vorsatz D, Gupta M, Yu S, Patel PL, Fragkias M, Li X, and Seto KC. Global scenarios of urban density and its impacts on building energy use through 2050. *PNAS* 2017;114 (34):8945–8950.
11. Berardi U. Clarifying the new interpretations of the concept of sustainable building. *Sustain. Cities Soc.* 2013;8:72–78.
12. Son H, Kim C, Chong WK, Chou JS. Implementing sustainable development in the construction industry: Constructors' perspectives in the US and Korea. *Sustain. Dev.* 2011;19:337–347.
13. Kibwami N, Tutesigensi A. Enhancing sustainable construction in the building sector in Uganda. *Habitat Int.* 2016;57:64–73.
14. Wang L, Toppinen A, Juslin H. Use of wood in green building: A study of expert perspectives from the UK. *J. Clean Prod.* 2014;65:350–361.
15. Chao C. Smart green buildings of tomorrow. *Indoor Built Environ.* 2013;22:595–597.
16. USGBC. *Green Building and LEED Core Concepts Guide*, 2nd ed. Washington, D.C. USA, 2011.
17. United Nations. Resolution adopted by the General Assembly. 60/1, Agenda items 46 and 120. World Summit Outcome, New York City, 2005.
18. Mateus R, Bragança L. Sustainability assessment and rating of buildings: Developing the methodology SBToolPT–H. *Build Environ.* 2011;46:1962–1971.
19. Bragança L, Mateus R, Koukkari H. Building sustainability assessment. *Sustainability* 2010;2:14.
20. Costanza R, Patten BC. Defining and predicting sustainability. *Ecol. Econ.* 1995;15:193–196.
21. ECBC. Energy conservation building code. https://beeindia.gov.in/content/ecbc (Accessed: February 11, 2017).
22. https://energy.gov/eere/buildings/building-energy-codes-program (Accessed: March 20, 2017).
23. https://www.iccsafe.org/
24. GBPN. http://www.gbpn.org/databases-tools/purpose-policy-tool-new-buildings (Accessed: July 10, 2017).
25. Young R. Global approaches: A comparison of building energy codes in 15 countries. *ACEEE* 2014;3:351–366.
26. https://cleanenergysolutions.org/resources/energy-efficiency (Accessed: February 11, 2017).
27. UNDP IEA. Modernising building energy codes: To secure our global energy future. IEA, 2013, pp. 22–23.

28. Laustsen J. Energy efficiency requirements in building codes, energy efficiency policies for new buildings. *Int. Energ. Agen.* 2008, pp. 27.

29. Binh K, Nguyen HA. Comparative review of five sustainable rating systems. *Proc. Eng.* 2011;21:376–386.

30. Kubba S. Introduction: The green movement—Myths, history, and overview. *Handbook of Green Building Design and Construction.* Boston, MA: Butterworth-Heinemann, 2012, pp. 1–19.

31. Gou ZH, Xie XH. Evolving green building: Triple bottom line or regenerative design? *J. Clean. Prod.* 2017;153:600–607.

32. Chen PH, Nguyen TC. Integrating web map service and building information modeling for location and transportation analysis in green building certification process. *Automat Constr.* 2017;77:52–66.

33. Yudelson J. Springer link (Online service). *Sustainable Retail Development New Success Strategies.* Dordrecht, the Netherlands: Springer Netherlands, 2010, pp. xxiii, 212.

34. Cole RJ. Transitioning from green to regenerative design. *Build. Res. Inf.* 2012;40:39–53.

35. Ahn YH, Pearce AR, Wang Y, Wang G. Drivers and barriers of sustainable design and construction: The perception of green building experience. *Int. J. Sustain. Build. Technol. Urban Dev.* 2013;4:35–45.

36. Butera FM. Climatic change and the built environment. *Adv. Build. Ener. Res.* 2010;4:45–75.

37. Darko A, Zhang C, Chan APC. Drivers for green building: A review of empirical studies. *Habitat Int.* 2017;60:34–49.

38. Darko A, Chan APC, Owusu-Manu D-G, Ameyaw EE. Drivers for implementing green building technologies: An international survey of experts. *J. Clean. Prod.* 2017;145:386–394.

39. ARUP. *SPeAR® Handbook 2012.* 2012.

40. Svec P, Berkebile R, Todd JA. REGEN: Toward a tool for regenerative thinking. *Build. Res. Inf.* 2012;40:81–94.

41. Plaut JM, Dunbar B, Wackerman A, Hodgin S. Regenerative design: The LENSES Framework for buildings and communities. *Build. Res. Inf.* 2012;40:112–122.

42. Cole RJ, Busby P, Guenther R, Briney L, Blaviesciunaite A, Alencar T. A regenerative design framework: Setting new aspirations and initiating new discussions. *Build. Res. Inf.* 2012;40:95–111.

43. Taleb H, Musleh MA. Applying urban parametric design optimisation processes to a hot climate: Case study of the UAE. *Sustain. Cities Soc.* 2015;14:236–253.

44. Aboul-Naga M, Al-Sallal KA, El Diasty R. Impact of city urban patterns on building energy use: Al-Ain city as a case study for hot-arid climates. *Architect. Sci. Rev.* 2000;43:147–158.

45. Dall'O' G, Galante A, Pasetti G. A methodology for evaluating the potential energy savings of retrofitting residential building stocks. *Sustain. Cities Soc.* 2012;4:12–21.

46. Sivakumar P, Palanthandalam-Madapusi HJ, Dang TQ. Control of natural ventilation for aerodynamic high-rise buildings. *Build. Simul. China* 2010;3:311–325.

47. Isaac M, van Vuuren DP. Modeling global residential sector energy demand for heating and air conditioning in the context of climate change. *Energ. Policy.* 2009;37:507–521.

48. Nemry F, Andreas U, Makishi C, Wittstock B, Braune A, Wetzel C et al. Environmental improvement potentials of residential buildings (IMPRO-building). *JRC Scientific and Technical Research Series. Luxembourg Office for Official Publications of the European Communities,* Luxembourg, UK: JRC Scientific and Technical Research Series, 2008.

49. Salvalai G, Malighetti LE, Luchini L, Girola S. Analysis of different energy conservation strategies on existing school buildings in the a Pre-Alpine Region. *Energ. Build.* 2017;145:92–106.

50. Aditya L, Mahlia TMI, Rismanchi B, Ng HM, Hasan MH, Metselaar HSC, Muraza O, Aditya HB. A review on insulation materials for energy conservation in buildings. *Renew. Sustain. Energ. Rev.* 2017;73:1352–1365.

51. Ashrafian T, Yilmaz AZ, Corgnati SP, Moazzen N. Methodology to define cost-optimal level of architectural measures for energy efficient retrofits of existing detached residential buildings in Turkey. *Energ. Build.* 2016;120:58–77.

52. Balaras CA. The role of thermal mass on the cooling load of buildings. An overview of computational methods. *Energ. Build.* 1996;24:1–10.
53. Kolokotroni M, Webb BC, Hayes SD. Summer cooling with night ventilation for office buildings in moderate climates. *Energ. Build.* 1998;27:231–237.
54. Shaviv E, Yezioro A, Capeluto IG. Thermal mass and night ventilation as passive cooling design strategy. *Renew. Energ.* 2001;24:445–452.
55. Yang L, Li YG. Cooling load reduction by using thermal mass and night ventilation. *Energ. Build.* 2008;40:2052–2058.
56. Geros V, Santamouris M, Karatasou S, Tsangrassoulis A, Papanikolaou N. On the cooling potential of night ventilation techniques in the urban environment. *Energ. Build.* 2005;37:243–257.
57. Givoni B. Comfort, climate analysis and building design guidelines. *Energ. Build.* 1992;18:13.
58. Ramponi R, Angelotti A, Blocken B. Energy saving potential of night ventilation: Sensitivity to pressure coefficients for different European climates. *Appl. Energ.* 2014;123:185–195.
59. Bellia L, Marino, C, Minichiello, F, Pedace, A. An overview on solar shading systems for buildings. *Energ. Proc.* 2014;62:9.
60. Bellia L, De Falco F, Minichiello F. Effects of solar shading devices on energy requirements of standalone office buildings for Italian climates. *Appl. Therm. Eng.* 2013;54:190–201.
61. Krone U, Ascione, F., Bianco, N., Tschirner, T., Böttcher, O. Prescriptive-and performance-based approaches of the present and previous German DIN 4108–2. Hourly energy simulation for comparing the effectiveness of the methods. *Energ. Proc.* 2015;75:10.
62. Shen C, Li XT. Solar heat gain reduction of double glazing window with cooling pipes embedded in venetian blinds by utilizing natural cooling. *Energ. Build.* 2016;112:173–183.
63. Al-Sallal KA, Al-Rais L, Bin Dalmouk M. Designing a sustainable house in the desert of Abu Dhabi. *Renew. Energ.* 2013;49:80–84.
64. Radhi H. Viability of autoclaved aerated concrete walls for the residential sector in the United Arab Emirates. *Energ. Build.* 2011;43:2086–2092.
65. Radhi H. Evaluating the potential impact of global warming on the UAE residential buildings—A contribution to reduce the CO_2 emissions. *Build. Environ.* 2009;44:2451–2462.
66. Banting HD, Li J, Missios P. *Report on the Environmental Benefits and Costs of Green Roof Technology for the City of Toronto.* Toronto, Canada: Deptartment of Architectural Science, Ryerson University, 2005.
67. Mentens J, Raes D, Hermy M. Green roofs as a tool for solving the rainwater runoff problem in the urbanized 21st century? *Landsc. Urban Plan.* 2006;77:217–226.
68. Stovin V, Dunnett N, Hallam A. Green roofs—Getting sustainable drainage off the ground. *6th International Conference of Sustainable Techniques and Strategies in Urban Water Mangement (Novatech 2007)*, Lyon, France, 2007, pp. 11–18.
69. Berndtsson JC, Bengtsson L, Jinno K. Runoff water quality from intensive and extensive vegetated roofs. *Ecol. Eng.* 2009;35:369–380.
70. Yang J, Yu Q, Gong P. Quantifying air pollution removal by green roofs in Chicago. *Atmos. Environ.* 2008;42:7266–7273.
71. Teemusk A, Mander U. Greenroof potential to reduce temperature fluctuations of a roof membrane: A case study from Estonia. *Build. Environ.* 2009;44:643–650.
72. Koehler M. Plant survival research and biodiversity: Lessons from Europe. *Greening Rooftops for Sustainable Communities*, Chicago, IL. Toronto, ON: The Cardinal Group, 2003, pp. 313–322.
73. Lazzarin RA, Castellotti F, Busato F. Experimental measurements and numerical modelling of a green roof. *Energ. Build.* 2005;37:1260–1267.
74. Del Barrio EP. Analysis of the green roofs cooling potential in buildings. *Energ. Build.* 1998;27:179–193.
75. Liu K, Baskaran B. Thermal performance of green roofs through field evaluation. *The First North American Green Roof Infrastructure Conference.* Chicago, IL: Awards and Trade Show, 2003, pp. 1–10.

76. Onmura S, Matsumoto M, Hokoi S. Study on evaporative cooling effect of roof lawn gardens. *Energ. Build.* 2001;33:653–666.
77. Herman R. *Green Roofs in Germany: Yesterday, Today and Tomorrow.* Chigago, IL: Greening Rooftops for Sustainable Communities, 2003, pp. 41–45.
78. Harzmann U. German green roofs. *Annual Green Roof Construction Conference,* Chicago, IL, 2002.
79. Dunnett N, Kingsbury N. *Planting Green Roofs and Living Walls.* Portland, OR: Oregon Timber Press, 2004.
80. Gaffin S. *Energy Balance Modelling Applied to a Comparison of White and Green Roof Cooling Efficiency.* Washington, DC: Greening Rooftops for Sustainable Communities, 2005.
81. *Energy Efficiency Factsheet—Reflective Roof Coatings.* Washington State University Cooperative Extension Energy Program, Olympia WA, 1993.
82. Lui K, Minor J. *Performance Evaluation of an Extensive Green Roof.* Washington, DC: Greening Rooftops for Sustainable Communities, 2005.
83. Santamouris M, Pavlou C, Doukas P, Mihalakakou G, Synnefa A, Hatzibiros A et al. Investigating and analysing the energy and environmental performance of an experimental green roof system installed in a nursery school building in Athens, Greece. *Energy* 2007;32:1781–1788.
84. Niachou A, Papakonstantinou K, Santamouris M, Tsangrassoulis A, Mihalakakou G. Analysis of the green roof thermal properties and investigation of its energy performance. *Energ. Build.* 2001;33:719–729.
85. Wong NH, Cheong DKW, Yan H, Soh J, Ong CL, Sia A. The effects of rooftop garden on energy consumption of a commercial building in Singapore. *Energ. Build.* 2003;35:353–364.
86. Grant DJDG. *Greater Manchester Green Roof Programme—Feasibility Study,* EDAW, UK, 2009.
87. Grant DJDG. *Greater Manchester Green Roof Programme—Guidance Document,* AECOM, UK, 2009.
88. Wilkinson SJ, Reed R. Green roof retrofit potential in the central business district. *Prop. Manag.* 2009;27:284–301.
89. Clark C, Adriaens P, Talbot FB. Green roof valuation: A probabilistic economic analysis of environmental benefits. *Environ. Sci. Technol.* 2008;42:2155–2161.
90. Williams NSG, Rayner JP, Raynor KJ. Green roofs for a wide brownland: Opportunities and barriers for roof top greening in Australia. *Urban Fore. Urban Green.* 2010;9:245–251.
91. Saiz S, Kennedy C, Bass B, Pressnail K. Comparative life cycle assessment of standard and green roofs. *Environ. Sci. Technol.* 2006;40:4312–4316.
92. Chang NB, Rivera BJ, Wanielista MP. Optimal design for water conservation and energy savings using green roofs in a green building under mixed uncertainties. *J. Clean. Prod.* 2011;19:1180–1188.
93. de Souza DM, Lafontaine M, Charron-Doucet F, Bengoa X, Chappert B, Duarte F et al. Comparative life cycle assessment of ceramic versus concrete roof tiles in the Brazilian context. *J. Clean. Prod.* 2015;89:165–173.
94. Hwang BG, Shah M, Binte Supa'at NN. Green commercial building projects in Singapore: Critical risk factors and mitigation measures. *Sustain. Cities Soc.* 2017;30:237–247.
95. Darko A, Chan AP, Ameyaw EE, He BJ, Olanipekun AO. Examining issues influencing green building technologies adoption: The United States green building experts' perspectives. *Energ. Build.* 2017;144:320–332.
96. Lim GH, Hirning MB, Keumala N, Ghafar NA. Daylight performance and users' visual appraisal for green building offices in Malaysia. *Energ. Build.* 2017;141:175–185.
97. Darko A, Zhang C, Chan AP. Drivers for green building: A review of empirical studies. *Habitat Int.* 2017;60:34–49.
98. Darko A, Chan APC, Owusu-Manu D, Ameyaw EE. Drivers for implementing green building technologies: An international survey of experts. *J. Clean. Prod.* 2017;145:386–394.
99. MacNaughton P, Satish U, Laurent JGC, Flanigan S, Vallarino J, Coull B et al. The impact of working in a green certified building on cognitive function and health. *Build. Environ.* 2017;114:178–186.
100. Dwaikat LN, Ali KN. Green buildings cost premium: A review of empirical evidence. *Energ. Build.* 2016;110:396–403.

101. Lau LC, Tan KT, Lee KT, Mohamed AR. A comparative study on the energy policies in Japan and Malaysia in fulfilling their nations' obligations towards the Kyoto *Protocol*. *Energ. Policy* 2009;37:4771–4778.

102. Kats G, Alevantis L, Berman A, Mills E, Perlman J. The costs and financial benefits of green buildings. A report to California's sustainable building task force. http://www.usgbc.org/Docs/News/News477.pdf, 2003. Accessed: March 18, 2017.

103. Madew R. The dollars and sense of green buildings, A report for the Green Building Council of Australia. http://www.gbca.org.au/resources/dollars-and-sense-of-green-buildings-2006-building-the-business-case-for-green-c/1002.htm, 2006. Accessed: March 18, 2017.

104. Service GPB. Assessing building performance: A post occupancy evaluation of 12 GSA buildings research. http://www.gsa.gov/graphics/pbs/GSA_AssessGreen_white_paper.pdf, 2008. Accessed: March 18, 2017.

105. Ries R, Bilec MM, Needy KL, Gokhan NM. The economic benefits of green buildings: A comprehensive case study. *Eng. Econ.* 2006;51:259–295.

106. Yudelson J. *The Green Building Revolution*. Washington, DC: Island Press, 2008.

2

An Introduction of Zero-Energy Lab as a Testbed: Concept, Features, and Application

Tingzhen Ming and Yong Tao

CONTENTS

2.1   Introduction..33
2.2   The Basic Information and Concept of the ZØE Lab 36
    2.2.1   Basic Information...36
    2.2.2   The Concept of the ZØE Lab..36
2.3   Important Features of the ZØE Lab 38
    2.3.1   Solar Features 38
    2.3.2   Geothermal Energy Utilization 41
    2.3.3   Wind Energy Utilization...43
    2.3.4   Rainwater Harvesting and Filtration System 44
    2.3.5   Indoor Environment and Thermal Comfort System 46
    2.3.6   Building Energy Monitoring and Control System....................50
2.4   Experiments of ZØE Lab Performances...52
    2.4.1   Overall Energy Consumption and Generation 52
    2.4.2   Energy Allocation 52
2.5   Conclusion 55
References...55

2.1 Introduction

Commercial and residential buildings are central in the aspects of the nation's energy savings, reduction of greenhouse gases (GHGs) and pollution emissions, and environment protection, as they consume almost 40% of the primary energy and approximately 70% of the electricity in the United States [1]. Obviously, the energy consumption by the building sector is continuously increasing due to the fact that more and more buildings are being constructed, more quickly than ever, all which brings up energy use. Existing data indicate that electricity consumption in the commercial building sector doubled between 1980 and 2000, and it is expected to increase another 50% by 2025 [2]. The trend of drastically increasing energy consumption in commercial and residential buildings will not stop in the long run unless there is a worldwide awakening of the people and governments, effective technological improvements, and preferential policies for the encouragement of renewable energy applications.

Technically, a direct and effective answer to save the buildings' energy is to construct zero-energy buildings (ZEB) or net-zero energy buildings (NZEB) [3]. A ZEB refers to a building with a net energy consumption of zero over a typical period of time. The energy consumption, that is, consumption of heat, electricity, and fuel, is greatly reduced in a ZEB, and the utilization of renewable energy technologies, such as solar photovoltaic (PV) panel, solar heat collector, wind turbine, and ground source heat pump, can compensate for this reduced energy demand. The negative environmental impact of current building practices can be overcome by low energy houses. ZEB can be the pinnacle of saving energy and the best standard for sustainable houses in the near future. Here comes the appealing concept of ZEB that is envisioned as the technology of tomorrow toward a green and sustainable future. Enormous efforts are being made to make the ZEB more practical and cost effective to make it a mainstream energy solution for future generations. These efforts are integrating various renewable energies [4–8], using insulated materials to avoid excessive energy dissipation [9–11], designing advanced HVAC systems [12–16], adopting various energy-saving facilities and apparatuses [17–20], and studying the relationship between occupant schedules and energy consumption in a building [21–26]. A detailed review on the development of two design strategies, that is, energy-efficient measures to minimize the need for energy use in buildings (especially for heating and cooling) and renewable energy technologies to meet the remaining energy needs, has been presented by Li et al. [27].

Recently, the research on ZEBs/NZEBs has aroused worldwide attention. Iqbal [28] presented a feasibility study of a wind energy conversion system based on a zero-energy home (ZEH) in Newfoundland. This study was based on annual recorded wind speed data, solar data, and power-consumed data in a typical R-2000 standard house in Newfoundland. National Renewable Energy Laboratory's software HOMER was used to select an optimal energy system. Da Graca et al. [29] explored the feasibility of solar NZEB systems for a typical single-family home in the mild southern European climate zone. Using a dynamic thermal simulation of two representative detached house geometries, solar collector systems were sized in order to meet all annual energy needs. The impact of building envelope, occupant behaviors, and domestic appliance efficiency on the final energy demand, and the solar NZEB system size was analyzed. After sizing a set of solar thermal (ST) and PV solar systems, an analysis was performed to identify the best system configuration from a financial and environmental perspective. The introduction in the analysis of a micro-generation government incentive scheme shows great potential for financially attractive NZEB homes in this climate zone. Nielsen and Moller [30], Marszal et al. [6], and Marszal and Heiselberg [31] presented a very detailed analysis on the application of NZEBs in Denmark from the points of heat reduction, on-site or off-site renewable energy supply options, and life cycle cost analysis. Zeiler and Boxem [32] reported a ZEB school which was built in the Netherlands and presented a detailed analysis on the pros and cons of the NZEB buildings. The authors advanced some suggested ways to find different solutions which increases the advantages of ZEB while at the same time improve some of the ZEB disadvantages. Silva et al. [33] presented a new prefabricated retrofit module solution for the facades of existing buildings and also the steps taken to optimize its performance, which included a judicious choice of materials, 3D modeling, cost-benefit analysis, and use of different simulation tools for performance optimization and prototyping. What is also shown is the implementation of the retrofit module within an integrated retrofit approach, whose final goal was to obtain a building with the minimum possible energy consumption and GHG emissions. The research above indicates that it is feasible to build ZEBs/ NZEBs toward significant energy savings and emission reduction.

Madeja and Moujaes [34] numerically investigated the energy consumption to compensate the heating and cooling loads of two residential homes and compared them with their experimental values. In their research, one home was classified as a ZEH and employed advanced construction features, which was designed to consume significantly less energy than a normal home. The baseline home was of the exact same dimensions and floor plan as the ZEH, but used more traditional construction practices. The results showed that the ZEH reduced the cooling load by about 76% and the heating load by about 12% experimentally. To investigate the energy performance of net-zero energy homes (NZEHs) and nearly net-zero energy homes (NNZEHs) in New England, Thomas and Duffy [35] gathered construction and occupational statistics on 20 homes, measured 12 months of energy consumption, cost, and production data, developed custom models to predict consumption and production, and compared measured performance to modeled predictions. They found that even in cold New England, these types of homes, using very diverse systems and designs, can meet or exceed their designed-for energy performance, though their actual performance varies widely. Bucking et al. [36] presented a theoretical analysis to estimate and reduce building energy consumption at the design stage using optimization algorithms coupled with the powerful tool of Building Performance Simulation (BPS). However, the application of optimization approaches to building design is not common practice due to time and computation requirements. Then, they proposed a hybrid evolutionary algorithm which used information gained during previous simulations to expedite and improve algorithm convergence using targeted deterministic searches. Wang et al. [10] presented a case study of zero energy (ZØE) house design in UK. They used EnergyPlus to facade design considering building materials, window sizes and orientations and TRNSYS to investigate the feasibility of ZEHs with renewable electricity, solar hot water system, and energy-efficient heating systems under Cardiff weather conditions. Scognamiglio and Rostvik [37] presented a very detailed analysis on the feasibility of combining solar PV and NZEB. They drew the conclusion that in a ZEB scenario, PV was very suitable for generating energy on-site and at-site. The authors envisioned possible formal results, opportunities and challenges, for the use of PV in ZEBs, as well as new research issues for the future relationships between PV and ZEBs from the architecture and landscape design point of view.

Thygesen and Karlsson [8] presented an analysis of three different combinations of PV-system, ST system, and solar assisted heat pump systems with regards to economics and energy. They found that a PV system in combination with a heat pump is a superior alternative to a ST system in combination with a heat pump.

Zhu et al. [11] experimentally compared a ZEH with a baseline house in suburban Las Vegas. The data showed that a radiant barrier and a water-cooled air conditioner were major contributors to the energy savings, while an insulated floor slab and thermal mass walls were not effective for energy conservation during cooling periods. PV roof tiles produce enough green power to cover the use in the ZEH, and the solar water heater can reach a peak efficiency of 80%. Obviously, a good design of the configuration of ZEBs/NZEBs and appropriate utilization of renewable energies according to the local weather conditions and topography are beneficial to achieve energy reduction.

This study is to present the concept, features, and performance of an NZEB served as a research lab in Discovery Park, University of North Texas (UNT). This building is designed to get a net ZØE consumption on the annual basis, and is named ZØE Lab. Renewable energy technologies including wind turbine and solar PV panels are used to generate energy for the building; advanced building and air conditioning technologies like Structural Insulated Panel (SIP) panel, low-emission glasses, and radiant floor are utilized to save energy. Detailed information of the construction, configuration, and performance through the real-time operation data will be presented in the following sections.

2.2 The Basic Information and Concept of the ZØE Lab

2.2.1 Basic Information

The ZØE Lab, which was completed in April 2012, is a unique kind of building in North Texas. It was designed specifically to test and demonstrate various alternative energy generation technologies in order to achieve a net-zero energy consumption on an annual base. The net-zero energy philosophy is based on a combination of different renewable energy technologies in a building, such as solar, geothermal, and wind systems that lead to the production of enough energy to power a highly reduced energy demand building and in many cases even create excess energy to return to the power grid, and thus the net energy consumption over a period (mostly a year) becomes zero.

The ZØE Lab has a floor area of 111.5 m² (1200 ft²) lying in the east of the Discovery Park campus of UNT (Figure 2.1 [38]). The ZØE's footprint shape is rectangular, and the building faces south. It is composed of two main activity areas: a research area and a living area, and with some auxiliary areas: an aisle, a control room, an electricity room, and a restroom. The research area occupies approximately 60% of the total area of the building, and is where almost all activities will take place. Students can utilize this area to monitor various data such as the indoor air temperature, air speed, humidity, electricity consumption, and so on. The living area has some electrical appliances and furniture, such as a refrigerator, a microwave, a bed, a desk, and several chairs. Thermal comfort studies are done in the living area. There are some experimental devices in the living area through which students can conduct research to monitor human behavior and comfort level to minimize energy consumption without reducing the comfort level of the occupants.

2.2.2 The Concept of the ZØE Lab

As shown in Figure 2.2 [38], the ZØE Lab was designed for building energy efficiency research, where various experiments can be conducted concerning energy consumption, energy generation, heat transfer, energy storage, thermal comfort, physical properties of materials, and envelope design. The unique characteristics of the ZØE Lab are:

1. *Integration of various renewable energy technologies*: As mentioned above, energy prices will continuously increase with decreasing fossil energies supply unless new fossil energy sources are explored, and the annual worldwide energy supply can overwhelm the annual worldwide energy consumption. Significant improvement in renewable energy technologies are crucial in future houses which rely on renewable energies. The ZØE Lab combines various technologies by using solar energy, wind energy, and geothermal energy, which are beneficial to the development of renewable energies. As a result, the ZØE Lab becomes a good demonstration and exploration toward the future zero energy buildings with better performances owing to successful application of renewable energy technologies.

2. *Multi-functions*: The ZØE Lab includes a research area and a living area to explore two kinds of human behaviors as the energy allocations in these two areas are totally different due to different functions. From the living area we can analyze the energy consumptions of refrigerator, microwave, water heaters, lights, and so on. We can also know more information about the water consumption in the kitchen and the restroom. As for the research area, the energy allocations of

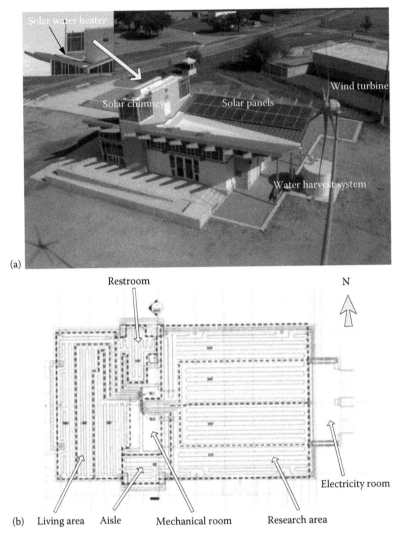

FIGURE 2.1
The ZØE in discovery park, UNT: (a) Bird's view; (b) plan including radiant floor zones.

personal computers, laptops, projector, meters, lights, radiant ground floor, and other experimental devices present a new figure for us to know more about the human behavior in a research lab.

3. *Flexibility*: A key concept of the ZØE Lab is flexibility which aims to perform various research experiments. Almost any part of the envelop is flexible to be replaced: the roof can be replaced by another roof with different materials and shapes, and the lateral walls can be replaced by different materials for further research on thermal isolation, energy storage, Retrofit, and/or facade design. The experimental devices inside the buildings can be reinstalled and moved in new places to analyze the energy consumption and temperature distributions concerning the occupant thermal comfort issue, and almost all the apparatus concerning various experiments, even the lamp bulbs, can be replaced by new products which represent the state-of-the-art technologies. Further, electricity generation systems

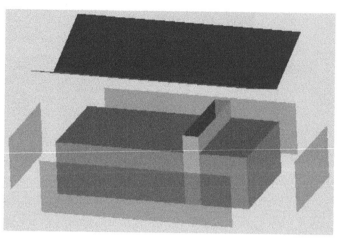

FIGURE 2.2
The concept of ZØE Lab.

from different companies can be replaced or new air conditioning system with higher efficiency can be added to achieve higher building performance. A good example is that the 5-blade wind turbine of the ZØE Lab has been replaced by a Be-Wind turbine for a new research on its output performance in June 2014.

2.3 Important Features of the ZØE Lab

2.3.1 Solar Features

1. *Solar PV panels*: As shown in Figure 2.1a, the roof of the ZØE, shaped as a "V", is divided into two parts by parallel walls that represent on the interior, the aisle, control room, and restroom, two being exterior walls that cross the V-shaped perpendicularly, dividing this roof into its east and west parts: east roof and west roof. At the north side of the east roof are installed 24 solar panels, with the total capacity being 5.6 kW. These solar panels mainly supply electric energy to the whole building, supporting all energy needed by the electricity-consuming facilities such as the light bulbs, the refrigerator, the meters and sensors, and the control and monitoring system. Considering the very strong solar radiation in the vicinity of noon time, the excessive electricity generated by the solar panels can be sent to the grid. Thereby, a connection of the ZØE Lab and the grid is an efficient way for energy consumption. The output powers may vary among different solar panels due to the afternoon shading effect of the wall on the panels' individual performance.

2. *Solar water heater*: A solar water heater was installed on the north part of the west roof (Figure 2.1a). This unique equipment is an evacuated tube collector solar water heater which provides the hot water for the ZOE Lab. The manifold of the solar water heater accommodates 20 KST0004 evacuated tubes. The gross area of the solar water heater is 2.84 m² (30.55 ft²) and the total fluid volume handled by

the solar water heater's manifold is 0.29 gal. More detailed information of the solar water heater can be seen in Table 2.1 [38]. The system works based on seven important components: solar collector, pump station, hot water storage tank, back-up heat source, hot water distribution system, solar controller, and expansion tank. The working principle of this system is based on solar energy. It so happens that, the evacuated tubes collect solar energy from the Sun and then this heat is transferred to propylene glycol mixture (PGM), a heat transfer liquid. Then, the PGM is pumped from the solar heater to the hot water storage tank where it transfers heat to the cold water. From the bottom of the tank, the cooled PGM is circulated back to the solar heater. During cloudy days, if the solar energy is not enough to supply the hot water requirement of the ZØE Lab, the back-up heat source is called into the loop. As ZØE is a south facing building, maximum solar energy utilization is possible and this orientation is useful for the solar heater.

3. *Low-e glass window*: The windows used in the ZØE Lab are Viracon VUE 11–50 low-emission (Low-e) windows. The glass used for the windows is high solar radiation reflecting. Basically, the glass used in these windows is insulated. The optical parameters of this kind of glass can be seen in Table 2.2 [38]. Here the VUE represents low-e coating applied to the windows' surface. Typically, this type of window is a double-pane separated by a spacer. In order to keep the interior of the window less moist, double sealants are used. The primary sealant is polyisobutylene (PIB) which prevents air from entering the airspace and absorbs moisture in the cavity, thus acting as a desiccant, and the secondary sealant used is Silicone,

TABLE 2.1

Basic Parameters of the Solar Water Heater

	English Units	**SI Units**
Aperture area	24.49 ft^2	2.27 m^2
Width of manifold	55.8″	1.42 m
Length of tube and manifold	78.9″	2 m
Depth	3.8″	0.096 m
Gross area	30.55 ft^2	2.84 m^2
Fluid volume	0.29 gallons	1.09 liters
Inlet/outlet dimension	0.75″	0.019 m
Weight while empty	110 lbs	49.89 kg
Recommended inclination	20°–70°	20°–70°
Rated flow rate	0.71 gpm	2556 m^3/hr
Maximum operating pressure	116 psi	8 bar
Vacuum	<10^{-8} bar	—
Temperature limitation	275°F	135°C

TABLE 2.2

Optical Properties of the Low-e Windows Used in the ZØE Lab

Absorptivity		**Transmissivity**	
Direct visible	0.71	Direct visible	0.29
Direct IR	0.81	Direct IR	0.115
Hemispherical dffuse	0.77	Hemispherical diffuse	0.096

being really resistant to UV radiations for a long time and, at the same time, it acts as a strong adhesive that holds the glass panes and the whole window unit together. Structural silicone has been found to be extremely resistant to UV light for longer duration and it can also withstand high temperatures and harsh atmospheric conditions.

4. *Reflected light color roof*: The roofs and walls of the ZØE Lab are built with steel faced Structural Insulated Panels (SIPs). Basically, the SIPs contain exterior and interior faces built of roll-formed steel and they are laminated with type-1 Expanded Polysterene. For the roofing purposes, 6" panels were used which have an R-value of 3.44 in winter and 3.84 in summer. These panels were made without using wood, which makes them very energy efficient because they prevent thermal bridging. These panels contain 99% insulation and 0% wood. According to Precision Foam Fabricators, which manufactured and installed these panels, the panels are capable of reducing building energy consumption by 50%. They also claim that these panels provide air-tight atmosphere at interiors so that energy loss is prevented to the exterior. The foam used in these panels spends a small amount of petroleum during manufacturing and there are no CFC gas emissions from the foam. These features make these panels very much energy efficient and environmentally safe. Also, due to these panels having air-tight construction, they have very high thermal performance. One more attribute of the panels is that no water is used during installation and construction, so it saves on water resources.

5. *Solar chimney for ventilation*: The solar chimney in the ZØE Lab is a south-facing chimney with dimensions of 4.12 m in height, 2.108 m in width, and 2.238 m in depth. The north wall, which is the absorber wall, runs 4.013 m in height and is composed entirely of gypsum board and is painted white for its full length. The other three walls (south, east, west) have two parts. The upper part has been made of Low-e glass windows for 1.87 m; the lower part is composed of three brick walls for 1.34 m. The inlet height of the chimney is 0.61 m and the outlet height is 0.305 m. The roof and outer walls of the solar chimney are made of steel-faced SIP panels. Outer and inner views of the chimney can be seen in Figure 2.3 [38]. A solar chimney is a technique providing necessary ventilation to a building using natural stack effect.

(a)   (b)

FIGURE 2.3
The solar chimney of ZØE Lab: (a) exterior view; (b) interior view from below.

The solar energy enters the glass windows and heats the wall which results in an increase of temperature, the air inside will be ultimately heated. Due to this increase in temperature, the density of air inside the chimney decreases and the air rises up and is vented out through the outlet of chimney and its place is taken by cooler air from the building by virtue of stack effect. This is known as natural ventilation which contributes significantly to reducing energy consumption in buildings.

2.3.2 Geothermal Energy Utilization

Building heating and cooling is provided by a primary and secondary systems which are both rejecting heat (in cooling mode) and absorbing heat (in heating mode) from a vertical bore hole geothermal ground loop. The primary system is a radiant floor heating and cooling system. The radiant hydronic water is heated or cooled by a water-to-water heat pump located in the first floor mechanical room. Secondary heating and cooling is through an overhead ducted air system served by a water-to-air heat pump located in the second floor mechanical room. The air system runs during occupied periods to ensure adequate ventilation to the occupants but the heating and cooling function runs as supplement of the radiant floor system when needed to achieve the setpoint temperatures above the dew point. Natural and mixed-mode ventilation is also available during mild weather through manual user changeover at the building automation system.

1. *Vertical ground loop heat exchanger*: Six boreholes (Figure 2.4 [38]) extend down 225 ft from the ground surface (finished grade). The bottommost portion of the U-tube assembly extends down 222 ft from the finished grade. The intended active

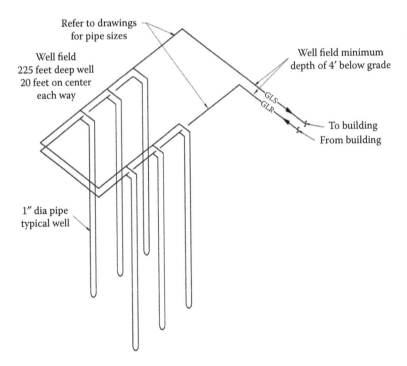

FIGURE 2.4
Vertical ground loop heat exchanger.

U-bend assembly length is 220 ft and allows a depth of 5 ft from the finished grade for back fill. The U-bend assembly extends 6 ft above the finished grade and is sealed and capped. The diameter of the boreholes and the U-bend are 3 and 1 in., respectively. High density polyethylene pipes (HOPE) are installed for the U-bend assembly. Pipes are manufactured and UV stabilizer in accordance with American Society for Testing and Materials (ASTM) standards. A thermally enhanced bentonite grout mixture is used to fill the annular space between the borehole wall and U-bend assembly. The grout has minimum solid content of 65%–70% and thermal conductivity of 1.73 W/(m.K). A pressure pump with a tremble pipe system is used to place the grouting materials.

2. *Water to water heat pump (WWHP)*: The WWHP is the primary source of the ZØE Lab, and it includes two types of water-source heat pumps: exposed vertical units (6 tons and smaller) and exposed floor-mounted water-to-water heat pumps. In heating mode, ground heat exchanger acts as a heat source and in cooling mode, it acts as a heat sink. The water-source heat pump is factory assembled and tested according to ASHRAE standards. The package includes a galvanized-steel cabinet, a chassis, a centrifugal-direct driven fan, a water circuit, refrigerant-to-air coils, a refrigerant circuit, a condenser water reheater, filters, and controller units. The heating and cooling capacities of this section are 8.3 MW and 2.33 Refrigeration tons, respectively. The radiant floor of ZØE building is schematically shown in Figure 2.1b. The radiant floor is composed of polyethylene tubes (total length: 573 m and inner diameter: 0.0127 m) with thermal conductivity of 0.381 W/(m.K). Maximum flow rate of water for the WWHP system is 0.00631 m³/s. Ground source WWHP and the connected radiant floor manifolds are shown in Figure 2.5 [38].

3. *Water to air heat pump (WAHP)*: Similar to the WWHP, a water to air heat pump (WAHP) is also coupled with vertical ground heat exchanger (Figure 2.6 [38]). The ground source WAHP of the ZØE Lab is shown in Figure 2.7 [38]. Cold water comes from the ground into the WAHP through the ground heat exchanger, gets hot by exchanging heat with the heat pump refrigerant, and returns to the ground. The refrigerant rejects heat into the underground cold water and is condensed. The condensed refrigerant moves through the expansion valve to the evaporator section and cools the indoor air by absorbing the heat and being evaporated.

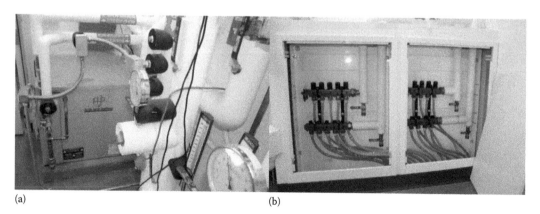

(a)   (b)

FIGURE 2.5
Ground source water-to-water heat pump: (a) heat pump; (b) radiant floor manifolds connected to WWHP.

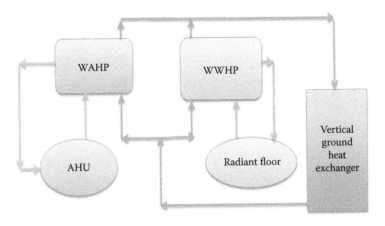

FIGURE 2.6
Overall HVAC system diagram for ZØE.

FIGURE 2.7
Ground source water to air heat pump of ZØE lab.

2.3.3 Wind Energy Utilization

According to NREL's wind report, the potential rank of wind energy in Texas is the top in the U.S. The annual average wind speed measured at 30 m high in Dallas-Fort Worth varies between 6.5 and 7 m/s. The wind profile measured in Denton shows similarity to that of Dallas-Fort Worth and serves as the reason for installing a wind turbine in ZØE Lab. Through monitoring the data of energy generation by the wind turbine, the performance of the wind turbine has been studied.

A horizontal axis type wind turbine, as shown in Figure 2.8 [38], is located a few meters south-east of the ZØE Lab. It has 3.5 kW rated power output with a 14.6 m hub height. It uses 5 downwind type blades which are made of fiberglass reinforced plastic. The swept area of the blades is 12.6 m². The rated wind speed of the turbine is 11 m/s. The cut-in

FIGURE 2.8
Wind turbine of the ZØE lab.

wind speed of the turbine is 3.2 m/s and the cut-out wind speed is 22 m/s. It is designed to have three operation modes: normal, high wind speed, and parked. The wind turbine has been operating in the normal mode with the wind speed being less than 15 m/s. When the wind speed exceeds 15 m/s, the operation mode will change to the high wind speed mode. In this mode, the brake is engaged to slow the turbine RPM down while still producing power in the 300–1000 W range. If the wind speed rises over 22 m/s, the system mode changes to parked mode. In parked mode, the wind turbine stops until the wind storm dies down.

2.3.4 Rainwater Harvesting and Filtration System

The rainwater harvesting and filtration system for the ZØE Lab is one unique feature that promises sustainability and possible off-the-grid applications pertaining to water usage (Figure 2.1a). With the inverted shape of the roof, the rainwater is directly channeled to the 3000-gallon rainwater storage tank by means of a slight slope of the roof. Figure 2.9 [38] shows the schematic diagram of the rainwater harvesting system of the ZØE Lab. At the edge of the roof is a funnel into which the rainwater falls (1). A 16 gal drip-off tank (2) collects the first 16 gal of rain that fall on the roof. This quantity of water is collected according to the size of the roof and is considered to be the dirtiest water which washes the dirt and debris from the roof. This dirty water is then dripped off to the exterior from a small nozzle at the bottom of the tank.

The remaining water associated with the rainfall is then fed into the storage tank using a calming or smoothing tube (3) which allows the water to enter from the bottom of the tank

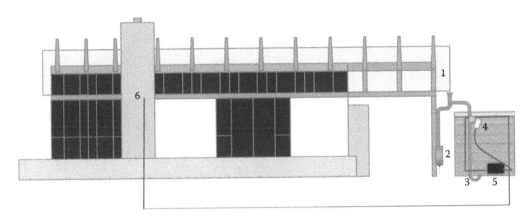

FIGURE 2.9
The schematic diagram of rainwater harvesting system of the ZØE lab: 1—roof funnel; 2—mesh filter; 3—pressure transducer; 4—main controller; 5—sediment filter; 6—carbon filters.

to prevent water agitation which can cause the settled dust and dirt to further contaminate the water. The rainwater is fed through a floating intake hose containing a 300-micron filter (4) which allows the cleaner water to be collected from the top of the tank. The rainwater is then supplied to the building by the means of a 20 gpm in-tank submersible pump (5) providing 60 psi pressure to the second floor of the building where the filtration system (6) is located.

The rainwater filtration system can be seen in Figure 2.10 [38]. The rainwater is pumped from the rainwater storage tank to the filtration system (1), and then it passes through a 100-micron mesh filter (2). Following the mesh filter is a pressure transducer (3) which is

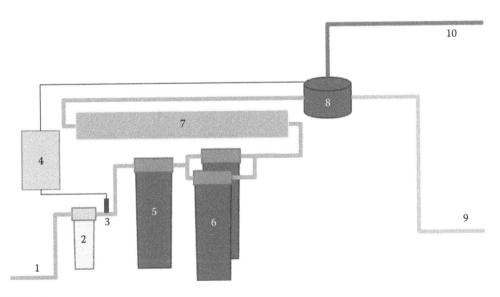

FIGURE 2.10
The schematic diagram of the rainwater filtration system: 1—roof funnel; 2—mesh filter; 3—pressure transducer; 4—main controller; 5—sediment filter; 6—carbon filters; 7—ultraviolet tube; 8—3-way valve; 9—utility water; 10—To building.

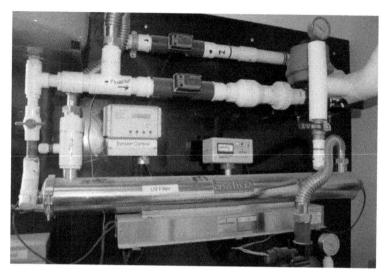

FIGURE 2.11
The water filtration system of the ZØE Lab.

the sensor that reports to the main controller (4) that controls the submerged pump in the tank. Next is a 5-micron sediment filter (5) followed by parallel-connected 5-micron carbon filters (6). Following the carbon filters, the rainwater passes through an ultraviolet tube (7) which kills any remaining bacteria in the water by exposure to an ultraviolet lamp. The water is then potable and is controlled by a 3-way valve (8) which is also controlled by the main controller (4). Local utility water (9) is also supplied to the 3-way valve (8) which can be programmed or user-controlled. The potable water is then supplied to the building (10). The control system of the water filtration system as shown in Figure 2.11 [38] is installed in the upstairs of the bedroom.

2.3.5 Indoor Environment and Thermal Comfort System

A method to distinguish human level of thermal comfort has been developed in the ZØE Lab by using a thermal camera, physiological sensors, and a surroundings sensor (Figure 2.12 [38]). The method has successfully collected data from hominal facial features, breathing rate, skin temperature, room temperature, blood volume pressure, relative humidity, and air velocity. Participants from all genders and races were involved in two sessions of a human thermal comfort experiment including a psychology survey session. The variables, such as room temperature and clothing are controlled to maintain steady test conditions. The region of interest was determined by body temperature and facial temperature as registered by the thermal imaging camera.

When adjusting a thermostat, a person is setting a goal for comfortable air temperature. Optimal thermal comfort requires the heat loss of the human body to be in balance with its heat production. The connection between a person's physical environment and the psychological state can be distinguished by specific levels of thermal comfort and the environment. To improve understanding about thermal comfort, in addition to ambient temperature all other affecting variables such as humidity level, physical activity, radiant temperature, air velocity, and clothing level should be considered.

FIGURE 2.12
Experimental setup of human thermal comfort in the ZØE Lab.

All activities are performed inside the living area of the ZØE Lab. The experiment is designed to detect changes in physiological signals from participants by inducing thermal comfort/discomfort conditions through low-to-medium-level physical activities. Our side goal is to detect a range of temperature body, heart rate pulse, and respiration changes in participants, which are associated with the comfort/discomfort conditions. The experiment consists of a thirty minutes session during which participants' physiological, verbal, and non-verbal responses were recorded using the following devices:

1. *Thermal camera*: A thermal camera (Figure 2.13 [38]) is used to detect sensible heat. It is a small and reliable infrared camera to be used in networked multi camera house installations. The set parameters of the thermal camera are described in Table 2.3 [38], and a case of experimental results can been seen in Figure 2.14 [38].

   The points of interest on this mapping region are nasal nostril (nose) area, face region, and periorbital region. Nostril is a significant area to compute breathing rate analysis. The thermal camera recognized the warm temperature in the breathing system which experiences significant temperature differences. Periorbital region is taken as an essential area. The periorbital featured a higher thermal signature than did the face region.

2. *Web camera and physiological sensors*: The camera, shown in Figure 2.15 [38], is a recording device which is provided to track the results within the participant body or face region in ongoing experiments. Wearable physiological sensors, including: volume pulse, skin conductance, temperature, and respiration sensors are shown in Figure 2.16 [38]. The system included four different sensors connected through a sensor box encoder from Thought Technology LTD.

FIGURE 2.13
Thermal camera.

TABLE 2.3

Parameters of the Thermal Camera

Emissivity	0.92	Camera model	Thermacam P/S/B
Reflective temperature	20.0°C	Camera serial No.	25100298
Distance	2 m	Lens	FOV 24
Atmospheric temperature	20	Filter	NOF
Ext. temperature	20	Frame rate	60 fps
Ext. transition	1	Company	Flir system
Relative humidity	50%	Series	Thermovision A40

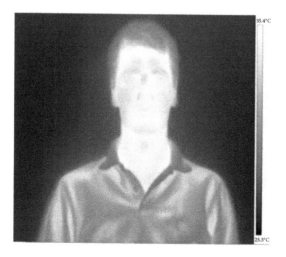

FIGURE 2.14
A thermal image from the camera.

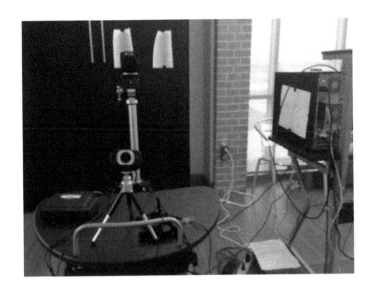

FIGURE 2.15
A web camera.

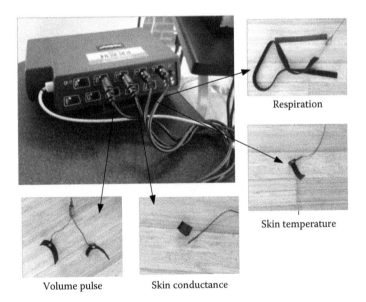

FIGURE 2.16
Physiological sensors.

3. *HOBO sensors*: HOBO sensors and wind meter were positioned in various positions of the living area, in order to record the basic indoor ambient parameters such as the relative humidity, lighting illuminate, ambient temperature, and wind speed.

The hominal thermal features appeared ubiquitously for any possible outcome. A summary of the clean data signal would be the first approach to determine

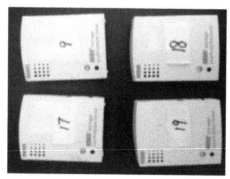 

HOBO sensors   Wind meter

FIGURE 2.17
HOBO sensors and wind meter on participant's space.

aspects of comfort. To overcome a challenge on the complexity of the data, each participant's signal is set into a group plan simultaneously. The boundary has been selected and uniform analysis has been performed (Figure 2.17 [38]). The objective is to achieve both physiological and psychological results in each variable from the data as well as from the correlation output.

2.3.6 Building Energy Monitoring and Control System

For monitoring and controlling the systems in the ZØE Lab, more than 100 sensors are installed in the building. Temperature, humidity, flow rate of air and water in HVAC systems, and irradiation from the sun have been measured since October 2012. The software used in ZØE is TAC VISTA which is a software to control, monitor, explore the data of a building, and issue warnings especially, the energy-related issues for the ZØE Lab. A snapshot of the software has been shown in Figure 2.18 [38]. The electricity consumption data are measured separately for plug-in equipment, lightings and HVAC systems as seen in Figure 2.19 [38]. The electricity consumption data for plug-in equipment are measured at Panel B, the lighting energy consumption is measured at Panel C, and energy consumption of HVAC system is measured at Panel C. From measuring Panel A, total energy consumption data in the ZØE Lab are obtained. The electricity generation data from the wind turbine and PV systems are collected separately. Therefore, it is possible to find a breakdown of energy consumption data and compare electricity consumption and generation data in real time.

This software is connected to two different serves. The whole capacity of TAC Vista server to collect data is 4 GB and it is the main server which carries all the work loads of the ZØE Lab. TAC Xenta server is the auxiliary server which helps reduce the work load of the permanent server. The TAC Xenta server has the logging ports in which all the logged data are kept at the first stage of logging procedure until it gets to the set up capacity. At that time the logged data in the TAC Xenta server will be uploaded to the TAC Vista server.

This program also controls the equipment in the ZØE Lab such as the WAHP, WWHP, energy recovery ventilator, radiant floor system, ventilation and exhaust fans, and all pumps. The control sequence of each equipment is programmable so the user can change the control logic used in the existing systems.

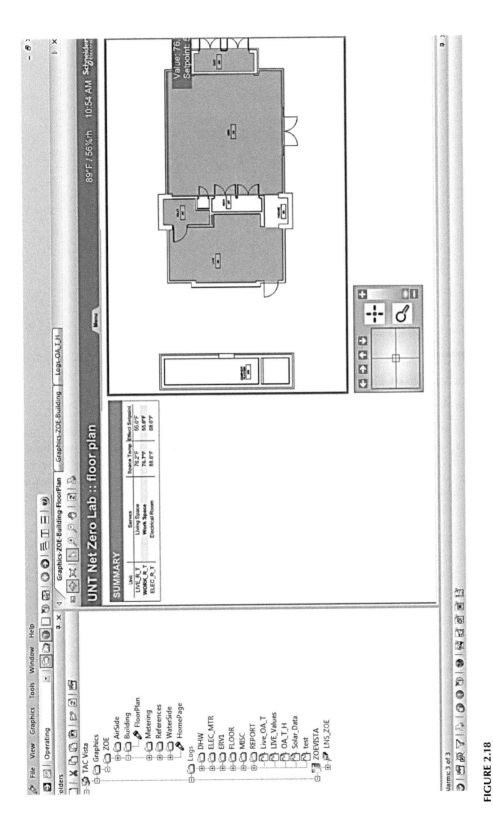

FIGURE 2.18
TAC Vista system to test the ZØE lab performance.

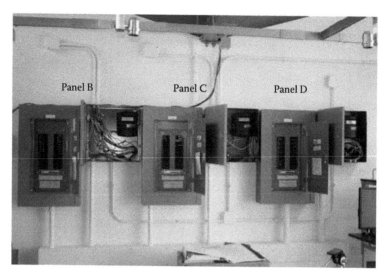

FIGURE 2.19
The monitoring and controlling the systems in the ZØE lab.

2.4 Experiments of ZØE Lab Performances

2.4.1 Overall Energy Consumption and Generation

As seen in the above section, to test the overall performance of the ZØE Lab, very detailed weather data such as the solar radiation and ambient air speed, temperature, and humidity are collected; temperatures underground with different depths, water flow rates through the WWHP and WAHP, and 16 points of temperatures at the ground floor surface have been monitored; and energy consumptions of each device and energy input from the grid have been recorded. The meters of panels B, C, and D are collecting the real-time energy consumption data of all the devices. The basic operation parameters of the HVAC systems, the ambient information, and the ground floor temperatures are shown and stored in Tac Vista.

Total energy consumption, energy from and to the grid, power generated by solar PV and wind turbine throughout one year (January 1, 2013 12:00 a.m. to January 1, 2014 12:00 a.m.) have been presented in Figure 2.20 [38]. It is found that the total energy used in the ZØE Lab is 26.9 MWh, Energy generated is 8.70 MWh, and the net energy bought from the grid is about 18.2 MWh. During this period, the wind turbine stopped at the beginning of June 2013 because of the breakdown of the wind turbine's speed control system. In addition, energy use varies with the season, where maximum energy consumption can be found in June 2013 as it is very hot during this time. However, in July and August 2013, frequent heavy rains decreased remarkably the energy consumption of the ZØE Lab. Energy generation varies slightly due to the insignificant solar radiation intensity variation along a year.

2.4.2 Energy Allocation

The energy consumption budget for different devices was done by recording the data of the panels of the ZØE Lab and also by retrieving the data from the computer program TAC

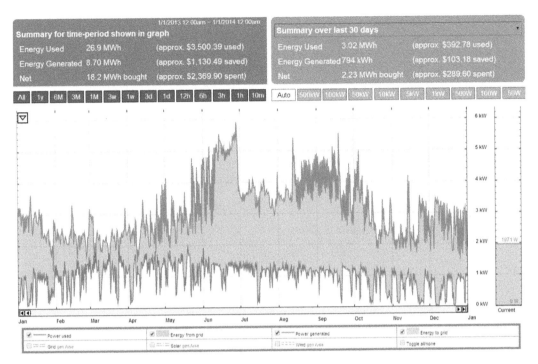

FIGURE 2.20
Energy consumption and generation of one year (January 1, 2013 12:00 a.m. to January 1, 2014 12:00 a.m.).

Vista. The panels are connected to different parts of the lab and the breakers inside the panels turn on certain components (inside lighting, receptacles, etc.). Charts were made in order to record the data from the panels. The charts had columns labels to the different panels and the rows were labels to specific times. At specific times the data were recorded from the panels. The times that were chosen were 7, 9, 11 a.m., and 4, 6, 8 p.m. Those times were chosen to give the most constant outside temperature reading. An Excel spreadsheet was completed in order to analyze the data recorded and see which components of the lab consumed the most energy.

The three main panels that were analyzed were panels B, C, and D. Panel B includes the receptacles of the lab. Panel C includes the lighting of the lab. Panel D includes the HVAC system. Detailed items of each panel are shown in Table 2.4 [38].

The results shown in Figure 2.21 [38] clearly indicate the energy consumption of various systems of the ZØE Lab. As expected, Panel D uses the majority of the consumed energy: well about 87% of the total energy consumed by the ZØE Lab. Out of the heat pumps, the WWHP consumes twice as much energy as the WAHP. It means the radiant floor uses more energy than the air system does. The ambient temperature in Denton, Texas, is very high through the cooling season of a year and the HVAC system is in operation all year round. Further, water flowing downward to 225 ft underground consumes a large part of the total energy. Thereby, future improvement can be achieved by using higher performances of WWHP and HVAC systems.

Panels B and C used roughly about the same amount of energy. For Panel B, the bathroom, refrigerator, and floor boxes consumed about 3%, 18%, and 10% of the energy, respectively. The receptacles expanded about 40%–50% of the energy from panel B. The receptacles and floor box energy would increase if electrical items are plugged into them. From this

TABLE 2.4

Detailed Items of the Panels

Panel B: Collect the Data from All Receptacles

1	Floor box in living area	2	Living area
3	Passage	4	Bathroom
5	Kitchen	6	Refrigerator
7	Work area	8	Electrical room
9	Floor box in work area	10	Work area
11	Mechanical room	12	Raceway
13	Communication port	14	Energy meter

Panel C: Collect the Data from All Lighting

1	Exterior	2	Busway
3	Living and work area	4	Energy meter

Panel D: Collect Data from Pumps and Else

1	GP-1	2	EF-2
3	GP-2	4	Solar control panel
5	WWHP	6	BP-1
7	Filter	8	BP-2
9	Motorized air louvers	10	WAHP

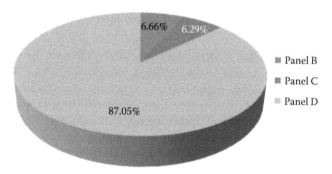

6.66%   6.29%

87.05%

■ Panel B
■ Panel C
■ Panel D

FIGURE 2.21
Energy allocation of the ZØE lab.

conclusion, it is seen that the other 20%–30% is used by the other components of panel B that were not analyzed. A way to reduce the energy consumed from panel B is to turn off electronics when not in use, especially the computers which are seldom shut down.

The results from panel C are not very conclusive except that it is evident that the living and work lighting uses the most energy. The busway and exterior lighting were inconsistent and gave values from 0.1% to 20%. A way to reduce the energy consumed from panel C is to add motion sensors to the living and work areas to turn off the light when it is not needed.

Detailed experimental results of heat pumps, solar PV, wind turbine, will be reported in next work. In addition, experiments of thermal comfort are also not included in this paper. Similarly, the results concerning the influences of facial features, breathing rate, skin temperature, room temperature, blood volume pressure, relative humidity, and air velocity on the human thermal comfort will be presented in future work.

2.5 Conclusion

Energy consumption in buildings accounts for a large part of the total human energy consumption, and new technologies in the utilization of renewable energies, advanced HVAC systems, and design of new and retrofit constructions are crucial in buildings' energy savings. In this work, the concept, unique features, and overall energy consumption of a ZØE Lab as a validation testbed have been presented which aim to demonstrate some basic characteristics of the future houses having lower energy consumptions.

1. The envelope of the ZØE Lab is flexible and each part of the walls and even the roof are dismountable, which is beneficial to the design of various experiments to achieve the best energy saving strategy. The ZØE Lab is multi-functional including the living area and research area, which is helpful to test the energy consumptions and thermal comfort performances of the building under different functions.

2. The ZØE Lab is built combining various renewable energy technologies using solar energy, wind energy, and geothermal energy to produce power and also reduce the energy usage from the grid. The net energy consumption of the ZØE Lab during a year (January 1, 2013 12:00 a.m. to January 1, 2014 12:00 a.m.) is 18.2 MWh; more solar panels, high performance wind turbines, and other solar energy storage systems are suggested to be used to supply more energy to the building.

3. The HVAC system accounts for about 87% of the total energy consumption of the ZØE Lab, adopting a new HVAC system with higher performance will greatly reduce the building energy consumption and then greatly improve the overall performance of the ZØE Lab.

References

1. Energy Efficiency and Renewable Energy, U.S. Department Of Energy, *2011 Buildings Energy Data Book*. D&R International, Ltd. http://buildingsdatabook.eren.doe.gov/docs/DataBooks/2011_BEDB.pdf (Accessed: March, 2012).
2. EIA, Annual Energy Review 2004. Washington, DC. www.eia.doe.gov/emeu/aer/contents.html (Accessed: August, 2005).
3. S. Pless, P. Torcellini, Net-zero energy buildings: A classification System based on renewable energy supply options, in Technical Report. NREL/TP-550-44586, June 2010, U.S. Department of Energy Office of Scientific and Technical Information, Oak Ridge, TN.
4. I. Visa, M.D. Moldovan, M. Comsit, A. Duta, Improving the renewable energy mix in a building toward the nearly zero energy status, *Energy and Buildings*, 68 (2014) 72–78.
5. J.P. Praene, M. David, F. Sinama, D. Morau, O. Marc, Renewable energy: Progressing towards a net zero energy island, the case of Reunion Island, *Renewable and Sustainable Energy Reviews*, 16 (2012) 426–442.
6. A.J. Marszal, P. Heiselberg, R.L. Jensen, J. Norgaard, On-site or off-site renewable energy supply options? Life cycle cost analysis of a net zero energy building in Denmark, *Renewable Energy*, 44 (2012) 154–165.
7. B.R. Lin, J.Y. Dong, New zero-voltage switching DC-DC converter for renewable energy conversion systems, *IET Power Electronics*, 5 (2012) 393–400.

8. R. Thygesen, B. Karlsson, Economic and energy analysis of three solar assisted heat pump systems in near zero energy buildings, *Energy and Buildings*, 66 (2013) 77–87.
9. L. Zhu, R. Hurt, D. Correia, R. Boehm, Detailed energy saving performance analyses on thermal mass walls demonstrated in a zero energy house, *Energy and Buildings*, 41 (2009) 303–310.
10. L.P. Wang, J. Gwilliam, P. Jones, Case study of zero energy house design in UK, *Energy and Buildings*, 41 (2009) 1215–1222.
11. L. Zhu, R. Hurt, D. Correa, R. Boehm, Comprehensive energy and economic analyses on a zero energy house versus a conventional house, *Energy*, 34 (2009) 1043–1053.
12. B.A. Sun, P.B. Luh, Q.S. Jia, Z. O'Neill, F.T. Song, building energy doctors: An SPC and kalman filter-based method for system-level fault detection in HVAC systems, *IEEE Transactions on Automation Science and Engineering*, 11 (2014) 215–229.
13. L.P. Wang, S. Greenberg, J. Fiegel, A. Rubalcava, S. Earni, X.F. Pang, R.X. Yin, S. Woodworth, J. Hernandez-Maldonado, Monitoring-based HVAC commissioning of an existing office building for energy efficiency, *Applied Energy*, 102 (2013) 1382–1390.
14. M. Maasoumy, A. Sangiovanni-Vincentelli, Total and peak energy consumption minimization of building HVAC systems using model predictive control, *IEEE Design & Test Of Computers*, 29 (2012) 26–35.
15. L. Perez-Lombard, J. Ortiz, J.F. Coronel, I.R. Maestre, A review of HVAC systems requirements in building energy regulations, *Energy and Buildings*, 43 (2011) 255–268.
16. I. Korolija, L. Marjanovic-Halburd, Y. Zhang, V.I. Hanby, Influence of building parameters and HVAC systems coupling on building energy performance, *Energy and Buildings*, 43 (2011) 1247–1253.
17. G.P. Moynihan, D. Triantafillu, Energy savings for a manufacturing facility using building simulation modeling: A case study, *EMJ-Engineering Management Journal*, 24 (2012) 73–84.
18. I. Kistelegdi, I. Haber, Building aerodynamic investigations for an energy-plus production facility in Sikonda (Hungary), *Bauphysik*, 34 (2012) 107–120.
19. J.A. Chica, I. Apraiz, P. Elguezabal, M.O. Rrips, V. Sanchez, B. Tellado, Kubik: Open building approach for the construction of an unique experimental facility aimed to improve energy efficiency in buildings, *Open House International*, 36 (2011) 63–72.
20. O.F. Osanyintola, C.J. Simonson, Moisture buffering capacity of hygroscopic building materials: Experimental facilities and energy impact, *Energy and Buildings*, 38 (2006) 1270–1282.
21. A. Roetzel, A. Tsangrassoulis, U. Dietrich, Impact of building design and occupancy on office comfort and energy performance in different climates, *Building and Environment*, 71 (2014) 165–175.
22. B. Lee, M. Trcka, J.L.M. Hensen, Building energy simulation and optimization: A case study of industrial halls with varying process loads and occupancy patterns, *Building Simulation*, 7 (2014) 229–236.
23. J. Widen, A. Molin, K. Ellegard, Models of domestic occupancy, activities and energy use based on time-use data: deterministic and stochastic approaches with application to various building-related simulations, *Journal Of Building Performance Simulation*, 5 (2012) 27–44.
24. C. Martani, D. Lee, P. Robinson, R. Britter, C. Ratti, ENERNET: Studying the dynamic relationship between building occupancy and energy consumption, *Energy and Buildings*, 47 (2012) 584–591.
25. O.G. Santin, L. Itard, H. Visscher, The effect of occupancy and building characteristics on energy use for space and water heating in Dutch residential stock, *Energy and Buildings*, 41 (2009) 1223–1232.
26. I. Richardson, M. Thomson, D. Infield, A high-resolution domestic building occupancy model for energy demand simulations, *Energy and Buildings*, 40 (2008) 1560–1566.
27. D.H.W. Li, L. Yang, J.C. Lam, Zero energy buildings and sustainable development implications—A review, *Energy*, 54 (2013) 1–10.
28. M.T. Iqbal, A feasibility study of a zero energy home in Newfoundland, *Renewable Energy*, 29 (2004) 277–289.

29. G.C. da Graca, A. Augusto, M.M. Lerer, Solar powered net zero energy houses for southern Europe: Feasibility study, *Solar Energy*, 86 (2012) 634–646.
30. S. Nielsen, B. Moller, Excess heat production of future net zero energy buildings within district heating areas in Denmark, *Energy*, 48 (2012) 23–31.
31. A.J. Marszal, P. Heiselberg, Life cycle cost analysis of a multi-storey residential net zero energy building in Denmark, *Energy*, 36 (2011) 5600–5609.
32. W. Zeiler, G. Boxem, Net-zero energy building schools, *Renewable Energy*, 49 (2013) 282–286.
33. P.C.P. Silva, M. Almeida, L. Braganca, V. Mesquita, Development of prefabricated retrofit module towards nearly zero energy buildings, *Energy and Buildings*, 56 (2013) 115–125.
34. R. Madeja, S. Moujaes, Comparison of simulation and experimental data of a zero energy home in an arid climate, *Journal of Energy Engineering-ASCE*, 134 (2008) 102–108.
35. W.D. Thomas, J.J. Duffy, Energy performance of net-zero and near net-zero energy homes in New England, *Energy and Buildings*, 67 (2013) 551–558.
36. S. Bucking, R. Zmeureanu, A. Athienitis, An information driven hybrid evolutionary algorithm for optimal design of a net zero energy house, *Solar Energy*, 96 (2013) 128–139.
37. A. Scognamiglio, H.N. Rostvik, Photovoltaics and zero energy buildings: A new opportunity and challenge for design, *Progress in Photovoltaics*, 21 (2013) 1319–1336.
38. T.Z. Ming, D. Chen, S.N. Toucieshki, S. Talele, G.T. Checketts, N. Hasib, C. Wicaksono et al., A zero energy lab as a validation testbed: Concept, features, and performance, *International Journal of Hydrogen Energy*, 40 (2015) 12854–12867.

3

Building Envelopes: A Passive Way to Achieve Energy Sustainability through Energy-Efficient Buildings

Manoj Kumar Srivastava

CONTENTS

3.1 Introduction 60
3.2 Energy Sustainability 60
3.3 Walls...61
    3.3.1 Passive Solar Walls..61
    3.3.2 Walls with Latent Heat Storage...62
    3.3.3 Lightweight Concrete..62
    3.3.4 Ventilated or Double Skin Walls...62
3.4 Windows and Doors (Fenestrations) 63
3.5 Types of Glazing Materials and Technologies 64
    3.5.1 Vacuum Glazing...64
    3.5.2 Switchable Glazing..64
    3.5.3 Suspended Particle Devices...64
    3.5.4 Electrochromic Devices...64
    3.5.5 Polymer Dispersed Liquid Crystal Devices 65
    3.5.6 Micro-Blinds 65
3.6 Frames..65
3.7 Roofs...66
    3.7.1 Ventilated Roofs...66
    3.7.2 Domed Roofs..66
    3.7.3 Reflective Roofs 66
    3.7.4 Green Roofs..67
    3.7.5 Photovoltaic Roof Envelopes 68
    3.7.6 Thermal Insulation of Roofs..68
    3.7.7 Evaporative Cooling Roofs...69
3.8 Thermal Mass 69
3.9 Building Simulation Software/Programs...69
3.10 Maintenance of Building Envelope 70
3.11 Conclusion...70
References..70

3.1 Introduction

In scientific and academic circles worldwide, the opportunity to develop the emerging discipline of sustainability science has never been greater. This new science has its origins in the concept of sustainable development proposed by the World Commission on Environment and Development (1987) (WCED), also known as the Brundtland Commission. Defining sustainable development as "development that meets the needs of the present without compromising the ability of future generations to meet their own needs," the WCED gained worldwide support for its argument that development must ensure the coexistence of economy and the environment. Today, "sustainability" is recognized the world over as a key issue facing twenty-first century society.

3.2 Energy Sustainability

In general, energy sustainability has become a global concern. There are many ways to address this global concern, but energy sustainability through energy-efficient buildings (EEBs) is presented in this study. In recent years, building designs with capability of mitigating energy consumption or "Green Buildings" have been initiated. The World Green Building Council (WGBC) has involved 80 nations across the globe in taking up green building initiatives to some degree. Different standardization mechanisms such as US-based Leadership in Energy and Environmental Design (LEED) and UK-based Building Research Establishment's Environmental Assessment Method (BREEAM) are developed to acquaint the builders and international society with how to get their building projects rated under these standards.

About 30%–40% of natural resources available in any industrialized country are exploited by the buildings. Almost half of the total energy consumed by a building is for cooling/heating the environment inside the buildings. Hence, the need for EEBs was deeply felt worldwide. There are both active and passive strategies to face this challenge. Active strategies include Heating, Ventilation, and Air-Conditioning (HVAC), and illumination, and passive strategies include concerns about building envelope. A building envelope is the surface that separates the indoor environment of a building from the outdoor environment. This includes walls, roofs, fenestrations, external shadings, thermal insulation, thermal mass, and so on. With specially designed building envelopes, EEBs can be constructed to attain energy sustainability. The buildings we expect today should have energy-efficient architecture along with environmentally friendly construction materials. The idea of sustainable buildings is comprised of conservation of water, energy, building material and land, along with appropriate care for environmental issues. Let us discuss the different types of building envelopes to improve the energy efficiency of a building. It has been found that about 31%–36% of energy savings can be ensured through specific designs of different building envelopes. An energy-efficient building envelope design saved as much as 35%

and 47% of total and peak cooling demands respectively [1]. During the preliminary stage of any building's design, an architect plans the shape of the building. As per the space available for the building and near surroundings, he plans the height and orientation of a building so that concept of photovoltaic (PV) building envelopes can also be realized. Building envelopes have a vital role to play in an EEB. Different building envelopes are discussed below.

3.3 Walls

Walls cover the largest area among different building envelopes. Hence, the designing of a wall envelope with appropriate energy-saving technology is of paramount importance and concern. Thermal and acoustic comforts within a building is also primarily due to the envelope's design. The thermal resistance (R-value) of a building envelope is crucial because it is one of the important parameters which controls the energy consumption of a building to a considerable extent. This is especially true in skyscrapers where wall to total envelope area is quite high. Walls with thermal insulation have a greater chance of condensation under high humidity conditions, which is aggravated in the winter months of cold climate regions. This moisture condensation on exterior walls damages the walls with undesirable microbial growth. This might reduce the life of a building. Walls are classified according to the material needed to construct them: there are wood-based walls, metal-based walls and masonry-based walls. However, modern wall designs with advanced technologies can improve the energy efficiency of a building considerably. Different advanced wall technologies are discussed below.

3.3.1 Passive Solar Walls

The walls that trap and transfer the solar energy efficiently to the building are called passive solar walls. These walls are made with a concrete thickness of 12 inches facade to absorb solar radiation, which is available in plenty toward the south direction in the northern hemisphere. A glazing is used for outer covering of the walls to provide a greenhouse effect.

There are four ways reported to attain energy efficiency inside buildings using passive solar walls. These are unventilated solar walls, Trombe walls, insulated Trombe walls, and composite solar walls. These walls (except unventilated solar walls) transfer heat inside the building. Heat transfer in these walls is both due to conduction and convection through circulating air. However, in the case of unventilated solar walls, heat transfers through conduction only. Choices of walls are made as per the climate of the area. For longer summer seasons, Trombe walls or unventilated solar walls are preferred; however, shorter heating seasons can be managed by composite solar walls or insulated Trombe wall designs. Overheating of the buildings during summer seasons can be regulated by solar shields [2]. Innovative designs of Trombe walls with integration of PV solar cells on Trombe walls can yield better results in EEBs.

Heat rejected by these solar cells and heat absorbed by the thermal mass will be utilized for indoor space heating using air circulated by a DC fan. This fan imposes forced convection and heats up the space. A new concept of Photovoltaic Trombe wall (PV–TW) for both space heating and cooling has also been reported [3]. It generates electricity along with heating of space with more aesthetic value. Though cooling of PV cells was not adequate, it was a double advantage design. Phase Change Material (PCM)-based Trombe walls have also been reported [4]. Experimental results suggest that PCM Trombe walls are better than concrete walls.

The trans wall is composed of water filled between two parallel glass panes supported by some rigid frame. A semi-transparent glass sheet is kept at the center of the panes, which will not only heat up the inside of a building but also illuminate it [5].

3.3.2 Walls with Latent Heat Storage

Impregnation of PCM enhances the thermal storage capacity of a wall. PCM can be impregnated in both gypsum and concrete walls. PCM impregnation by weight works efficiently. Impregnation of 30%–60% by weight is possible. Researchers found that PCM-based walls lowered the maximum temperature of the indoors by up to 4°C–4.2°C and reduced the heating demand at night [6,7]. Choice of PCM should be such that they have melting/freezing temperature in the practical range and must have a high latent heat of fusion and high thermal conductivity. PCM should also have desirable environmental properties to decrease environmental hazards.

3.3.3 Lightweight Concrete

Concrete walls were popular because of its cost effectiveness and less skilled labor requirement for its construction. These walls are thermal resistant, but their thermal resistance can further be improved by using lightweight aggregates such as pumice, diatomite expanded clay, and so on. In recent years, people have also reported low thermal conducting aggregates such as polystyrene, vermiculate, and so on [8]. Mixing aluminium powder in the base constitution material also produces LWC material by generating very small air bubbles. LWC walls are common in countries where thermally insulated walls are not popular.

3.3.4 Ventilated or Double Skin Walls

In these walls, air gap is maintained between two walls envelopes. The cavity thus created is ventilated with both artificial and forced ventilations. These walls also help in passive cooling of a building. It has been found that, energy savings increase with an increase in width of air gap up to an optimal value of 0.15 m, beyond which, it again starts diminishing.

3.4 Windows and Doors (Fenestrations)

Fenestrations mean an opening in a building envelope, which includes windows, doors, ventilators, and so on. Fenestration design not only gives the aesthetics of a building but also plays vital role in thermal comfort and illumination level of a building. Fenestration technology has been advanced significantly in recent years. These technologies include solar-controlled window panes, low emissive paints, PV glazing, aerogel-filled fenestrations, and so on.

Actually while designing the windows of a building, the following parameters should be taken into account:

1. Thermal conductivity
2. Solar heat gain coefficient (SHGC or g-value)
3. Orientation
4. Climatic conditions

Windows with low U-value and high solar energy transmittance are used to heat up the indoors with solar energy. Low emissive coatings on panes with selective transparency toward visible sun radiations, helps us to illuminate the inside of a building during the day [9]. Low emissive coatings are of two types: hard and soft coatings. Thermal emissivity is measured with reference to perfect black body, which has an emissivity of 1. Thermal emissivity of various surfaces is listed below.

Materials	Emissivity
Silver (polished)	0.02
Marble	0.56
Asphalt	0.88
Brick	0.90
Concrete (rough)	0.91
Glass (uncoated)	0.91

As given above, glass window panes have reasonably large e-value. To control emissivity, thin film coatings are applied on glass window panes. Fluorinated tin oxide (SnO_2:F) using pyroelectric Chemical Vapour Deposition (CVD) and silver coating with anti-reflection layer using Magnetron sputtering are broadly used to reduce emissivity of window panes. Silver based coatings are required to be protected by the insulated glazing. Hence, better control of thermal energy inside the building can be ensured with these coatings. Heat radiated from indoors during winter is trapped inside while heat during summer does not emit from exterior thus keeping inside cool. While glass with different emissivities can be made by different doping, they cannot be used in window panes because these glasses generally have higher near-infrared (NIR) transmission leading to excess heat loss or gain in a building through fenestrations. There are some drawbacks of low emissive panes also such as low e-value glass windows may block radio signals which leads to poor reception of signals of TV mobile, and so on.

3.5 Types of Glazing Materials and Technologies

Glazing of a building plays an important role in designing an EEB. With the advancements in the field of EEBs, various types of glazing are reported and used. Followings are the glazing used with state-of-the-art features.

3.5.1 Vacuum Glazing

In vacuum glazing, two glass panes of 3–4 mm thickness are spaced about 0.2 mm and space between the panes is evacuated to reduce the heat transfer by conduction and convection. This reduces the U-value as low as 1 watt/m² K. Generally inner glass pane or both is coated with low emissive thin film; though, maintaining a vacuum inside is a challenge. A detailed study of vacuum glazing in presented [10]. A programmed analysis of double vacuum triple layer glazing is also reported [11], which claims the U-value as low as 0.2 watt/m² K.

3.5.2 Switchable Glazing

In switchable glazing, the light and heat transmission can be controlled by voltage, heat, and light falling on the glazing. These glass panes are called smart glasses. These glasses can change from transparent to translucent as per the intensity of solar radiations falling on them. Smart glasses technology includes electrochromic, thermochromic, photochromic, suspended particle microblinds, and polymer-dispersed liquid crystals. Lighting inside the building can be controlled up to an extent where curtains and light screens may not be needed at all. With these smart glasses, entry of ultraviolet radiations can be blocked up to 99%

3.5.3 Suspended Particle Devices

In suspended particle devices (SPDs), nanoscale particles are suspended in a liquid and inserted between two transparent sheets. When there is no external voltage applied, these suspended nanoparticles are randomly oriented and block the light. When voltage is applied, alignment of these particles allows the passage of light and variation in panel voltage regulates the intensity of light entering the building.

This regulated transmission of heat and light through windows reduces the need of air conditioning during summers and heat blowers during winters. The optimization of light inside the building helps a lot in designing EEBs.

3.5.4 Electrochromic Devices

There are certain materials which change their opacity under the application of electric potential. These materials transmit their state from colored to translucent and transparent states. A burst of electricity is required to change their opacity. Once the change is initiated, no electricity is required to maintain the changed state. Changing of state starts from the edges and may take a few minutes to complete depending on the size of the window. Recently, reflective electrochromic materials such as transition metal hydrides are reported. These become reflective rather than absorptive and behave like a mirror.

3.5.5 Polymer Dispersed Liquid Crystal Devices

Under this technology, a liquid crystal mixed with liquid polymer after curing or solidification, sandwiched between to transparent glass or plastic sheets. The curing or solidification of polymer causes the formation of droplets of liquid crystals inside the solid polymers. The size of droplets is affected by curing conditions, which in turn affects the transparency of smart windows. Now a potential difference is applied using electrodes pasted on the sheets. After the application of potential difference, randomly aligned liquid crystals start aligning themselves. As a result, scattering losses by randomly oriented liquid crystals in droplets reduces. Hence, transparency of the smart window increases and the window acts as a transparent glass. With no voltage, the randomly oriented liquid crystals scatter the light and window panes become translucent.

3.5.6 Micro-Blinds

Another contemporary and promising technology in the field of regulated solar radiation windows is mini and micro blinds. Micro-blinds are composed of rolled extremely thin metal blinds on the glass, which are not visible as by naked eyes. Deposition of these micro-blinds is done using lithographic processes. Glass window pane is pasted with conducting transparent oxide (TCO) layer. A thin insulator is deposited between the rolled metal layers and TCO layer for electric disconnection. When a voltage is applied between the rolled metal and TCO layer, rolled metal layer stretches out and restricts the incoming light through these panes. However, without this potential difference the rolled metal layers allows light to pass through. Hence, controlled illumination inside a building can be achieved. Micro-blinds are durable and cost-effective to fabricate. These blinds also have fast switching speeds.

Nanocrystal-impregnated window panes also have the ability to control both light and heat NIR, ensuring optimum illuminance and heat conditions inside a building. Technology uses a small potential difference (~2.5 volt) to switch over from an IR transmitter and NIR blocking transmitting and NIR blocking states. Nanocrystals of tin-doped indium oxide imbedded in glassy matrix of niobium oxide form a composite material is reported. The structural changes after voltage regulates the thermal and light conditions inside a building [10].

3.6 Frames

Frames are the spacers of modern fenestrations and play a significant role in controlling the U-value of different types of windows as reported [11]. Optimum design of appropriate frames is yet to be exploited commercially.

3.7 Roofs

Roofs play a significant role to control the thermal condition of the indoors for comfort because roofs account for a large amount of heat loss or gains as compared to the walls of a building. This is especially true for buildings with larger roof area such as sports complexes, auditoriums, exhibition halls, and so on. In the United Kingdom, the U-value of roofs had been improving continuously from 1.4 W/m²K in 1965 to 0.35 W/m²K in 1985 and now 0.25 W/m²K or less is required for all new buildings.

Hence, techniques used in the architecture of roofs are of great concern to regulate the climatic conditions of the inside of a building.

3.7.1 Ventilated Roofs

In literature, both slanted and flat ventilated roofs are reported. As we know, the roof design should respect the comfort of the users of the building. The roofs of a building have a greater role to play as compared to other building envelopes. In winter, the roofs should heat the indoors, while during summers, it should restrict the heat gain from surroundings, so that indoor temperature can be maintained to reduce the energy consumption in air-conditioners. Multilayer roofs to maintain air flow and thermal insulation are reported [12]. Ventilation in roofs has proven to provide climatic control inside a building up to certain extent [12,13]. There are other passive ventilation methods also reported like special roof systems using natural ventilation to control the thermal condition inside a building. Double skin roofs are created by adding a metallic screen on an existing sheet metal roof for passive cooling of building in tropical climate [14]. The effect of construction parameters were also analyzed at different temperature and humidity conditions of the surroundings and it was found that ventilated roofs are more remarkable than conventional roofs [15,16].

3.7.2 Domed Roofs

In the Middle East, domed roofs were largely found in the vernacular architectural of buildings where climate is hot and arid [17]. Ventilated and domed roofs absorb more heat because of larger surface area, but they speedily radiate also after sunset. It is found that for better thermal performances, a vaulted roof should be such that rim angle is more than 100°. North-south orientation of vaulted roofs is more advantageous than east–west orientation. The vault or domes cover more than 70% of thermal stratification into the volume keeping the lower part of the building cool.

3.7.3 Reflective Roofs

The solar reflective roofs are also called cool roofs. These roofs have high solar reflectance (SR), hence, the indoors of the building remain cooler comparatively. SR is also called reflectivity or albedo. Two surface properties of roofs can affect the thermal climate inside a building; one is SR and other IR emittance or emissivity. Standard conventional roofs have SR up to 20% which can be increased to 60% using different metallic reflective coatings, for example, aluminum coating can raise the SR value up to 50%. SR and infrared emittance properties of typical roof types along with temperature rise are listed below [18].

Roof Surface Type	Solar Reflectance	Infrared Emittance	Roof Surface Temperature Rise (°C)
Ethylene propylene diene monomer (EPDM)–black	0.06	0.86	46.1
EPDM–white	0.69	0.87	13.9
Thermoplastic polyolefin (TPO)–white	0.83	0.92	6.11
Bitumen–smooth surface	0.06	0.86	46.1
Bitumen–white granules	0.26	0.92	35
Built-up roof (BUR)–dark gravel	0.12	0.90	42.2
BUR–light gravel	0.34	0.90	31.7
Asphalt shingles–generic black granules	0.05	0.91	45.6
Asphalt shingles–generic white granules	0.25	0.91	35.6
Shingles–white elastomeric coating	0.71	0.91	12.2
Shingles–aluminum coating	0.54	0.42	28.3
Steel–new, bare, galvanized	0.61	0.04	30.6
Aluminum	0.61	0.25	26.7
Siliconized polyester–white	0.59	0.85	20.6

Source: http://eetd.lbl.gov/coolroofs/.

The choice of reflective coatings should require that it has larger IR emittance also. Otherwise, the indoors will remain warmer during nights. Though, there is a drawback of these cool roofs as they become colder during winter season. But it has been analyzed and found that energy savings in air-conditioning during summer is dominant over energy consumption during winter to heat up. Hence such coated roofs are more useful where winter does not occur effectively. Studies conducted in California reveals that the peak temperature of roof surface was reduced by 33–42 K using this technique.

3.7.4 Green Roofs

Green roofs are a passive cooling technique in which vegetation on roofs is used to restrain the solar radiation to reach on the roof surface. Green roofs are maintained through two to three layers which include water-proofing membrane, soil, or growing a medium vegetation layer itself. Sometimes a drainage layer is also designed. Green roofs not only keep the indoors cool but also offer many other environmental benefits such as rain water management [19,20], improved air quality [21], durability of roof [22], and reduction of urban heat island effect [23]. Other than these, green roofs also add the aesthetics of a building and can be used as a beautiful place for leisure time.

The major classification of green roofs includes extensive and intensive green roofs. Extensive green roofs have a thin solid or substrate layer; hence, big plants cannot be grown. It is used to maintain a grass lawn and will be lightweight. However, intensive green roofs have a deep solid layer, hence, shrubs and other like plants can be grown on these roofs.

It is reported by Herman in 2003, that about 14% of roofs were green roofs in Germany [24]. Extensive green roofs are preferred to use over intensive green roofs because these are relatively maintenance free. There is a challenge to maintain the substrate or soil for a longer period of time for a maintained intensive roof as reported [25].

3.7.5 Photovoltaic Roof Envelopes

PVs are one of the most widely and efficiently used renewable energy sources available. This is an elegant way of producing electricity directly from solar radiations without any damage to our environment. Solar cells are used in photovoltaics to produce electricity from solar radiation with poor maintenance cost and long life.

PVs provide clean energy solution onsite and especially very useful for our buildings. Today, PV elements are becoming integrated parts of a building, often serving as an exterior weather skin that not only fulfils the energy requirements but also protects building envelopes, produces thermal insulation and adds in aesthetics of the building also.

A building integrated photovoltaics (BIPV) system consists of integrated a photovoltaic module into building envelopes that includes both roofs and facades.

A PV panel is composed of an array of individual solar cells called modules connected electrically and mechanically. BIPV systems are effectively cheaper over a longer period as compare to PV envelopes which need separate mounting systems. A complete BIPV system includes the PV module which may be either thick or thin film-based, a power storage with a number of batteries or utility grid system, and equipment to convert DC output from power storage into AC compatible source with utility grid. A charge controller to regulate power into and out of the battery storage is also needed.

PVs can be integrated in different ways with building envelopes. These may be incorporated into awnings and sawtooth designs on a building facade, which not only adds the aesthetics but also offers passive shading. Using PVs for skylight systems can be both cost effective and elegant design features.

3.7.6 Thermal Insulation of Roofs

Thermal insulation is a material or combination of materials to reduce heat flow by conduction, convection, and radiation. It retards heat flow because of bad heat conductivity or high thermal resistance. This insulation reduces the energy consumption to a considerable extent. A simple and effective way to improve energy efficiency of a building is to make use of thermally insulated envelopes.

The selection of insulating material is achieved by looking at two parameters: thermal conductivity and thermal inertia of the material. Environmental and health impacts of particular insulating materials must be taken into account to choose appropriate insulating material.

Thermally insulated roofs can control thermal conditions inside a building because approximately 60% of the thermal leakage of a building takes place through roofs, whether it is winter or summer seasons. Thermally insulated walls have potential to save energy load of a building during both summer and winter seasons. Thermal insulation of roof can also be understood in terms of radiant transmittive barrier. When this is this barrier is supported with a reflective surface on the roofs, it is called reflective-transmittive barrier because it also reflects solar infrared radiation. On the other hand, roof insulation is also achieved using polystyrene, fiber glass, rock and mineral, wool, and so on. These insulating materials are mostly used in arid climatic conditions like Middle East and Asia. Building with thermally insulated roofs can save energy load up to 50% as compared to the buildings without roof insulation [26].

It has been reported that reflective surface roofs using polyurethane, polystyrene, polyethylene, sand and rubber, along with two different reflective surfaces of aluminium and galvanised steel can reduce heat flux more than 80% through these reflective thermally insulated roofs [27]. It has also been noted that aluminum as a reflecting surface and polystyrene and polyurethane are better option as compared to other reflective and insulating materials available. It is also studied and found that in the tropical climate of Sri Lanka, use of cellular polystyrene (thickness 25 mm) as insulating material can reduce the soffit temperature up to 10°.

3.7.7 Evaporative Cooling Roofs

An evaporative cooling roof is cooled using the latent heat evaporative technique. Shallow ponds on the flattop roofs are used in both fixed top roof insulation and moving top thermal insulation. During summers, the top movable insulation covers the pond during daytime and is exposed to the environment during night to cool inside the building. The above process is reversed during winter season to open the pond during day and close the pond during night to retain the heat inside the building. It has been observed that using roof pond, the indoor temperature of a building can be reduced up to 20°C [26].

Another technique is also available under evaporative cooling in which water soaked into bags are spread on the rooftops to cause evaporative cooling specially in region of hot and arid climates. The method has also reported to reduce the indoor temperature up to 15°C [26]. These water-based insulation techniques are discouraged as they consume an abundance of water.

3.8 Thermal Mass

Thermal mass refers to the materials which have high heat capacity, that is, they can absorb large amounts of heat without much rise in their temperature, store it and release afterwards. Thermal mass includes building components such as walls, partitions, roofs, ceilings, floors, and the furniture of a building. The absorption and progressive release of heat by these thermal masses can control the indoor temperature of a building. This reduces the peak energy load of a building and mean radiant temperature [27,28]. For effective thermal absorption, day-to-day ambience temperature variations should exceed 100 C. Thermal mass optimization can be achieved by thermophysical properties of building materials, building orientation, thermal insulation, ventilation and auxiliary cooling techniques and occupancy patterns. The effects of thermal mass and night ventilation on building cooling loads are mathematically modelled [29]. As reported, in New York, energy savings up to 18%–20% inside commercial building of area 27000 ft^2 were achieved.

3.9 Building Simulation Software/Programs

In the modern world, to get best out of available resources, computer assistance is becoming inevitable. In the field of EEB design computational simulations for appropriate selections of different building envelopes can be programmed. A building code can be generated with ideal optimization of various building envelopes that can be compared with practical situations. Crawley et al. [30] performed an up-to-date comparison of the features and capabilities of 20 major building energy simulation codes. This programmed simulation-based study can help the designer of a building to design an EEB. The modeling tools can also be used to predict a cost- effective and energy-efficient retrofit to an existing building. Several building energy modeling codes have been developed across the world. The accuracy of these simulations will depend upon certain feedbacks such as orientation, climatic condition of the place, geographical morphology, construction details, types of the building (residential or commercial), and so on.

3.10 Maintenance of Building Envelope

Maintenance of building envelopes is inescapable for indoor comfort living because it is the envelope of a building, which separates indoor from outdoor climatic conditions. These envelopes are prone and are affected by temperature, humidity, air movement, rain, snow, solar radiation, and various other natural factors of the surroundings. Generally, the building envelope repair and replacement costs contribute 20%–30% of the overall building repair and maintenance life cycle costs [31]. Commonly, a building envelope is largely affected and damaged by water runoff. Whenever water runs down over building envelope components, it can leave behind contaminants that react with or stick to the surface of the exposed envelope components, thus causing a temporary or permanent damage to the building envelope [32]. It has been reviewed that windstorms affect the high rise building envelopes to a very large extent [33]. Diagnosis of repair/replacement of any building envelopes should be done in a regular manner to yield best results of any EEB.

3.11 Conclusion

The present article discussed various building envelope components from an energy efficiency and savings perspective. Improvements to building envelope fall under the passive energy strategies for EEBs. Passive energy-efficiency strategies are actually governed by different geographical and meteorological factors and, therefore, require a broader understanding of the climatic factors by a building designer. For example, application of thermal mass as an energy-saving method is more effective in places where the atmospheric air temperature differences between the days and nights are high. For moderate temperature differences, thermal mass technique is not fruitful. Therefore, building energy modeling and simulation computer codes play an important role in choosing the best energy efficiency options for a given location.

References

1. Chan KT, Chow WK. Energy impact of commercial-building envelopes in the sub-tropical climate. *Applied Energy* 1998;60(1):21–39.
2. Zalewski L, Lassue S, Duthoit B, Butez M. Study of solar walls—Validating a simulation model. *Building and Environment* 2002;37(1):109–121.

3. Jie J et al. Modelling of novel Trombe walls with PV cells. *Buildings and Environment* 2005;42:1544–1552.
4. Tyagi VV, Buddhi D. PCM thermal storage in buildings: A state of art. *Renewable and Sustainable Energy Reviews* 2007;11(6):1146–1166.
5. Nayak JK. Transwall versus Trombe wall: Relative performance studies. *Energy Conversion and Management* 1987;27(4):389–393.
6. Athienitis AK, Liu C, Hawes D, Banu D, Feldman D. Investigation of the thermal performance of a passive solar test-room with wall latent heat storage. *Building and Environment* 1997;32(5):405–410.
7. Kuznik F, Virgone J. Experimental assessment of a phase change material for wall building use. *Applied Energy* 2009;86(10):2038–2046.
8. Al-Jabri KS, Hago AW, Al-Nuaimi AS, Al-Saidy AH. Concrete blocks for thermal insulation in hot climate. *Cement and Concrete Research* 2005;35(8):1472–1479.
9. Robinson PD, Hutchins M. G. Advanced glazing technology for low energy build-ings in the UK. *Renewable Energy* 1994;5(1–4):298–309.
10. Llordes A et al. Polyoxometalates and collidal nano-crystals as building blocks for metal oxide nno composite films. *Journal of Material Chemistry* 2011;21:11631–11638.
11. Ciampi M et al. Energy analysis of ventilated and micro-ventilated roofs. *Solar Energy* 2005;79:183–192.
12. Manca O, Mangiacapra A, Marino S, Nardini S. Numerical investigation on thermal behaviors of an inclined ventilated roof. *Presented at 12th Biennal Conference on Engineering Systems Design and Analysis ESDA2014*, Copenhagen, Denmark, June 25–27, 2014.
13. Tong Y, Zhai J, Wang C, Zhou B, Niu X. Possibility of using roof openings for natural ventilation in a shallow urban road tunnel. *Tunnelling and Underground Space Technology* 2016;54:92–101.
14. Biwole PH, Woloszyn M, Pompeo C. Heat transfer in a double-skin roof ventilated by natural convection in summer time. *Energy and Buildings* 2008;40:1487–1497.
15. Villi G, Pasut W, De Carli M. CFD modelling and thermal performance analysis of a wooden ventilated roof structure. *Building Simulation* 2009;2:215–228.
16. Dimoudi A, Lykoudis S, Androutsopoulos A. Thermal performance of an innovative roof component. *Renewable Energy* 2006;31:2257–2271.
17. Tang R, Meir IA, Wu T. Thermal performance of non air-conditioned buildings with vaulted roofs in comparison with flat roofs. *Building and Environment* 2006;41(3):268–276.
18. LBNL's cool roofing materials database. http://eetd.lbl.gov/coolroofs/ (Accessed: April 15, 2017).
19. Mentens J, Raes D, Hermy M. Green roofs as a tool for solving the rainwater runoff problem in the urbanized 21st century? *Landscape and Urban Planning* 77 (2006) 217–226; Castleton HF et al. Green roofs; building energy savings and the potential for retrofit. *Energy and Buildings* 2010;42:1582–1591.
20. Stovin V, Dunnett N, Hallam A. Green roofs—Getting sustainable drainage off the ground. *6th International Conference of Sustainable Techniques and Strategies in Urban Water Mangement* (Novatech 2007), Lyon, France, 2007, pp. 11–18.
21. Yang J, Yu Q, Gong P. Quantifying air pollution removal by green roofs in Chicago. *Atmospheric Environment* 2008;42 (31):7266–7273.
22. Teemusk A, Mander U. Green roof potential to reduce temperature fluctuations of a roof membrane: A case study from Estonia. *Building and Environment* 2009;44 (3):643–650.
23. Banting D et al. Report on the environmental benefits and costs of green roof technology for the City of Toronto, 2005, Hitesh Doshi Department of Architectural Science, Ryerson University, Toronto, Ontario.
24. Herman R. Green roofs in Germany: Yesterday, today and tomorrow. *Green-ing Rooftops for Sustainable Communities*, Chicago, IL, 2003, pp. 41–45.
25. Williams NSG et al. Green roofs for a wide brown land: Opportunities and barriers for rooftop greening in Australia. *Urban Forestry & Urban Greening* 2010;9(3):245–251.
26. Sanjay M, Prabha C. Passive cooling techniques of buildings: Past and present—A review. *ARISER* 2008;4(1):37–46.

27. Antinucci M, Asiain D, Fleury B, Lopez J, Maldonado E, Santamouris M, Tombazis A, Annas S. Passive and hybrid cooling of buildings—State of the art—PB—Taylor & Francis. *International Journal of Solar Energy* 1992;11(3):251.
28. Balaras CA. The role of thermal mass on the cooling load of buildings. An overview of computational methods. *Energy and Buildings* 1996;24(1):1–10.
29. Yang L, Li Y. Cooling load reduction by using thermal mass and night ventilation. *Energy and Buildings* 2008;40(11):2052–2058.
30. Crawley DB, Hand JW, Kummert M, Griffith BT. Contrasting the capabilities of building energy performance simulation programs. *Building and Environment* 2008;43(4):661–673.
31. Genge GR. Repair of faults in building envelopes. *Journal of Building Physics* 1994;1 8(1):81–88.
32. Burton B. The building envelope: A maintenance challenge. *Journal of Building Physics* 1992;16(2):134–139.
33. Minor JE. Windborne debris and the building envelope. *Journal of Wind Engineering and Industrial Aerodynamics* 1994;53(1–2):207–227.

4

Passive and Low Energy Buildings

Lu Aye and Amitha Jayalath

CONTENTS

4.1 Introduction...73
4.2 Design Strategies and Performance Parameters of Passive and Low Energy
Buildings..74
    4.2.1 Walls...74
    4.2.2 Glazing...76
    4.2.3 Roof..77
    4.2.4 Thermal Insulation, Thermal Mass and Phase Change Materials...................80
    4.2.5 Ground Cooling...83
    4.2.6 Night Ventilation..84
4.3 Embodied Energy..84
4.4 Conclusions..85
References..86

4.1 Introduction

The quest for environmentally friendly renewable energy resources has gained remarkable global interest in recent years due to the rapid depletion and adverse environmental impacts of fossil fuels. The building and construction industry is a prime consumer of the world's fossil energy resources and accounts for nearly 40% of total energy consumption, of which more than 27% is related to residential energy consumption [1]. The highest proportion of energy consumed in a residential building is on space heating, which accounts for more than 54% with the increased demand for thermal comfort in buildings [1]. These facts emphasize the need for energy conservation in buildings, which has gained significant attention from research communities and governments. World Green Building Council (WGBC) is taking up green building initiatives in an effort to reduce the energy footprint in buildings and currently 82 nations across the globe share the same interest under WGBC [2]. Leadership in Energy and Environmental Design (LEED) is a green building certification program developed by the U.S. Green Building Council to access design, construction, operation, and maintenance of green buildings [3].

Enhancement in energy efficiency in buildings can be realized with active, passive, and combined strategies. Active and passive measures can be used to improve Heating, Ventilation, and Air-Conditioning (HVAC) systems, lighting arrangements, building envelope elements, and so on. However, in recent years, passive strategies to enhance thermal and energy performance in building envelope has gained a renewed interest and continued to thrive with new technologies. Passive buildings fall under low-energy building where special design criteria is in place to reduce the operational energy consumption in a building. Building envelope is a prime component in building structure and separates the indoor and outdoor environment and reduces the heat transfer. Various components including walls, roof, glazing, insulation, shading, thermal mass, and so on, factor into building envelope and passive measures can be used to reduce building energy consumption. Moreover, passive and low-energy cooling technologies can provide comfort in non-air conditioned buildings and decrease the cooling load of the buildings by heat gain prevention, heat modulation, and heat dissipation techniques. A proper architectural design of building envelope along with passive cooling strategies can have a significant impact on the energy consumption and the greenhouse gas (GHG) emissions of the building sector.

4.2 Design Strategies and Performance Parameters of Passive and Low Energy Buildings

Main building envelope components and some passive strategies to improve energy performance in buildings are discussed in the following sections.

4.2.1 Walls

Walls comprise the outermost fabric of a building envelope and are expected to provide thermal and acoustic comfort for the inhabitants. Wall area accounts for the largest portion of the building envelope especially in high rise buildings. The surface area of wall elements acts crucially in thermal transmission, and consequently energy consumption in buildings. Improved thermal performance of walls with advanced construction methods and higher thermal resistance (R-value) values is very useful in increasing energy efficiency and comfort levels in buildings.

Passive solar walls absorb solar radiation and transmit part of the thermal energy into the building reducing energy consumption [4]. The absorption of solar energy is facilitated by the glazing on the outer covering of the wall, which creates a greenhouse effect. The composite solar wall comprises of transparent outer cover, air gap, storage wall, ventilated air gap, and insulation layer with two vents [5]. The storage wall absorbs solar radiation and heat-up due to greenhouse effect. The absorbed energy of the wall is transmitted toward the inner surface through conduction and then into the interior through radiation and convection. The insulation layer provides good heat resistance and limits the overheating during summer times. The Trombe wall is more efficient compared with composite solar wall, where the heated air is swiftly supplied by heating the outside ventilated air layer as presented in Figure 4.1. However, due to removal of insulation layer, the control over the absorbed solar energy is reduced and there is a risk of overheating due to direct solar supplies. Both these configurations have the drawback of reversed heat circulation when the storage wall is colder than the ventilated air. Furthermore, unwanted heat gain during summertime can be minimized using a solar shield [5]. Jie et al. [6] have proposed a novel Trombe wall with PV modules that was attached to the back of the glass panel. Both the heat generated by the PV modules and the absorbed solar energy of the storage

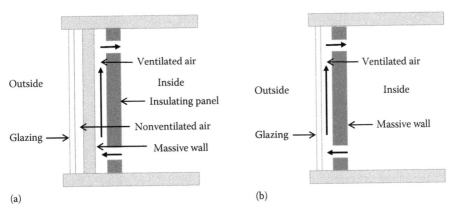

FIGURE 4.1
(a) Composite solar wall and (b) Trombe wall. (Adapted from Zalewski, L. et al., *Build. Environ.*, 37, 109–121, 2002.)

wall was used for space heating. Removal of excess heat reduces the temperature of PV modules and improves the electrical efficiency.

Double skin façades (DSF) also can be used for passive cooling in buildings. Two layers of building element, where usually a glazing surface is placed in front of a regular building façade with a cavity in between, constitute a DSF. The cavity in between can be air tight or generally naturally/mechanically ventilated [3] to reduce overheating during summer and for energy savings in winter. DSF improves thermal comfort for occupants and provides insulation and solar radiation control [7]. The transparent properties of DSF maintains the visibility of the surrounding contrast to the conventional curtain wall façades, yet allows a large amount of solar radiation to enter the building with reduced glare [8]. However, appropriate ventilation air velocities are recommended to maintain to reduce the build-up of elevated temperature gradients during summer periods [9]. Moreover, narrow cavity depth, high transmittance of the glazing material at the exterior layer, and increased height of the façade cavity is preferred in DSF design for improved heat gain and buoyancy force for natural ventilation [10].

The cavity depth and the type of glazing material used can affect the thermal performance of DSF. A parametric and optimization study on DSF design performed by Jie et al. [6] showed that the window glazing type on the outside surface of the inner layer has the highest impact. The total energy consumption was reduced with the reduction of the cavity depth. The optimal DSF design resulted in about 5.6% reduction in energy consumption, however the results depend on the climate, environment, energy sources used, and the building design. Limited work has been done related to DSF with integrated solar PV, which can be used to produce electricity. A theoretical study on PV-integrated DSF with motorized shading in the middle of the cavity showed 21% reduction in electricity generation with 25% increment in PV area [8].

Lightweight cement-based materials with lower material density, lower thermal conductivity, and ease of construction have also attracted a considerable interest as an alternative for traditional concrete constructions. Lightweight concrete (LWC) is any concrete with a density less than 2000 kg m^{-3}. Lightweight aggregates such as pumice, dolomite expanded clay, and foamed slags have been used in manufacturing of LWC. Autoclave aerated concrete (AAC) is a non-combustible cement-based material, a sub-category of LWC, manufactured with Portland cement, fly ash, quick lime, gypsum, water, and aluminium powder. The inherent nature of high porosity of AAC results in superior thermal properties, thus, with application in buildings, a reduction in heating and cooling loads can be realized. Furthermore, AAC improves the acoustic properties of the building along with reduced construction time and ease of transportation.

4.2.2 Glazing

Windows constitute a main element in building envelope and play an important role in thermal losses in cold climates and unnecessary heat gains in warm climates. It provides an aesthetic value to building design and has a vital impact on lighting levels and thermal comfort in buildings. The impact of windows does not solely depend on the thermal conductivity (U-value) or the solar heat gain coefficient (SHGC) of glazing material but also on the climate condition, building design, orientation, and shading.

Heat transfer reduction through glazing system is achieved using multiple layers of glass with air or inert gas filled in between to reduce conduction and convection. Though an increase in number of glazed layers aid reduction of heat transfer, increased cost, weight, and reduction of light transmittance due to increased thickness are some downsides [11]. However, replacement of glazing systems with improved glazed properties, addition of shading devices, and suitable orientation of glazed area with respect to the sun can lead to energy savings in buildings [12,13]. Furthermore, different materials and technologies have been developed to achieve high performance solar gain control solutions with improved thermal isolation properties for glazing systems in buildings. Aerogel is a mesoporous solid with a combination of low thermal conductivity and high transmittance of daylight and solar energy [14]. Aerogels have the lowest thermal conductivity among solids; ~0.010 $Wm^{-1}K^{-1}$ at room temperature, depending on the pressure [15] and with volume porosity of greater than 50%. Silica aerogels have interesting optical properties where they exhibit high transmittance in the range of visible light: wavelength between 380 and 780 nm. Thus, work has been done to develop high insulating windows with granular or monolithic aerogel. Monolithic translucent aerogel in 10 mm thick packed bed has a solar transmittance of 0.88. At present, two commercial products of aerogel-based daylight systems is available with visible light transmission of 0.40 and heat transmittance coefficient between 0.6 and 0.3 W $m^{-2}K^{-1}$ for layers of 30 and 60 mm, respectively [14]. Optical characteristics of glazing systems with granular and monolithic aerogel were analyzed by Buratti and Moretti [15] that represented a specific alternative to conventional windows. Monolithic aerogel showed best performance compared with granular system for both light transmittance and thermal insulation. A reduction in heat losses and light transmission was observed compared to double glazing with low-emittance layer.

Heat transfer through glazed surfaces by conduction and convection is nearly eliminated with vacuum glazing where center-of-glass U-value is as low as 1 W $m^{-2}K^{-1}$ [2]. Low transmittance coating is generally applied to one or both glass panes to reduce the radiative heat transfer. Maintaining vacuum for longer periods is one of the challenges faced and the glazing edge seal should be able to maintain low vacuum pressure such as 0.1 Pa for its lifetime [16]. Furthermore, a support pillar array can be used to hold the two glazed surfaces apart while maintaining the low vacuum as shown in Figure 4.2 [17].

To further reduce the heat transfer of vacuum glazing, triple vacuum glazing has been studied where theoretical studies make thermal conductance predictions of less than 0.2 W $m^{-2}K^{-1}$. The space between two glass layers can be reduced to achieve a compact glazing system. The additional vacuum cavity in triple glazing results in increased thermal

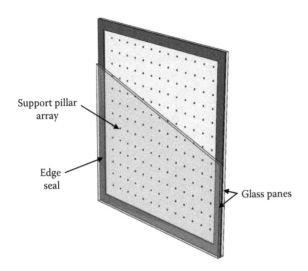

FIGURE 4.2
Schematic diagram of a vacuum glazing.

resistance compared to double vacuum glazing and the increased distance between inner and outer glass panes aids in reducing deflection [18].

Though external shading devices are considered as the best option for solar control, this approach is less desirable especially in high-rise buildings due to maintenance, structural limitations, and wind loading [16]. However, reflection of solar radiation can be achieved using switchable reflective glazing. The optical properties of the reflective glazing systems can be changed either using low DC voltage or switchable reflective light shelves [2] and currently this technology is at its early stage of development.

Commercial vacuum glazing produced by Nippon Sheet Glass (NSG) in Japan is available in the current market. The three different glazing systems produced by NSG result in U-value range from 0.7 to 1.5 W m^{-2}K^{-1}. Though the theoretical prediction for best performance of a vacuum glazed system is at 0.5 W m^{-2}K^{-1}, neither laboratory nor the market has been able to achieve this at present [11]. Furthermore, numerical simulations of triple vacuum glazing were studied by Fang et al. [19] and have achieved 0.26 W m^{-2}K^{-1} thermal transmission at the center-of-glazing area.

4.2.3 Roof

Roof is a main element in building envelope and the roof surface area accounts for 20%–25% of total urban surfaces [20]. It is susceptible for solar radiation; thus it affects the indoor occupant comfort as well as the outdoor air temperature in urban areas. The heat gains or loss through the roof depends on the surface area, orientation, and the thermal properties of roof material. Thermal performance of the roof is vital in improving the thermal performance of a building, thus regulations have been put in place in building constructions. For example, in accordance with the UK building regulations, the upper limit for the roof U-value is 0.25 W m^{-2}K^{-1} required for new buildings in UK [2].

Some passive techniques have been used to reduce the solar absorptivity of the roof, which includes solar reflective/cool roofs, green roofs either partly or fully covered with vegetation, roof insulation, evaporative roofs, and photovoltaic (PV) roofs. Cool roofs are characterized by the high solar reflectance and high thermal emittance. Most of the incident solar radiation during the day can be reflected using high solar reflectance coatings on roof, thus reducing the roof surface temperatures and heat transfer in to the building [21]. High thermal emittance allows to dissipate the heat stored in the roof structure during nighttime. Conventional roofing materials only have solar reflectance values of 0.05–0.25 compared with 0.6 or more with reflective roof coatings [2]. Though metal roofs are highly reflective, they tend to get hot since they cannot emit the absorbed solar energy due to low infrared emittance values. Special roof coatings need to be used to improve the infrared emittance of metal roofs. Aluminum and white elastomeric coatings can be used to increase the solar reflectance of roof materials and the layer thickness can also affect the improved values. A study on the effects of cool roof on energy use in six different buildings in California was performed with white elastomeric coating or PVC single-ply membrane roof [19]. Results showed that the cool roof can reduce daily peak roof temperature by 33°K–42°K. In studied institutional and commercial buildings, savings on average air conditioning energy use of 17%–52% was realized with reflective roofs. As an investigation of strategies to mitigate urban heat island effects in metropolitan Hyderabad, data gathered on cool roofs in India was evaluated [22]. The application of white coatings on previously black roofs and concrete roofs yielded 26% and 19% reduction in cooling energy use, respectively, and the reduction in annual direct GHG emissions was estimated as 11–12 CO_2-e m^{-2} of flat roof area.

A green roof fully or partially covers the building roof with vegetation to protect the building environment as shown in Figure 4.3. The composite layer system consists of waterproof medium, soil layer, and the vegetation layer. A green roof reduces the thermal loads entering the building due to solar radiation and affects the convective and radiative heat transfer. This is due to absorption of solar radiation by the plants to

FIGURE 4.3
Roof top garden, Burnley Campus, The University of Melbourne (U.O.M. 2017 [17]).

support their life cycle and the added insulation by the soil layer. The installation of the rooftop garden yielded 15% savings on annual energy consumption and 17%–79% in the space cooling load. Based on a study on the effect of green and cool roofs on energy performances in Mediterranean area, 12% reduction in energy demand for non-insulated buildings was realized with green roof [20]. However, cool roofs were the most effective solution for reducing cooling demand and increase energy savings for center and southern Mediterranean basin.

The moisture content in the soil layer can affect its insulating properties, the wetter the medium, the poorer the insulating value. Based on a numerical study performed on a hypothetical commercial building in Singapore, the dry clay soil resulted in 16%–64% reduction in peak roof thermal transfer, whereas clay soil with 40% moisture content reduces only 2%–22% [23]. Figure 4.4 shows the energy exchanges in a traditional roof, wet green roof, and a dry green roof. The wet green roof has almost double the amount of evapotranspiration compared with dry green roof. Thus, in hot and dry climates, the wet soil of green roof is advantageous and helps to remove heat from the building through evapotranspiration and act as a passive cooler [2].

Integration of PV in to building envelope is another passive technique to improve the thermal performance while facilitating energy production. The building envelope, particularly the roof, is ideal for installation of PV especially in countries where restrictions on land usage are in place. PV roofing systems based on standard modules can be mounted on sloped or flat roofs with different anchoring systems (aluminum frames) and laminates (frameless modules) [24]. The PV roof tiles replace the roofing material and are installed directly on to the structure. Furthermore, thin film system of PV component developed in the U.S. can be rolled on to a roof like a carpet. Due to its flexibility, it can be used with many roofing materials and sandwiched PV-roofing materials with added roof insulation and for electricity production [24]. A novel transparent roof made by solid Compound Parabolic Concentrator (CPC) PV/T/D (Photovoltaic/Temperature/ Daylighting) system is designed to achieve excellent light control during daytime and to improve thermal environment in the building [25]. The PV/T/D system can substantially reduce the light intensity entering the room at noon, while producing electricity and hot air by surplus light.

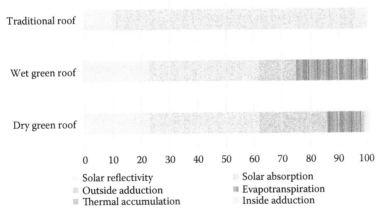

FIGURE 4.4
Comparison of the energy exchanges of the dry or wet green roof with a traditional roof, summer season. (Adapted from Sadineni, S.B. et al., *Renew. Sustain. Energ. Rev.*, 15, 3617–3631, 2011.)

4.2.4 Thermal Insulation, Thermal Mass and Phase Change Materials

The presence of glass surfaces and the insulating capacity of the outer cladding are the main reasons for heat loss and gain within the building envelope. Thermal insulation in buildings facilitates the thermal comfort for occupants. Insulation reduces unwanted heat loss or gain through conduction, convection, and radiation. The suitable use of insulation can decrease the energy demands of heating and cooling systems, thus downsizes the requirement for HVAC system during design stage [2]. The current product range on insulation varies among paints, coatings, thin films, or rigid panels. Some of these conventional insulation materials include mineral fiber blankets or loose fill (fiberglass and rock wool), rigid boards or sprayed in place insulation (polyurethane and extruded polystyrene), and reflective materials (aluminum foil, ceramic coatings). The selection of suitable thermal insulation material depends on the thermal conductivity and the thermal inertia. The increase in temperature and the moisture content can have detrimental effects on the performance of thermal insulation. Flame retardancy is also another factor in selecting insulation type. The flammability behavior of the insulation materials is investigated based on test parameters like heat release rate (HRR), ignition time, peak heat release rate (PHRR), and smoke and carbon monoxide yield.

For better thermal management and for GHG emissions reduction, traditional insulating materials were being used in thick and multiple layers. This introduced additional cost, reduced space, and possible heavier load bearing components. Essentially, the focus has been and still will be to achieve the highest possible thermal insulation values, that is, the lowest thermal conductivity for the materials and the lowest thermal transmittance, U-value, for the structures and buildings [26].

Vacuum insulation panels (VIP) provide much higher thermal insulation compared with traditional insulation materials as shown in Figure 4.5. In a VIP, better thermal performance is achieved through applying a vacuum to an evacuated foil encapsulated porous material. Though a vacuum can be used to reduce the thermal conductivity of traditional insulation, it depends on the properties of evacuated material. Fumed silica (SiO_x) is one of the most commonly used core material in VIPs and helps to maintain the

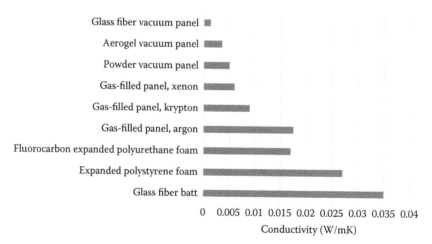

FIGURE 4.5
Thermal conductivities of advanced and conventional insulation materials. (Adapted from Baetens, R. et al., Energ. Build., 42(2), 147–172, 2010.)

vacuum throughout its operational life. Thermal conductivity values of 0.004 W m^{-1}K^{-1} can be achieved using a VIP. However, the formation of thermal bridges in panel envelope and the degradation of thermal performance through its lifetime are some drawbacks of VIPs. During operation, gas and water vapor can penetrate through sealed edges of the VIP [27]. Furthermore, VIPs cannot be cut at the site and panels are fragile in handling. Thus, research work has been brought up to further reduce the thermal conductivity and to increase the serviceable lifetime of VIP. Development of core materials with lower conductivity and small pore sizes, high quality air, and moisture tight envelopes are some of them.

Nanoscale materials provide a far better solution in building insulation due to their high surface-to-volume ratio, which enable them to trap still air within a thin layer of material. These trapped air pockets provide the resistance to heat flow while consuming less space compared to conventional insulation materials. The requirement for a vacuum and airtight envelope can be eliminated by using mesoporous materials with a pore size of 50 nm or less or microporous materials with a pore size of 2 nm or less [14].

Some commercial available insulation products with nanomaterials include Aerogel and Nansulate®. Silica aerogel consists of cross-linked nanostructured internal structure of SiO$_2$ chains with large number of air-filled pores. The average pore diameter varies from 10 to 100 nm, which will take up to 99% of the total aerogel volume. High porosity makes aerogel the lightest solid in the world. It also results in remarkable physical, thermal, acoustic properties, and low mechanical strength in the solid [28]. In the building industry the main markets for aerogel is thermal and acoustic insulation where the demand is expected to reach up to $646 M by 2013 [14].

Nansulate® coating is a patented insulation technology that incorporates a nanocomposite called Hydro-NM-Oxide, a product of nanotechnology. It's an excellent insulator due to its low thermal conductivity and the nanomaterial used. Nansulate® is a nanocomposite insulation material which is composed of 70% percent "Hydro-NM-Oxide" and 30% acrylic resin and performance additive. Nansulate® can be directly applied to the existing buildings without incurring any post-construction addition with conventional insulating, thus it creates tremendous energy saving with existing buildings. Nansulate® coating thickness on an applied surface is around 200 microns DFT (approximately three coats coverage) with very high savings on space reduction [29].

Thermal energy storage (TES) systems are considered as an effective method of improving thermal mass effect of building envelopes. Thermal mass is a property of construction materials to store heat that provides inertia against temperature fluctuations. An integration of higher thermal mass of construction materials with appropriate thermal design of buildings with insulation can manipulate indoor temperatures for enhanced thermal comfort with minimal usage of mechanical heating and cooling. Dynamic thermal masses, such as phase change materials (PCMs) with high storage density per unit mass, can be used to improve the thermal mass effect of building envelopes even with less structural mass attached. Integration of PCMs with buildings aims to reduce heating and cooling loads by reducing demand on heating, ventilating, and air-conditioning (HVAC) systems, improving thermal comfort, shifting peak load, and stabilizing the indoor temperature in a building by reducing the temperature fluctuations due to external weather conditions.

PCMs are latent heat storage materials that involve a phase transition, and the amount of thermal energy stored depends on the enthalpy of phase change. The endothermic and exothermic process around a narrow temperature range, with high storage density for unit mass compared with sensible storage materials like concrete, makes them preferable for building applications. For instance, a PCM wall thickness of 25 mm will have equivalent thermal mass compared to a 420 mm thick concrete wall [30]. A PCM with a phase transition temperature within the operating temperature zone of a building can improve the thermal comfort levels for occupants and reduce demand for heating and cooling [31–34].

Wallboards provide the most cost-effective replacement of standard thermal mass; and due to low thermal inertia of wallboards, addition of PCM can substantially improve the thermal storage capacity of the material. Furthermore, different construction materials used in buildings also has great potential in developing high performance thermal storage materials. Encapsulation of PCMs is preferred as it can hold the liquid phase of the PCM and minimize interactions with the outer environment [35]. Recent developments of microencapsulated PCMs provides more efficient heat transfer [36] by creating finely dispersed PCMs with high surface area as shown in Figure 4.6 [37].

A study on gypsum plaster wallboards with embedded PCM has revealed reduction in peak indoor temperature and reduced cooling demand and increased thermal comfort in lightweight buildings [38]. Wallboards with aluminium honeycomb matrix filled with 60% PCMs was studied by Evola et al. [39]. It was stated that appropriate PCM phase transition range should be selected to achieve a satisfactory comfort condition and the utilization of total latent heat storage potential is affected by convective heat transfer of wall surface, climate condition, amount of direct solar radiation, and so on. Microencapsulated PCM substituted for fine aggregates in cementitious mortar and concrete was used to improve the thermal performance of the material [37]. This study showed that as the volume fraction of PCM in substitution was increased, the thermal conductivity and diffusivity was decreased while the heat capacity was increased. An optimum amount of 20% replacement based on volume basis in cementitious mortar and concrete was suggested for acceptable strength properties along with improved thermal performance in buildings.

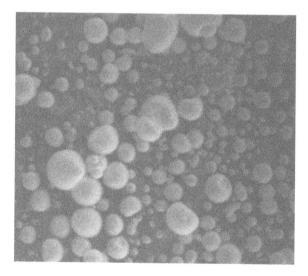

FIGURE 4.6

SEM micrograph (using BSE detector) of Micronal DS5040X and their agglomerates. (Jayalath, A. et al. *Constr. Build. Mater.*, 120, 408–417, 2016.)

4.2.5 Ground Cooling

The temperature of the ground at a depth of a few meters remains constant all year and can be used as a natural heat sink to dissipate excess heat in summer and as a heat source for heating purposes in winter. The most common technique is to use air tunnels known as earth to air heat exchangers (EATHE) to couple the building with the ground [40]. The ambient air is drawn into earth–air tubes and during the passage through ground, heat exchange takes place between soil and ventilation air as shown in Figure 4.7. Due to the low thermal conductivity of soil, the temperature of the soil is lower in summer and higher in winter, thus the ventilation air passed through the earth to air exchangers can be directly used for space heating and cooling in buildings [41].

Some of the factors affecting the effectiveness of ground cooling systems are the air temperature at the outlet, geological properties of the soil (e. g. thermal conductivity, density diffusivity), pipe characteristics related to heat exchange (e.g., pipe depth, diameter, air flow rate, pipe material), and site climatic conditions. Various experimental and numerical studies have investigated the performance of EATHE systems in different climatic conditions. Ground cooling system was experimentally tested for semi-arid climate in Morocco for cooling purposes during the hot season [42]. Earth to air heat exchanger performed well as a semi-passive system for air cooling where blown air temperature in the building was maintained at 25°C with relative humidity around 40% while outdoor temperatures reached 40°C. The reduction of indoor air temperature displayed an exponential drop as a function of the pipe length. Numerical simulations with TRNSYS resulted in specific heat capacity of 55 W m^{-2} for three pipes with 26°C outlet temperature. The feasibility of using earth to air heat exchanger for energy and GHG emission reduction in Mediterranean climate was investigated using a multipurpose building [43]. Energy reductions of about 29% in winter conditions were realized compared with 36%–46% reduction in summertime. The influence of air flow rate on the performance of EATHE system is highlighted where maximizing airflow rates during summer and minimizing during winter is recommended. Based on a comprehensive review on earth to air heat exchanger systems, it is stated that a well-designed EATHE system can reduce the electricity consumption of a typical building by 25%–30% and can be regarded as a feasible option to replace conventional air-conditioning [41].

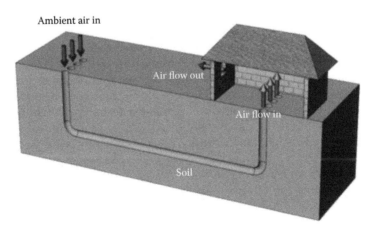

FIGURE 4.7
Schematic view of a building combined with an earth-air tube heat exchanger system.

4.2.6 Night Ventilation

Night ventilation is one of the most attractive passive cooling techniques where cold night air is used to cool down the heat absorbed by the building and to reduce daytime temperature increase. This can be induced naturally using stack or wind forces or with the support of a fan to achieve sufficient airflow rates. The effectiveness of night ventilation depends mainly on the relative differences between the indoor and outdoor temperature and air flow rate of the ambient air. Thermal capacity of the building and the suitable coupling of thermal mass with the air flow also can affect the effectiveness of the technique [44].

A linear relationship was found between cooling needs of the building and the contribution of the night ventilation based on the energy data from 214 air-conditioned residential buildings in Greece [44]. It was found that the higher the cooling demand of the building, the higher the potential contribution of the night ventilation technique. A reduction of cooling loads up to 40 kWh m^{-2} per year with an average contribution close to 12 kWh m^{-2} was realized for residential buildings. The impact of urban environment on the performance of night ventilation system under air-conditioned and free-floating operations was evaluated for ten urban canyons in Greece [45]. The study shows that the efficiency of night ventilation system is significantly affected by the air temperature and the wind velocity in controlled indoor environment. Moreover, night ventilation is more efficient in climates with large diurnal temperature range (above 15°C) and where the nighttime temperatures in summer are below 20°C [46]. Night ventilation techniques have been tested for both residential and office buildings, and it has shown that regardless of the climate condition, it improves thermal comfort and reduces air conditioning demand in office buildings [40]. For optimization and decision-making during design stages, numerical simulation of night ventilation techniques is important. However, simulation of overall system is difficult due to the involvement of large number of uncertainties related to cooling potential and effectiveness [40].

4.3 Embodied Energy

While the main aim of reducing energy demand is fulfilled in passive and low-energy buildings, increased use of raw materials to achieve passive status (insulation, improved glazing, and so on) require significant amount of energy to manufacture. During the lifecycle of a building, energy is consumed directly and indirectly. Direct consumption arises through construction, operation, refurbishing, and demolishing, while the indirect consumption is through the production of the building materials discussed earlier. The second form of energy, "embodied energy" is needed to be accurately considered during lifecycle energy analysis in passive buildings. Lifecycle energy analysis estimates total energy inputs, outputs, and flows through the lifetime of a building and necessarily include operational energy and embodied energy [47]. Accurate inclusion of embodied energy provides a holistic view and assists in forming regulation for better sustainability.

Several studies analyze the relative importance of operational and embodied energy and their relative share in total lifecycle energy analysis in passive and low-energy buildings [48,49]. It has identified that embodied energy dominates in lifecycle energy analysis of passive and low-energy buildings compared with the operational energy in the past [47,48]. The share of embodied energy has increased up to 26%–57% in low-energy

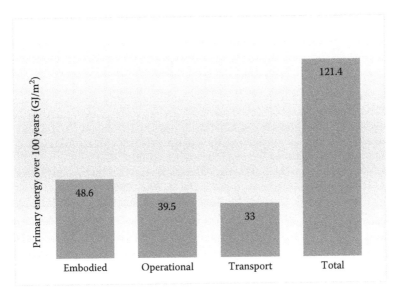

FIGURE 4.8
Life cycle energy demand of the base case passive house, by use. (Adapted from Stephan, A. et al., Appl. Energ., 112, 23–34, 2013.)

buildings compared to 6%–20% in conventional buildings. In passive buildings, embodied energy shows a closer share and accounts for 11%–33% [47]. Another study identifies 55% share of embodied energy in lifecycle energy demand in passive buildings as shown in Figure 4.8 [48] when only embodied and operational energy are considered. Different methods used in calculation of embodied energy could be the cause for the mismatch and future research with extended analysis on embodied energy is recommended.

4.4 Conclusions

This chapter reviewed passive building envelope components and low-energy cooling techniques toward improving thermal performance and reducing energy consumption in buildings. However, the thermal performance of these methodologies are highly dependent on the climatic conditions, thus an appropriate design strategy needs to be used with proper understanding on climatic factors. The passive measures discussed here bear great potential toward energy-efficient buildings (EEBs) in future and require further studies for optimization and to quantify the realized benefits in financial and quantitative terms. However, the increase in embodied energy due to implementation of passive measures in buildings needs to be considered in evaluating the benefits. Thus, future studies need to focus on the net energy benefits realized from passive and low-energy buildings and inclusion of lifecycle energy analysis in the framework of energy efficiency regulations has been highlighted in some current research studies. Furthermore, government incentives and rebate schemes can be used to promote the customer awareness and market penetration of these technologies.

References

1. IEA, *Energy Efficiency Requirements in Building Codes, Energy Efficiency Policies for New Buildings*, Paris, France, 2008.

2. S.B. Sadineni, S. Madala, R.F. Boehm, Passive building energy savings: A review of building envelope components, *Renewable and Sustainable Energy Reviews* 15(8) (2011) 3617–3631.

3. H. Omrany, A. Ghaffarianhoseini, A. Ghaffarianhoseini, K. Raahemifar, J. Tookey, Application of passive wall systems for improving the energy efficiency in buildings: A comprehensive review, *Renewable and Sustainable Energy Reviews* 62 (2016) 1252–1269.

4. J. Shen, S. Lassue, L. Zalewski, D. Huang, Numerical study on thermal behavior of classical or composite Trombe solar walls, *Energy and Buildings* 39(8) (2007) 962–974.

5. L. Zalewski, S. Lassue, B. Duthoit, M. Butez, Study of solar walls—Validating a simulation model, *Building and Environment* 37(1) (2002) 109–121.

6. J. Jie, Y. Hua, H. Wei, P. Gang, L. Jianping, J. Bin, Modeling of a novel Trombe wall with PV cells, *Building and Environment* 42(3) (2007) 1544–1552.

7. J. Joe, W. Choi, Y. Kwak, J.-H. Huh, Optimal design of a multi-story double skin facade, *Energy and Buildings* 76 (2014) 143–150.

8. M.A. Shameri, M.A. Alghoul, K. Sopian, M.F.M. Zain, O. Elayeb, Perspectives of double skin façade systems in buildings and energy saving, *Renewable and Sustainable Energy Reviews* 15(3) (2011) 1468–1475.

9. J. Darkwa, Y. Li, D.H.C. Chow, Heat transfer and air movement behaviour in a double-skin façade, *Sustainable Cities and Society* 10 (2014) 130–139.

10. S. Barbosa, K. Ip, Perspectives of double skin façades for naturally ventilated buildings: A review, *Renewable and Sustainable Energy Reviews* 40 (2014) 1019–1029.

11. P.C. Eames, Vacuum glazing: Current performance and future prospects, *Vacuum* 82(7) (2008) 717–722.

12. C.A. Balaras, K. Droutsa, A.A. Argiriou, D.N. Asimakopoulos, potential for energy conservation in apartment buildings, *Energy and Buildings* 31(2) (2000) 143–154.

13. R.J. Fuller, M.B. Luther, Thermal simulation of an Australian university building, *Building Research and Information* 30(4) (2002) 255–263.

14. R. Baetens, B.P. Jelle, A. Gustavsen, Aerogel insulation for building applications: A state-of-the-art review, *Energy and Buildings* 43(4) (2011) 761–769.

15. C. Buratti, E. Moretti, Glazing systems with silica aerogel for energy savings in buildings, *Applied Energy* 98 (2012) 396–403.

16. A.S. Bahaj, P.A.B. James, M.F. Jentsch, Potential of emerging glazing technologies for highly glazed buildings in hot arid climates, *Energy and Buildings* 40(5) (2008) 720–731.

17. U.O.M. The Melbourne newsroom, Green thumbs on high: Gardens on the roof and champion tree climbers at Burnley Open Day. http://newsroom.melbourne.edu/news/green-thumbs-high-gardens-roof-and-champion-tree-climbers-burnley-open-day, 2017 (Accessed July 17, 2017).

18. H. Manz, S. Brunner, L. Wullschleger, Triple vacuum glazing: Heat transfer and basic mechanical design constraints, *Solar Energy* 80(12) (2006) 1632–1642.

19. Y. Fang, T.J. Hyde, N. Hewitt, Predicted thermal performance of triple vacuum glazing, *Solar Energy* 84 (2010) 2132–2139.

20. M. Zinzi, S. Agnoli, Cool and green roofs. An energy and comfort comparison between passive cooling and mitigation urban heat island techniques for residential buildings in the Mediterranean region, *Energy and Buildings* 55 (2012) 66–76.

21. H. Akbari, R. Levinson, L. Rainer, Monitoring the energy-use effects of cool roofs on California commercial buildings, *Energy and Buildings* 37(10) (2005) 1007–1016.

22. T. Xu, J. Sathaye, H. Akbari, V. Garg, S. Tetali, Quantifying the direct benefits of cool roofs in an urban setting: Reduced cooling energy use and lowered greenhouse gas emissions, *Building and Environment* 48 (2012) 1–6.

23. N.H. Wong, D.K.W. Cheong, H. Yan, J. Soh, C.L. Ong, A. Sia, The effects of rooftop garden on energy consumption of a commercial building in Singapore, *Energy and Buildings* 35(4) (2003) 353–364.
24. A.S. Bahaj, Photovoltaic roofing: Issues of design and integration into buildings, *Renewable Energy* 28(14) (2003) 2195–2204.
25. C. Feng, H. Zheng, R. Wang, X. Yu, Y. Su, A novel solar multifunctional PV/T/D system for green building roofs, *Energy Conversion and Management* 93 (2015) 63–71.
26. B.P. Jelle, A. Gustavsen, R. Baetens, The path to the high performance thermal building insulation materials and solutions of tomorrow, *Journal of Building Physics* 34(Compendex) (2010) 99–123.
27. J. Fricke, H. Schwab, U. Heinemann, Vacuum insulation panels—Exciting thermal properties and most challenging applications, *International Journal of Thermophysics* 27(4) (2006) 1123–1139.
28. R. Baetens, B.P. Jelle, J.V. Thue, M.J. Tenpierik, S. Grynning, S. Uvsløkk, A. Gustavsen, Vacuum insulation panels for building applications: A review and beyond, *Energy and Buildings* 42(2) (2010) 147–172.
29. Nansulate, Industrial Nanotech. http://www.industrial-nanotech.com/, (Accessed: April 28, 2017).
30. S.N. Al-Saadi, Z. Zhai, Modeling phase change materials embedded in building enclosure: A review, *Renewable and Sustainable Energy Reviews* 21 (2013) 659–673.
31. S.E. Kalnæs, B.P. Jelle, Phase change materials and products for building applications: A state-of-the-art review and future research opportunities, *Energy and Buildings* 94 (2015) 150–176.
32. D. Buddhi, V.V. Tyagi, PCM thermal storage in buildings: A state of art, *Renewable and Sustainable Energy Reviews* 11(Copyright 2007, The Institution of Engineering and Technology) (2007) 1146–1166.
33. V.V. Tyagi, S.C. Kaushik, S.K. Tyagi, T. Akiyama, Development of phase change materials based microencapsulated technology for buildings: A review, *Renewable and Sustainable Energy Reviews* 15(2) (2011) 1373–1391.
34. K. Pielichowska, K. Pielichowski, Phase change materials for thermal energy storage, *Progress in Materials Science* 65 (2014) 67–123.
35. M.N.A. Hawlader, M.S. Uddin, M.M. Khin, Microencapsulated PCM thermal-energy storage system, *Applied Energy* 74(Compendex) (2003) 195–202.
36. A. Jamekhorshid, S.M. Sadrameli, M. Farid, A review of microencapsulation methods of phase change materials (PCMs) as a thermal energy storage (TES) medium, *Renewable and Sustainable Energy Reviews* 31 (2014) 531–542.
37. A. Jayalath, R. San Nicolas, M. Sofi, R. Shanks, T. Ngo, L. Aye, P. Mendis, Properties of cementitious mortar and concrete containing micro-encapsulated phase change materials, *Construction and Building Materials* 120 (2016) 408–417.
38. P. Schossig, H.M. Henning, S. Gschwander, T. Haussmann, Micro-encapsulated phase-change materials integrated into construction materials, *Solar Energy Materials and Solar Cells* 89(2–3) (2005) 297–306.
39. G. Evola, L. Marletta, F. Sicurella, A methodology for investigating the effectiveness of PCM wallboards for summer thermal comfort in buildings, *Building and Environment* 59 (2013) 517–527.
40. M. Santamouris, D. Kolokotsa, Passive cooling dissipation techniques for buildings and other structures: The state of the art, *Energy and Buildings* 57 (2013) 74–94.
41. M. Kaushal, Geothermal cooling/heating using ground heat exchanger for various experimental and analytical studies: Comprehensive review, *Energy and Buildings* 139 (2017) 634–652.
42. M. Khabbaz, B. Benhamou, K. Limam, P. Hollmuller, H. Hamdi, A. Bennouna, Experimental and numerical study of an earth-to-air heat exchanger for air cooling in a residential building in hot semi-arid climate, *Energy and Buildings* 125 (2016) 109–121.
43. F. Ascione, D. D'Agostino, C. Marino, F. Minichiello, Earth-to-air heat exchanger for NZEB in Mediterranean climate, *Renewable Energy* 99 (2016) 553–563.

44. M. Santamouris, A. Sfakianaki, K. Pavlou, On the efficiency of night ventilation techniques applied to residential buildings, *Energy and Buildings* 42(8) (2010) 1309–1313.

45. V. Geros, M. Santamouris, S. Karatasou, A. Tsangrassoulis, N. Papanikolaou, On the cooling potential of night ventilation techniques in the urban environment, *Energy and Buildings* 37(3) (2005) 243–257.

46. B. Givoni, Performance and applicability of passive and low-energy cooling systems, *Energy and Buildings* 17(3) (1991) 177–199.

47. P. Chastas, T. Theodosiou, D. Bikas, Embodied energy in residential buildings-towards the nearly zero energy building: A literature review, *Building and Environment* 105 (2016) 267–282.

48. A. Stephan, R.H. Crawford, K. de Myttenaere, A comprehensive assessment of the life cycle energy demand of passive houses, *Applied Energy* 112 (2013) 23–34.

49. I. Sartori, A.G. Hestnes, Energy use in the life cycle of conventional and low-energy buildings: A review article, *Energy and Buildings* 39(3) (2007) 249–257.

5

Energy-Efficient Building Construction and Embodied Energy

R. Singh and Ian J. Lazarus

CONTENTS

5.1   Introduction 89
5.2   Status of Green Building Construction 90
5.3   Significance of Embodied Energy 91
5.4   Embodied Energy of Construction Materials and Buildings 92
5.5   Life Cycle Energy of Buildings 98
5.6   Conclusion 103
References 104

5.1 Introduction

It has been recognized that climate change and global warming, major concerns for the sustainability of the society, are mainly the result of greenhouse gas (GHG) emissions from anthropogenic and natural activities [1]. Continuous industrialization, rapidly growing urbanization and improved living standards have increased energy consumption in recent years. The energy, for the increased demand, is being supplied primarily from the combustion of fossil fuels, which resulted into emission of GHGs, predominantly carbon dioxide. The building sector is one of the sectors that consume a significant amount of thermal and electrical energy for indoor thermal comfort and appliances. The sector also consumes vast quantities of materials, and each material must be extracted, processed, and finally transported to the building construction site. The building sector also affects the environment significantly through materials consumption [2], water use [3], land use [4], but mostly by its energy consumption as the built environment demands about 30%–40% of global energy demand [5]. Although the energy consumed during the various activities in buildings from construction to their operation are critically important for the modern lifestyle and development of the human, it also puts at risk the quality and sustainability of the biosphere due to undesirable results or second order effects [6,7]. Therefore, the growing environmental awareness has aroused great attention globally. Environmental certification is one of the important responses to negative impact of buildings on the environment [8]. Energy-efficient buildings (EEBs) and green buildings, which are built bearing energy efficiency [9], water conservation [10], use of environmental friendly construction material and their conservation [11] and efficient construction processes in mind, are better for the environment as well as sustainability of the society. Also, in EEB and green building construction, the workplace satisfaction of the occupants [12,13] is considered in addition to indoor thermal comfort, psychological [14,15], and behavioral [16] benefits. A substantial growth in the environmental certifications have been realized in the twenty-first century both for residential and nonresidential buildings [17]. The EEB construction not only improves energy efficiency [9,13] but also significantly preserves natural resources [18,19], mitigates environmental hazards [18,20], and safeguards the ecosystem [21]. The environmental advantage is undeniably the most obvious benefit of EEB and green buildings, but there exist other advantages as well. For instance, there is some financial gain to be made by constructing an EEB despite that there is usually an extra upfront cost for the energy-efficient construction compared to conventional buildings [22,23]. According to Lau et al. [24], over 55% energy cost can be saved in office building if green building concepts are used in the construction. Ross et al. [23] also indicated significant money can be saved using green building concepts over conventional building design concepts.

It has to be noted here that reducing energy consumption in the operational energy is not the only way of improving energy efficiency of buildings, but energy consumption and GHGs emissions can also be minimized using appropriate construction materials. The energy consumed and emission of pollutants such as carbon dioxide (CO_2) in producing construction materials and building construction may be regarded as being embodied within materials. In another words, the embodied energy of primary production can be viewed as the quantity of energy required to process and supply the construction materials to the construction site [25]. Inevitably, the building sector, accounting for approximately 30%–40% of the world's energy consumption and 30% of raw material use, becomes a major target for environmental improvement. The sector is also responsible for 25% of total solid waste generation, 25% of water use, 12% of land use, and 33% of the related global GHG emissions [26,27]. Moreover, the current annual material consumption about 60%, particularly in housing, is expected continue to increase with the rapidly growing population [28]. Further, inclusion of embodied energy of construction materials and transport requirement with the mobility of building users makes the building sector the more dominant driver of energy use and GHG emissions [29]. Therefore, in the light of energy crises and climate change issues, it is urgent and crucial that the building sector should be given adequate attention to produce buildings with a minimum negative environmental impact.

In recent years, reducing energy (operational and embodied) use and GHG across a building's life has been the focus in the building research. In this chapter, we aimed to discuss the process of estimating embodied energy consumption in buildings and total energy in their life cycle. As embodied energy and operational energy are crucial to rate the buildings for their greenness and sustainability, the status of green building construction around the globe to date has to be included. Eventually, embodied energy fraction in passive, low-energy, and nearly zero buildings is compared over conventional construction.

5.2 Status of Green Building Construction

As a result of reduction in energy, material, water consumption, and least waste generation in green buildings construction, an average 15% appreciation has been estimated in the resale price of the building [30], and significant emphasis on the green building construct

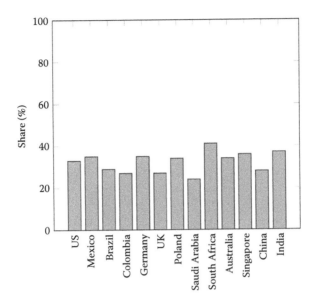

FIGURE 5.1
Status of green building construction percentage in different countries. (From Analytics DD, *World Green Building Trends 2016*, Dodge Data & Analytics, New York, 2016.)

and certification has been given under the leadership of different organizations and certification agencies around the world. Vyas and Jha [31] presented a study that focused on developing appropriate constructs to benchmark green building attributes in construction such that, with limited funds, the sustainable performance of the building is improved. The status of the share of green buildings in the total building infrastructure until 2015 has been estimated and shown in the following Figure 5.1 [32]. The status of green construction indicates how fast the concept is being adopted by the society to ensure the sustainability. The importance of the embodied energy in the green building construction is discussed in the other section of this chapter.

5.3 Significance of Embodied Energy

As indicated in previous section, construction materials play a significant role in the EEB/green building construction and embodied energy of materials is given adequate attention in EEB and green building rating and certification processes and kept under major elements (including sustainable site planning, water and wastewater management strategies, energy performance optimization, and quality of indoor environment, etc.). The share of different elements considered in the certification process is shown in Figure 5.2 [33,34]. Recent studies also indicated the significance of the embodied energy in total life cycle energy (LCE) of the buildings [35,36]. The material energy efficiency during the construction is ensured by choosing low embodied energy materials. Therefore, the embodied energy of the materials is assessed before adoption. Also, recycling and reusability of the construction materials after

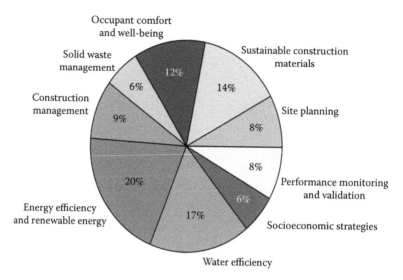

FIGURE 5.2
Weight given to various criteria in GRIHA, a green building certification tool, practiced in India. (From TERI, *GRIHA V 2016*, http://www.grihaindia.org/files/GRIHA_V2015_May2016.pdf, 2016.)

the useful life of the material is taken into account during the time of planning. Usually, in a lifespan of 50 years, embodied energy accounted for 45% of the total energy need, however, the percentage may vary with building typology, climatic conditions of the building location, and local availability of the desirable construction materials. The recycling potential was estimated between 35% and 40% of the embodied energy [37]. The embodied energy is expended once in the initial construction stage of a building, whereas operational energy accrues over the effective life of the building. Through material substitution, the embodied energy can be reduced by approximately 17% or increased by about 6% [19]. A literature survey assessed the proportion of embodied energy between 9% and 46% of the overall energy used over the building's lifetime for low-energy buildings and between 2% and 38% in conventional buildings. The survey was conducted for a total of 60 cases, which include residential and nonresidential buildings from nine countries (i.e., Sweden, Germany, Australia, Canada, and Japan) [38]. The embodied energy of construction materials depends on the production process of the material [39], while the embodied energy of buildings depends on ten parameters: system boundaries, primary and delivered energy, methods of embodied energy analysis, age of data sources, geographic location of study area, technology of manufacturing processes, source of data, completeness of data, temporal representativeness, and feedstock energy consideration [40].

5.4 Embodied Energy of Construction Materials and Buildings

Embodied energy: Embodied energy is defined as the energy consumed during the manufacturing phase of a material. The embodied energy of a building includes energy contents of all the materials used in the building and technical installations, as well as energy consumed at the time of erection/construction and renovation of the building. One of the major objectives in carrying out embodied energy analysis for building construction is to compare the total embodied energy content for different building materials, components, and designs. In the analysis, two major components are considered: initial and recurring embodied energies. The initial embodied energy is the sum of the energy required for extraction and manufacturing of a material and energy required in the transportation of the material to the construction site; the recurring embodied energy represents the sum of the embodied energies of the material used for maintenance, repair, restoration, refurbishment, or replacement during the service life of the building.

The embodied energy content for a building is generally estimated using the bottom up approach (also known as process-based approach) [41–43]. In the bottom–up approach, the embodied energy databases for construction materials as well as drawings, specifications, and/or data from the actual buildings are used in the calculation. The approach is generally quite accurate and reliable when information on building quantities, final drawings, and environmental impact databases for construction products are available. Distance of the building construction site from the construction material suppliers and waste management operators also plays a significant role in accurate estimation of the embodied energy of a building.

The quantity and type of construction materials and their embodied energy intensity factors are required to quantify the total embodied energy content of a given building design/building/element. The quantity of the materials is estimated from the technical drawings of the existing/newly proposed building to be evaluated. The initial embodied energy content is estimated using the following model:

$$E_{emb,initial,i} = E_{Extraction,i} + E_{manufacture,i} \qquad (5.1)$$

$$E_{emb,initial,i} = \sum_{1}^{i} (\alpha_i m_i) \qquad (5.2)$$

where:
$E_{emb,initial,i}$ is the initial embodied energy of the ith type of building material (in MJ)
$E_{emb,initial}$ is the initial embodied energy of the whole building (in MJ)
α_i is the embodied energy intensity factor for the ith type of building material (in MJ/kg)
m_i is the mass of the ith type of building material (in kg)
m_i should include not only the quantities of material in-place but also the wastages incurred during construction

The data of embodied energy intensities used in the calculation are not available precisely for each location/country in public domain. Therefore, the embodied energy intensities values for different materials, based on the literature, are listed in Table 5.1 and Figure 5.3.

Wastage of construction materials during the construction is very common and should be taken into account in the total embodied energy calculation. The values of waste

TABLE 5.1

Embodied Energy Intensities for Common Building Materials

Material	Energy Intensity	Material	Energy Intensity
Aggregate (general)	0.10 [44]	Insulation	
Virgin rock	0.04 [44]	Cellulose	3.30 [44]
River	0.02 [44]	Fiberglass	30.30 [44]
Aluminum (virgin)	191.00 [44]	Polyester	53.70 [44]
	236.8 [39]	Glass wool	14.00 [44]
Aluminum (recycled)	8.10 [44]		
Steel	42.0 [39]		
Steel, Virgin	32 [45]	Gravel	0.2 [45]
Asphalt (paving)	3.40 [44]	Paint	90.40 [44]
			93.3 (New Zealand) [45]
Cement	7.80 [44]		90.4 (China) [45]
	5.85 [39]	Solvent based	98.10 [44]
	6.15 [46]		
Cement mortar	2.00 [44]	Lime	5.63 [39]
Water	0.00814 [46]	Water based	76.00 [44]
		Aggregate	0.124 [46]
Ceramic		Plasterboard	6.10 [44]
Brick and tile	2.50 [44]	Plastics	
Brick (glazed)	7.20 [44]	PVC	70.00 [44]
Clay tile	5.47 [44]	Polyethylene	87.00 [44]
Concrete		Polystyrene	105.00 [44]
Block	0.94 [44]		117 (New Zealand) [45]
Brick	0.97 [44]		105 (China) [45]
Paver	1.20 [44]	Sealants and adhesives	
Precast	2.00 [44]	Phenol formaldehyde	87.00 [44]
Ready mix, 17.5 MPa	1.00 [44]	Urea formaldehyde	78.20 [44]
30 MPa	1.30 [44]	Steel (recycled)	10.10 [44]
	1.08 [46]	Reinforcing, section	8.90 [44]
Roofing tile	0.81 [44]	Reinforcement	8.08 [46]
Glass	25.8 [39]	Steel (virgin, general)	32.00 [44]
Float	15.90 [44,45]	Galvanized	34.80 [44]
Toughened	26.2 (New Zealand) [45]	Stainless	11.00 [44]
	26.4 (China) [45]	Timber (softwood)	
Laminated	16.30 [44]	Plywood	10.4 (New Zealand) [45]
Gypsum	8.64 [44]		18.9 (China) [44]
		Rough saw	5.18 [44]

factors depend on the type of building materials and are listed in Table 5.2. The actual waste factor may vary from these values according to the construction practices at different locations. Some effective measures, for example, prefabricated components and dimensionally coordinated modular flats can lead to smaller waste factors. Therefore, some of the countries already adopted these effective measures to minimize the waste

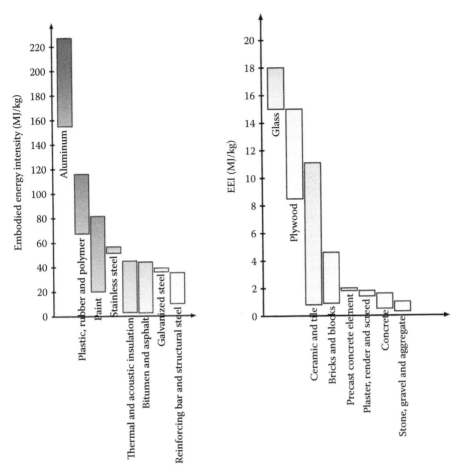

FIGURE 5.3
Embodied energy intensities for different types of building materials. (From Atmaca, A., and Atmaca, N., *Energ. Build.*, 102, 417–431, 2015; Baird, G. et al., *IPENZ Trans.*, 24, 46–54, 1997; Hammond, G.J.C., Inventory of Carbon and Energy (ICE) Version 2.0, *Bath DoMEUo*, Circular Ecology and University of Bath, UK, 2011; Huberman, N., and Pearlmutter, D., *Energ. Build.*, 40, 837–848, 2008; Kofoworola, O.F., and Gheewala, S.H., *Energ. Build.*, 41, 1076–1083, 2009.)

during the construction. Moreover, the life span of some building materials may not be equal to that of the building itself and may need to be replaced several times during the whole life of building. Also, furniture, paint, curtains, and few other components are replaced simply because of owner's choice. To ensure the high accuracy in the calculation of exact total embodied energy in the building lifetime, these factors should be taken into account in the estimation. The replacement factor for such components/materials can be calculated using the following formula:

$$\text{Replacement factor} = \frac{\text{Life span of building}}{\text{Average life span of component/material}} \qquad (5.3)$$

TABLE 5.2

The Value of Waste Factors for Different Types of Materials Used in the Construction of Buildings

Material	Waste Factor (%)	Material	Waste Factor (%)
Aluminium	2.5	Polystyrene	5
Coatings (paints and lacquers	5	Polythene	5
Concrete (reinforced)	2.5	Polyvinyl chloride (PVC)	5
Concrete (plain)	2.5	Steel	5
Copper	2.5	Tiles and clinkers	2.5
Glass	0	Timber (planed)	2.5
Gypsum wallboard	5	Timber (rough saw)	2.5
Mineral wool	5	Timber (shingles and shavings)	2.5

Source: Atmaca, A., and Atmaca, N., *Energ. Build.*, 102, 417–431, 2015.

The replacement factors for some common building materials and elements are given in Table 5.3.

Further, raw materials or semifinished products are generally first transported from source site to the construction site. The transportation of the construction materials may involve various modes of transportation to reach the user's site. Considerable energy is consumed in transportation of the construction materials and depends on the distance of the construction site from the supplier site as well as the mode of transportation used. Based on the literature [44], Table 5.4 gives general information about the energy use in different modes of transportation. The energy consumed in transportation of some of the common construction materials via road is listed in Table 5.5.

TABLE 5.3

The Replacement Factors of Typical Building Materials and Elements

Building Components	Replacement Factor	Building Components	Replacement Factor
Structural elements (column, beams, etc.)	1.0	Plastic carpeting	2.4
Flooring	1.0	Painting and wall papering	5.0
Wall and roofing tiles	1.3	Others	1.2
Ext./Int. walls	1.0–2.4	Doors	1.3–2.0
Windows	1.3–2.0	Paints and Coat.	5.0–15
Ceiling finishes	2.0–4.0	Acoustical tiles	2.5–3.75
Floor finishes	3.0–4.0	Vinyl flooring	2.5–4.16

Source: Chen, T.Y. et al., *Energy.*, 26, 323–340, 2001; Atmaca, A., and Atmaca, N., *Energ. Build.*, 102, 417–431, 2015.

TABLE 5.4

The Relatively Small Values among the Data of Energy Use in Different Modes of Transportation

Method of Transportation	Energy Use (MJ/(kg-km))
Deep-sea transport	0.216
Truck (road)	2.275
Coastal vessel	0.468
Class rail roads	0.275

Source: Atmaca, A., and Atmaca, N., *Energ. Build.*, 102, 417–431, 2015.

TABLE 5.5

Energy in Transportation of Building Materials

Method of Transportation	Energy Use (MJ/(km))
Sand (m³)	1.75
Crushed aggregate (m³)	1.75
Burnt clay bricks (m³)	2.00
Portland cement (tonnes)	1.00
Steel (tonnes)	1.00

Source: Reddy, B.V.V., and Jagadish, K.S., *Energ. Build.*, 35, 129–137, 2003.

The following revised mathematical formula can be used to estimate the total embodied energy of buildings, more precisely and accurately by including transporting, installing, and finishing the building materials and components during initial erection as well as renovation of the building [44]:

$$E_{emb,i} = E_{emb,initial,i} + E_{transport,i} + E_{installing,i} + E_{finishing,i} \qquad (5.4)$$

In recent past, significant emphasis has been given put on the assessment of the embodied energy in different countries and almost all types of buildings were investigated to identify the alternative low-energy construction materials to make buildings green and sustainable. Shukla et al. [52] evaluated the embodied energy for an adobe house and Debnath et al. [53] estimated for the load-bearing single-storey and multistorey concrete-structured buildings. The embodied energy of the load-bearing wall building was estimated lower than concrete-structured buildings [53]. Venkatarama Reddy and Jagdish [39] estimated the embodied energy of residential buildings using different types of masonry materials and roofing systems. In their study, soil–cement block masonry construction was found to be more energy efficient than burnt clay brick and concrete block masonry constructions. The soil cement block masonry and stabilized mud block (SMB) filler slab construction led to 62% lower embodied energy compared to RC framed structure building. These studies are confined to only analyzing and reducing the embodied energy of buildings.

In a Canadian office building, the embodied energy in a steel structure was estimated 1.61 times greater than that in a concrete structure, which was 1.27 times greater than that of a wooden structure [54]. Goggins et al. [46] assessed the embodied energy for a typical RC building structure in Ireland. The embodied energy of a typical 30 MPa concrete mix in Ireland was calculated to be 1.08 MJ/kg. Through a case study of a 3-story office block in Galway city, they showed that a 30% reduction in the embodied energy of the slab panels can be achieved by replacing 50% of the cement content with ground granulated blast slag (GGBS). Chang et al. [55] calculated the embodied energy equal to 309,965 GJ (6.3 GJ m^{-2}) for an education building in China. The intensity was higher than the mean value of residential buildings (5.5 GJ m^{-2}) but much lower than commercial buildings (9.2 GJ m^{-2}) in the U.K., Australia, and Japan. Reddy et al. [56] investigated the embodied energy of a school building built using cement stabilized rammed earth (CSRE). The embodied energy was estimated at 1.15 GJ/m^2, which is lower for the CSRE construction than the embodied energy value of 3–4 GJ/m^2 for conventional burnt clay brick load-bearing masonry construction. Chel and Tiwari [57] investigated thermal performance of an existing eco-friendly and low-embodied energy vault roof passive house, mud-house, located at Indian Institute of Technology (IIT) Delhi, India. The embodied energy analysis resulted in energy payback period of 18 years for the studied house. The period could be significantly high for RCC construction as the embodied energy per unit floor area of RCC construction estimated 3702.3 MJ/m^2 is quite high compared to the mud-house (for which the embodied energy was estimated 2298.8 MJ/m^2). Mandley et al. [58] estimated significant reductions potential for the UK building sector in annual material and embodied energy consumption in the short to midterm. Their projections indicated 4.7% and 6.4% by 2020 and 9.3% and 28.6% by 2030 savings in resource and embodied energy, respectively. Most of these studies ignored the transportation and wastage factors in the analysis.

5.5 Life Cycle Energy of Buildings

The total energy inputs, outputs, and flow through the life cycle of the building are estimated using the detailed life cycle energy analysis (LCEA) approach (see Figure 5.4). The boundaries of the system are expanded in order to take both the operating energy and the embodied energy into account [59]. That means that the LCE of a building includes the energy usages in production of building material, construction, operation, maintenance, disassembly, and waste management [2,60].

The operating energy is the energy used in the operation of a building, for example, energy consumed in space cooling, heating, ventilation, lighting, and to run domestic appliances. Out of all these operations, energy for space cooling and heating has the largest share and depends on the heat gain or loss of the building. Both operational and LCE can be minimized by minimizing the building envelope's heat loss/heat gain through

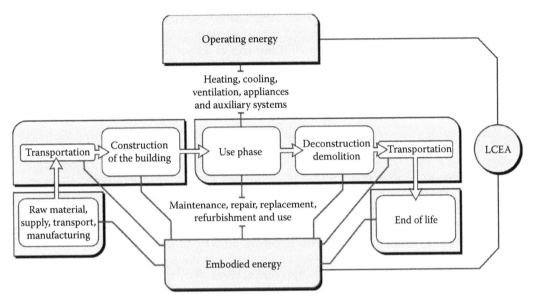

FIGURE 5.4
Boundaries of the system, inputs outputs and flows in Life Cycle Energy Analysis (LCEA) of buildings. (From Chastas, P. et al., *Build. Environ.*, 105, 267–282, 2016.)

appropriate selection of construction materials, used in different layers. The low thermal conductivity and considerable heat capacity at parity of all other conditions are essentially required to keep heat loss/heat gain. Moreover, the LCE can be reduced further if embodied energy of such materials is low. The LCE of the selected building is evaluated based on an assumed life span of 50–75 years using the following formula:

$$LCE = \sum m_i e_i + E_c + \overbrace{\sum m_i e_i \left[\left(\frac{L_b}{L_{mi}} \right) - 1 \right]}^{\text{Recurring embodied energy}} + \underbrace{E_o A L_b}_{\text{Operational energy}} + E_D + E_T \qquad (5.5)$$

Embodied energy

The first term in the right-hand side of the above equation represents building embodied energy, which includes E_c, the energy used at site for erection/construction of the building. The second term is used for recurring embodied energy includes embodied energy for replacement construction and finishing materials to rehabilitate the building and energy used in regular annual maintenance. The terms L_b and L_{mi} represent life span of the building and life span of the material (1), respectively. The third term in the above equation represents the operational energy, which is required for maintaining comfort conditions and day-to-day maintenance of the buildings. The thermal comforts are maintained by using heating, ventilation, and air-conditioning (HVAC) and day-to-day requirements, such as domestic hot water, lighting, and for running appliances. The operational energy varies largely depending on the level of required thermal comfort, climatic conditions, and operating schedules. At the end of service life of buildings, the buildings are demolished, and waste materials are required to be transported to the landfill site. Some energy is used in demolishing and waste transportation. The last two terms in the above equation are used for energy used in demolishing and transportation. In some cases, the energy used for on-site construction and demolition at the end of its service life are ignored as they contribute little (nearly 1%) to LCE.

The LCE use of buildings has been quantified, in few studies, across different scales of the built environment. Utama and Gheewala [61] evaluated LCE of a residential apartment in Jakarta, Indonesia, with two envelope materials: (a) double walls having external walls made from clay bricks, inner walls with gypsum plaster board, and an air gap in between and (b) single walls with clay bricks. Double walls had resulted in better energy performance (40% less) than single walls. Previously, Utama and Gheewala [62] also analyzed clay and cement-based single landed houses in Indonesia and observed that energy consumption of a clay house was lower than a cement house. Norman et al. [63] compared the LCE use of low and high density residential buildings around Toronto, Canada, including transport requirements. However, their study significantly underestimated embodied energy requirements. Recently, Fuller and Crawford [64] evaluated the life cycle, energy use, and GHG emissions for different housing patterns in and around Melbourne, Australia. They relied on the comprehensive hybrid analysis to quantify the embodied energy and their study has not evaluated specific energy reduction strategies and financial feasibility. Medgar and Martha [65] presented a life cycle assessment of a single-family house modeled with two types of exterior walls: wood-framed and insulating concrete form (ICF). The house was modeled in five different climates in the US. The energy use was estimated greater for the wood house over the ICF house in most of the studied cases in the selected climate. Xing et al. [66] presented the life cycle assessment for an office building constructed in China using steel and concrete. They observed that the embodied energy and environmental emissions of the steel-framed building were superior to the concrete-framed one. However, operational energy use and associated emissions were larger for the steel-framed building as a result of the higher thermal conductivity of steel than concrete. The LCE -consumption and environmental emissions of the steel-framed building were estimated slightly higher. Aste et al. [67] revealed that high thermal insulation together with the considerable thermal mass increases energy performance of buildings. Citherlet and Defaux [68] analyzed and compared a family house by changing its insulation thickness and type. The study confirmed that good insulation provides a significant reduction of energy (about 50%). Mithraratne and Vale [69] recommended provision of higher insulation to a timber-framed house as an energy-saving strategy for low-energy housing in the New Zealand context. Recently, T. Ramesh et al. [70] evaluated LCE demand for a residential building of usable floor area about 85.5 m² for five different Indian climates. The study was performed for conventional (fired clay) and other alternative wall materials including hollow concrete, soil cement, fly ash, and aerated concrete by varying the thickness of wall and insulation (expanded polystyrene) on wall and roof. The LCE savings in the building, with and without insulation, on wall and roof in five climates is shown in Figure 5.5. It is clearly shown in the Figure 5.5 that applying insulation on the roof is more beneficial than on the walls. Moreover, increasing the thickness of insulation yielded higher savings in all the climates except in moderate climate. Hence, on the roof, adding more insulation thickness above 5 cm would not be a prudent decision in the moderate climate.

In general, the construction materials represent more than 50% of the embodied energy in the building [71]. In this sense, the use of alternative materials, such as hollow

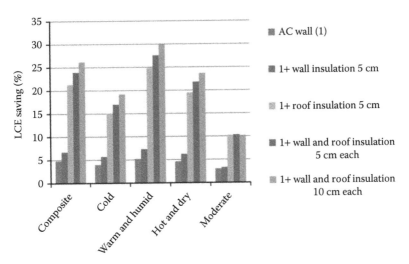

FIGURE 5.5
Life cycle energy savings (%) for five climatic conditions of India. (From Ramesh, T. et al., *Appl. Energ.*, 89, 193–202, 2012.)

concrete blocks, stabilized soil blocks or fly-ashes instead of high-embodied energy materials such as reinforced concrete, could save 20% of the cumulative energy over a 50-year life cycle [50]. Moreover, recycling of building materials [37,72] can significantly reduce the embodied energy in the building. For example, the use of recycled steel and aluminum confers more than 50% savings in embodied energy [44]. Ramesh et al. [73] showed that the embodied energy represents a share between 10% and 30% of the total LCE consumption in commercial buildings and between 5% and 60% in residential buildings (see Figure 5.6). Their conclusion indicated that the embodied energy of buildings cannot be neglected, especially because the operational energy is being continuously reduced via multipronged efforts related to technology and policy aspects, such as improvement of HVAC performance, utilization of new and renewable energy, adoption of the zero-energy building (nZEB) design concepts, and implementing of green building policies.

Stephan et al. [74,75] and Stephan and Crawford [76] have evaluated the overall LCE profile for various residential buildings in Australia, Belgium, and Lebanon. In their study, they have included embodied, operational, and transport requirements, however, financial requirements have been ignored. Most recently, Stephan and Stephan [77] quantified the LCE and cost requirements associated with 22 different energy reduction measures targeting embodied, operational, and user-transport requirements for an apartment building in Sehaileh, Lebanon. The embodied, operational, and transport energy requirements were estimated over 50 years using a comprehensive approach. Recently, 90 case studies of residential buildings evaluated between 1997 and 2016, around the world, were reviewed. The building life span of the studied cases ranges between 30 and 100 years with a most common value of 50 years. The share of embodied energy in the total LCE of the examined buildings ranges between 5% and 100%. The percentage of embodied energy in conventional buildings ranges between 5% and 36%; in low-energy buildings the share varies between 10% and 83%, and for nearly nZEB the share of embodied energy

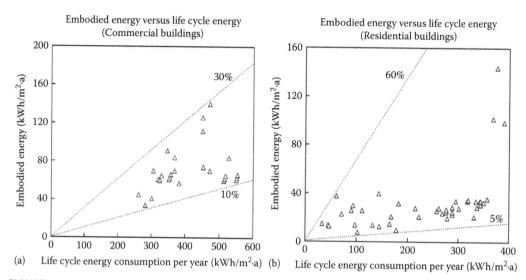

FIGURE 5.6
Relation between embodied and total energy: (a) Office buildings; (b) residential buildings. (From Stephan, A. et al., *Energ. Build.*, 55, 592–600, 2012.)

ranges between 69% and 100% for a nearly zero balance [59]. They have identified a significant gap of 17% as the difference between the minimum share of embodied energy of an nZEB and the maximum share of embodied energy of a low-energy building. The gap appeared to increase for the passive and conventional buildings, respectively. It can also be observed from Figure 5.6 that a low-energy building is not more energy efficient than a passive one

Most recently, a study for low-rise apartment residential buildings in Iskandar Malaysia Johor Bahru clearly indicated that IBS has a better advantage in terms of reducing embodied energy (MJ) and GWP (kg CO_2-Equiv.) toward a low carbon development [78]. In this case, concrete and steel were major contributors to the embodied energy and emissions to the environment. Atmaca and Atmaca [47] assessed the LCE and carbon dioxide emission for two actual residential buildings constructed in Gaziantep, Turkey. Their results indicated that the operation phase was dominant in both urban and rural residential buildings and contributed 76%–73%of the primary energy requirements and 59%–74% of CO_2 emissions, respectively. The embodied energy of the buildings was estimated in a range of 24%–27% of the overall life cycle energy consumption. Su and Zhang [79] have created a model for assessment of embodied energy consumption and carbon emissions in three steel constructed residential buildings in China. The model included the materials production phase, transportation phase, construction phase, recycle and demolition phases, and upstream of energy. The direct materials and energy consumption of these three residential buildings with different volumes were investigated on site. The results showed that the embodied energy consumption of steel members, concrete, and cement account for more than 60% of the total energy consumption of all building components. They also concluded that the embodied energy and environment issues of the building components of the steel-construction buildings are sensitive to building height rather than building volumes (Figure 5.7).

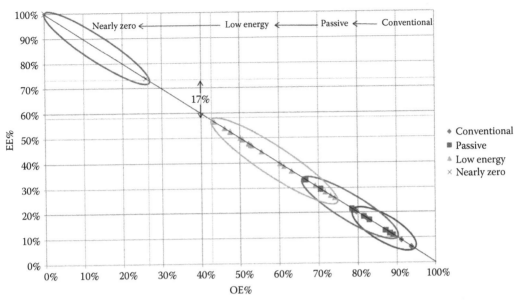

FIGURE 5.7
Total share of embodied (EE%) and operating energy (OE%) in the Life Cycle Energy Analysis (LCEA) of the 90 case studies of residential buildings. (From Chastas, P. et al., *Build. Environ.*, 105, 267–282, 2016.)

5.6 Conclusion

The building sector has been identified as one of the major energy and resource-consuming sectors to deal with the growing global warming, climate change issues, and rapidly depleting fossil fuels. The buildings are responsible for 30%–40% of global energy consumption, 30% raw material, and 25% water consumption, 12% land use and 25% solid waste generation. Also, 33% of global GHG emissions come from the building sector. These statistics have attracted scientists, policy makers, politicians, and other stakeholders' attention to tackle the above issues adopting sustainable solutions for building applications. Green building design strategies and integrating renewable energy technologies have been developed and are being utilized in the recent years. As a result, significant progress has been observed in the area of green building development in both developing as well as developed countries. In the EEB and green building concepts, a significant weight has been given to the sustainable construction materials, energy efficiency, and renewable energy. The embodied energy of building, which is a result of construction material consumption in the construction of buildings, is one of the aspects that help in green building and sustainable development. The chapter explores the estimation processes of embodied energy and LCE assessment in detail and how these methods have used to benefit the green building development. It has also been observed that shares of embodied energy in the total LCE vary between 5% and 36% for conventional buildings, between 10% and 83% for low-energy buildings and between 69% and 100% in the case of nearly nZEB. Hence, the variation in the embodied energy shares clearly indicates that selection of building materials, and accurate and detailed assessment of embodied energy and LCE are crucial, and adequate attention has to be given at the planning, designing, and construction stages of buildings.

References

1. Dakwale VA, Ralegaonkar RV, Mandavgane S. Improving environmental performance of building through increased energy efficiency: A review. *Sustain. Cities Soc.* 2011;1:211–218.
2. Adalberth K. Energy use during the life cycle of single-unit dwellings: Examples. *Build. Environ.* 1997;32:321–329.
3. Zeng Y, Yang C, Zhang J, Pu W. Feasibility investigation of oily wastewater treatment by combination of zinc and PAM in coagulation/flocculation. *J Hazard Mater.* 2007;147:991–996.
4. Levin H. Systematic evaluation and assessment of building environmental performance (SEABEP). Proc. Second International Conference on Buildings and the Environment, CSTB and CIB, 1997;2:3–10.
5. GhaffarianHoseini A, Dahlan ND, Berardi U, GhaffarianHoseini A, Makaremi N, GhaffarianHoseini M. Sustainable energy performances of green buildings: A review of current theories, implementations and challenges. *Renew. Sustain. Energ. Rev.* 2013;25:1–17.
6. Hammond GP. Energy, environment and sustainable development: A UK perspective. *Process. Saf. Environ.* 2000;78:304–323.
7. Yu H, Tian X, Yuan A. Green design in low carbon environment. *International Symposium on Water Resource and Environmental Protection*, Xian, China, 2011.
8. Zuo J, Zhao ZY. Green building research-current status and future agenda: A review. *Renew. Sustain. Energ. Rev.* 2014;30:271–281.
9. Turner C, Frankel M. *Energy Performance of LEED for New Construction Buildings.* Vancouver, WA: New Buildings Institute, 2008.
10. USGBC. *US Green Building Council USGBC History.* Washington, DC: USGBC, 2009.
11. Ren HY, Shun WA, Chuan WA. General analysis on green building materials and development in China. *Appl. Mech. Mater.* 2015;744–746:1427–1430.
12. Kim SK, Hwang Y, Lee YS, Corser W. Occupant comfort and satisfaction in green healthcare environments: A survey study focusing on healthcare staff. *J. Sustain. Dev.* 2015;8:156–173.
13. Thatcher A, Milner K. Changes in productivity, psychological wellbeing and physical wellbeing from working in a "green" building. *Work* 2014;49:381–393.
14. Allen JG, MacNaughton P, Satish U, Santanam S, Vallarino J, Spengler JD. Associations of cognitive function scores with carbon dioxide, ventilation, and volatile organic compound exposures in office workers: A controlled exposure study of green and conventional office environments. *Environ. Health Perspect.* 2016;124:805–812.
15. Armitage L, Murugun A, Kato H. Green offices in Australia: A user perception survey. *J. Corp. Real Estate* 2011;13:169–180.
16. Khashe S, Heydarian A, Gerber D, Becerik-Gerber B, Hayes T, Wood W. Influence of LEED branding on building occupants' pro-environmental behavior. *Build. Environ.* 2015;94:477–488.
17. USGBC. Green-building-facts.
18. Coelho A, de Brito J. Influence of construction and demolition waste management on the environmental impact of buildings. *Waste Manage.* 2012;32:532–541.
19. Thormark C. The effect of material choice on the total energy need and recycling potential of a building. *Build. Environ.* 2006;41:1019–1026.
20. Wang WM, Zmeureanu R, Rivard H. Applying multi-objective genetic algorithms in green building design optimization. *Build. Environ.* 2005;40:1512–1525.
21. Bianchini F, Hewage K. How "green" are the green roofs? Lifecycle analysis of green roof materials. *Build. Environ.* 2012;48:57–65.
22. Langdon D. Cost and benefit of achieving green. Build Efficiency Initiative, 2007. (https://www.usgbc.org/Docs/Archive/General/Docs2583.pdf). Accessed on 27/05/2017.
23. Ross B, Lopez-Alcala M, Small AA. Modeling the private financial returns from green building investments. *J. Green Build.* 2007;2:97–105.

24. Lau LC, Tan KT, Lee KT, Mohamed AR. A comparative study on the energy policies in Japan and Malaysia in fulfilling their nations' obligations towards the Kyoto Protocol. *Energ. Policy* 2009;37:4771–4778.

25. Slesser M. *Dictionary of Energy*, 2nd ed. New York: Macmillan, 1998.

26. UNEP. Buildings and climate change: Summary for decision-makers. In: *Programme UNE* (Ed.), Paris, France: United Nations Environment Programme (UNEP), 2008.

27. UNEP. Common carbon metric for measuring energy use and reporting greenhouse gas emissions from building operations. In: *Programme UNE* (Ed.), 2009.

28. Chani PS. Comparative analysis of embodied energy rates for walling elements in India. *IE(I) J. Arch. Eng.* 2005;84:47–50.

29. Anderson JE, Wulfhorst G, Lang W. Energy analysis of the built environment-A review and outlook. *Renew. Sustain. Energ. Rev.* 2015;44:149–158.

30. Nayara Kasai CJCJ. Barriers to green buildings at two Brazilian engineering schools. *Int. J. Sustain. Built. Environ.* 2014;3:87–95.

31. Vyas GS, Jha KN. Benchmarking green building attributes to achieve cost effectiveness using a data envelopment analysis. *Sustain. Cities Soc.* 2017;28:127–134.

32. Analytics DD. *World Green Building Trends 2016*, Dodge Data & Analytics, New York, 2016.

33. Oh JK, Lee JK, Han B, Kim SJ, Park KW. TiO_2 rutile nanowire electrodes for dye-sensitized solar cells. *Mater. Let.* 2012;68:4–7.

34. TERI, *GRIHA V 2016*, http://www.grihaindia.org/files/GRIHA_V2015_May2016.pdf, 2016.

35. Crawford RH, Treloar GJ. Validation of the use of Australian input output data for building embodied energy simulation. *Eighth International IBPSA Conference*, Eindhoven, the Netherlands, 2003.

36. Pullen SF, Holloway D, Randolph B, Troy P. Energy profiles of selected residential developments in Sydney with special reference to embodied energy. *Australian and New Zealand Architectural Science Association, 40th Annual Conference Challenge for Architectural Science in Changing Climate*, Adelaide, Australia, 2006.

37. Thormark C. A low energy building in a life cycle—Its embodied energy, energy need for operation and recycling potential. *Build. Environ.* 2002;37:429–435.

38. Sartori I, Hestnes AG. Energy use in the life cycle of conventional and low-energy buildings: A review article. *Energ. Build.* 2007;39:249–57.

39. Reddy BVV, Jagadish KS. Embodied energy of common and alternative building materials and technologies. *Energ. Build.* 2003;35:129–137.

40. Dixit MK, Fernandez-Solis JL, Lavy S, Culp CH. Identification of parameters for embodied energy measurement: A literature review. *Energ. Build.* 2010;42:1238–1247.

41. Heinonen J, Junnila S. A carbon consumption comparison of rural and Urban lifestyles. *Sustain. Basel.* 2011;3:1234–1249.

42. Peuportier BLP. Life cycle assessment applied to the comparative evaluation of single family houses in the French context. *Energ. Build.* 2001;33:443–450.

43. Junnila S. Life cycle assessment of environmentally significant aspects of an office building. *J. Real Estate Res.* 2004;2:81–97.

44. Chen TY, Burnett J, Chau CK. Analysis of embodied energy use in the residential building of Hong Kong. *Energy* 2001;26:323–340.

45. Jiao Y, Lloyd CR, Wakes SJ. The relationship between total embodied energy and cost of commercial buildings. *Energ. Build.* 2012;52:20–27.

46. Goggins J, Keane T, Kelly A. The assessment of embodied energy in typical reinforced concrete building structures in Ireland. *Energ. Build.* 2010;42:735–744.

47. Atmaca A, Atmaca N. Life cycle energy (LCEA) and carbon dioxide emissions (LCCO(2)A) assessment of two residential buildings in Gaziantep, Turkey. *Energ. Build.* 2015;102:417–431.

48. Baird G, Alcorn A, Haslam P. The energy embodied in building materials updated New Zealand coefficients and their significance. *IPENZ Trans.* 1997;24:46–54.

49. Hammond G JC. Inventory of Carbon and Energy (ICE) Version 2.0, *Bath DoMEUo*, Circular Ecology and University of Bath, UK, 2011.

50. Huberman N, Pearlmutter D. A life-cycle energy analysis of building materials in the Negev desert. *Energ. Build.* 2008;40:837–848.

51. Kofoworola OF, Gheewala SH. Life cycle energy assessment of a typical office building in Thailand. *Energ. Build.* 2009;41:1076–1083.

52. Shukla A, Tiwari GN, Sodha MS. Embodied energy analysis of adobe house. *Renew. Energ.* 2009;34:755–761.

53. Debnath A, Singh SV, Singh YP. Comparative assessment of energy requirements for different types of residential buildings in India. *Energ. Build.* 1995;23:141–146.

54. Cole RJ, Kernan PC. Life-cycle energy use in office buildings. *Build. Environ.* 1996;31:307–317.

55. Chang Y, Ries RJ, Lei SH. The embodied energy and emissions of a high-rise education building: A quantification using process-based hybrid life cycle inventory model. *Energ. Build.* 2012;55:790–798.

56. Reddy BVV, Leuzinger G, Sreeram VS. Low embodied energy cement stabilised rammed earth building-A case study. *Energ. Build.* 2014;68:541–546.

57. Chel A, Tiwari GN. Thermal performance and embodied energy analysis of a passive house—Case study of vault roof mud-house in India. *Appl. Energ.* 2009;86:1956–1969.

58. Mandley S, Harmsen R, Worrell E. Identifying the potential for resource and embodied energy savings within the UK building sector. *Energ. Build.* 2015;86:841–851.

59. Chastas P, Theodosiou T, Bikas D. Embodied energy in residential buildings-towards the nearly zero energy building: A. literature review. *Build Environ.* 2016;105:267–282.

60. Gustavsson L, Joelsson A. Life cycle primary energy analysis of residential buildings. *Energ. Build.* 2010;42:210–220.

61. Utama A, Gheewala SH. Indonesian residential high rise buildings: A life cycle energy assessment. *Energ. Build.* 2009;41:1263–1268.

62. Utama A, Gheewala SH. Life cycle energy of single landed houses in Indonesia. *Energ. Build.* 2008;40:1911–1916.

63. Norman J, MacLean HL, Kennedy CA. Comparing high and low residential density: Life-cycle analysis of energy use and greenhouse gas emissions. *J. Urban Plan D-ASCE* 2006;132:10–21.

64. Fuller RJ, Crawford RH. Impact of past and future residential housing development patterns on energy demand and related emissions. *J. Hous. Built. Environ.* 2011;26:165–183.

65. Marceau, ML, VanGeem MG. Comparison of the life cycle assessments of an insulating concrete form house and a wood frame house. *J ASTM Int.* 2006;3:1–11.

66. Xing S, Xu Z, Jun G. Inventory analysis of LCA on steel- and concrete-construction office buildings. *Energ. Build.* 2008;40:1188–1193.

67. Aste N, Adhikari RS, Buzzetti M. Beyond the EPBD: The low energy residential settlement Borgo Solare. *Appl. Energ.* 2010;87:629–642.

68. Citherlet S, Defaux T. Energy and environmental comparison of three variants of a family house during its whole life span. *Build. Environ.* 2007;42:591–598.

69. Mithraratne N, Vale B. Life cycle analysis model for New Zealand houses. *Build. Environ.* 2004;39:483–492.

70. Ramesh T, Prakash R, Shukla KK. Life cycle energy analysis of a residential building with different envelopes and climates in Indian context. *Appl. Energ.* 2012;89:193–202.

71. Asif M, Muneer T, Kelley R. Life cycle assessment: A case study of a dwelling home in Scotland. *Build. Environ.* 2007;42:1391–1394.

72. Blengini GA. Life cycle of buildings, demolition and recycling potential: A case study in Turin, Italy. *Build. Environ.* 2009;44:319–330.

73. Ramesh T, Prakash R, Shukla KK. Life cycle energy analysis of buildings: An overview. *Energ. Build.* 2010;42:1592–1600.

74. Stephan A, Crawford RH, de Myttenaere K. Towards a comprehensive life cycle energy analysis framework for residential buildings. *Energ. Build.* 2012;55:592–600.

75. Stephan A, Crawford RH, de Myttenaere K. A comprehensive assessment of the life cycle energy demand of passive houses. *Appl. Energ.* 2013;112:23–34.
76. Stephan A CR. A multi-scale life-cycle energy and greenhouse-gas emissions analysis model for residential buildings. *Architect. Sci. Rev.* 2014;57:39–48.
77. Stephan A, Stephan L. Life cycle energy and cost analysis of embodied, operational and user-transport energy reduction measures for residential buildings. *Appl. Energ.* 2016;161:445–464.
78. Wen TJ, Siong HC, Noor ZZ. Assessment of embodied energy and global warming potential of building construction using life cycle analysis approach: Case studies of residential buildings in Iskandar Malaysia. *Energ. Build.* 2015;93:295–302.
79. Su X, Zhang X. A detailed analysis of the embodied energy and carbon emissions of steel-construction residential buildings in China. *Energ. Build.* 2016;119:323–330.

6

Building Integrated Photovoltaic: Building Envelope Material and Power Generator for Energy-Efficient Buildings

Karunesh Kant, Amritanshu Shukla, and Atul Sharma

CONTENTS

6.1 Introduction 110
   6.1.1 Green Building Rating System in India 111
      6.1.1.1 Green Rating for Integrated Habitat Assessment 112
      6.1.1.2 Indian Green Building Council 112
      6.1.1.3 Bureau of Energy Efficiency 113
6.2 Photovoltaic Technology 114
   6.2.1 Single-Crystal Silicon Solar Cell 114
   6.2.2 Gallium Arsenide Solar Cell 114
   6.2.3 Thin-Film Solar Cell 114
   6.2.4 Amorphous Silicon Solar Cell 115
   6.2.5 Cadmium Telluride Solar Cell 116
   6.2.6 Organic Solar Cells 116
   6.2.7 Efficiency of Different Solar Cells 117
6.3 Building Integrated Photovoltaic 119
   6.3.1 Building Integrated Photovoltaic Products 119
      6.3.1.1 Building Integrated Photovoltaic Foil Products 119
      6.3.1.2 Building Integrated Photovoltaic Tile Products 119
      6.3.1.3 Building Integrated Photovoltaic Module Products 119
      6.3.1.4 Solar Cell Glazing Products 120
      6.3.1.5 Building Attached Photovoltaic Products 120
6.4 Methods of Installation 120
   6.4.1 Roof Integration of Photovoltaic 121
   6.4.2 Facade Integration of Phototvoltaic 124
6.5 Challenges of Building Integrated Photovoltaic 125
   6.5.1 Price 125
   6.5.2 Performance 125
   6.5.3 Codes and Standards 126
   6.5.4 Market Limitations 126
6.6 Role of Building Integrated Photovoltaic for Developing Smart Cities 127
6.7 Summary 128
References 128
Endnote 129

6.1 Introduction

Energy demand is projected to upsurge significantly in the upcoming years as the consequence of population growth and economic growth (EIA 2007). Numerous people in the world are currently experiencing dramatic shifts in lifestyle as their financial prudence make the transition from subsistence to an industrial or service base. The largest rises in energy demand will come to pass in developing nations where the proportion of global energy consumption is expected to upsurge from 46% to 58% between 2004 and 2030 (EIA 2007). Per capita, energy consumption figures are still likely to remain well below those in Organization for Economic Cooperation and Development (OECD) countries. Figure 6.1 shows the consumption and production of energy in different sources. The total energy consumed in the various sector is around 8679 (Mton) and building sector consumes 2911 (Mton), which is about 33.5% of total energy consumption. The direct use of renewable energy is around 29.6% of total energy consumed in buildings. The use of renewable is growing due to is availability and because it is environmental friendly. Research and development on energy proficiency and the use of renewable energy in the building sector have until recently principally focused on small-scale residential buildings and have been attained in terms of reducing energy use in this type of buildings.

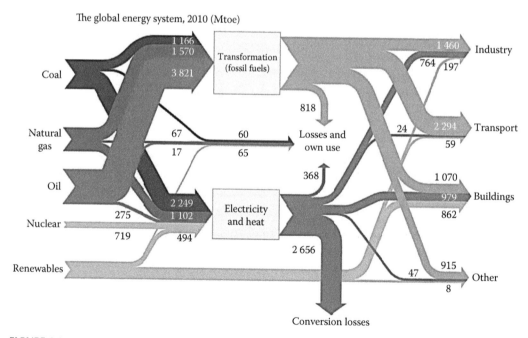

FIGURE 6.1
Total global energy consumption in 2010. (From Absi Halabi, M. et al., *Renew. Sustain. Energ. Rev.*, 43, 296–314, 2014.)

A "sustainable" or "green" building consumes main resources such as water, materials, energy, and land more efficiently than conventional buildings that are just built to code. With more natural lighting and enriched air quality, green buildings usually contribute to enhanced human comfort, health, and productivity. The United States Green Building Council (USGBC), a general non-profit membership association, established the Leadership in Energy and Environmental Design (LEED)™ system to offer a guideline and rating system for green buildings. If building green is cost effective, an extensive shift to green building offers a potentially auspicious way to help address a range of challenges facing Massachusetts, including (Kats and Capital 2003).

- Address growing costs of transmission and distribution congestion. The progress of Time of Use rates (TOU) by Massachusetts utilities and the creation of congestion pricing in the form of locational minimal pricing allows building holders to capture some of the benefits associated with lower overall and lower peak energy use in green buildings (Kats and Capital 2003).
- Reduce or slow rise in electricity and gas prices through expanded green construction and building retrofits and abridged energy demand.
- Improve the quality of the educational environment and improve school test scores.
- Improve competitiveness by providing work and living environments characterized by superior health and comfort in these environments.

Following are some relevant attributes common in green buildings that promote improved work environments.

- On average 25%–30% more energy efficient.
- Much lower source emissions from measures such as enhanced sitting (e.g., avoiding locating air intakes next to outlets, such as parking garages, and avoiding recirculation), and better building material source controls (e.g., required attention to storage). Certified and Silver level green buildings achieved 55% and Gold level *LEED* buildings achieved 88% of possible *LEED* credits for use of the following: (Kats and Capital 2003) less toxic
- Significantly better lighting quality including more daylighting (half of 21 *LEED* green buildings reviewed provide daylighting to at least 75% of building space [Kats and Capital 2003]), improved daylight harvesting and use of shading, greater occupancy control over light levels and less glare.
- Commonly enhanced thermal comfort and better ventilation—especially in buildings that use under floor air for space conditioning.
- Commissioning or use of measurement confirmation and CO_2 monitoring to ensure better performance of systems such as ventilation, heating, and air-conditioning.

6.1.1 Green Building Rating System in India

Whether buildings are really green is to be decided in contradiction of the predefined rating systems. There are three main Rating systems in India (Chaudhari et al. 2013).

6.1.1.1 Green Rating for Integrated Habitat Assessment

Green Rating for Integrated Habitat Assessment (*GRIHA*) is India's own rating system jointly developed by TERI and the Ministry of New and Renewable Energy, Government of India. The *GRIHA* rating system consists of 34 criteria categorized in to four different sections. Some of them are as follows: (1) site selection and site planning, (2) conservation and efficient utilization of resources, (3) building operation and maintenance, and (4) innovation.

Commonwealth Games Village, New Delhi; Fortis Hospital, New Delhi; Centre for Environmental Sciences & Engineering (CESE) Building, IIT Kanpur; Suzlon One Earth, Pune; and many other buildings have received *GRIHA* rating.

6.1.1.2 Indian Green Building Council

The LEED is the rating system developed by the US Green Building Council (USGBC), the organization promoting sustainability through Green Buildings. LEED is a framework for assessing building performance against set criteria and standard points of references. The benchmarks for the LEED Green Building Rating System were developed in the year 2000 and are currently available for new and existing constructions (Figure 6.2).

Confederation of Indian Industry (CII) formed the Indian Green Building Council (*IGBC*) in the year 2001. It is the nonprofit research institution having its offices in CII–Sohrabji Godrej Green Business Centre, which is itself a *LEED* certified Green building. *IGBC* has established the following green building rating systems for different types of building

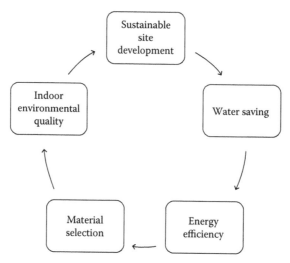

FIGURE 6.2
LEED India approach for green building.

TABLE 6.1

Some Examples of LEED-Rated Building in India

S. No	Green Buildings	Rating Received
1	ABN Amro Bank N.V., Ahmedabad	LEED "Platinum" rated
2	American Embassy School, Delhi	LEED "Gold" rated
3	Anna Centenary Library Building, Chennai	LEED "Gold" rated
4	Biodiversity Conservation India Ltd (BCIL)—Bangalore	LEED "Platinum" rated
5	Birla International School, Jaipur	LEED "Gold" rated
6	CII—Sohrabji Godrej Green Business Centre	LEED "Platinum" rated
7	ITC Green Centre—Gurgaon	LEED "Platinum" rated
8	Olympia Technology Park—Chennai	LEED "Gold" rated
9	Rajiv Gandhi International Airport—Hyderabad	LEED "Silver" rated
10	Suzlon Energy Limited—global headquarter in Pune	LEED "Platinum" rated

in line and conformity with US Green Building Council. To date, the following Green Building rating systems are available under *IGBC* (Table 6.1).

- LEED India for New Construction
- LEED India for Core and Shell
- IGBC Green Homes
- IGBC Green Factory Building
- IGBC Green SEZ
- IGBC Green Townships

6.1.1.3 *Bureau of Energy Efficiency*

The Bureau of Energy Efficiency (BEE) developed its own rating system for the buildings based on a 1 to 5-star scale. More stars mean more energy efficiency. BEE has developed the Energy Performance Index (EPI). The unit of Kilowatt hours per square meter per year is considered for rating the building and especially targets air conditioned and non-air-conditioned office buildings. The Reserve Bank of India's buildings in Delhi and Bhubaneswar, the CII Sohrabji Godrej Green Business Centre and many other buildings have received BEE 5 star ratings. Indians were aware of Green Building concepts from the beginning. Conventional homes with baked red color roof tiles and clay made walls is a really good example of energy-efficient structures that are used to keep cool during summers and warm during the winters. Most of the rural India is still attached to this building technology with naturally available materials like clay, wood, jute ropes, and so on. Today we have advanced technologies that create smarter systems to control temperature, lighting systems, power and water supply and waste generation. Green buildings might be a bit heavy on the purse but are good for the environment. In this rapidly changing world, we should adopt the technology that helps us to save precious natural resources. This would lead us to truly sustainable development.

6.2 Photovoltaic Technology

Semiconductor materials are used to fabricate solar cells and most of them are crystalline and thin film, which vary from each other in terms of light absorption effectiveness, energy conversion efficiency, manufacturing technology, and cost of production.

6.2.1 Single-Crystal Silicon Solar Cell

These types of silicon cells are the oldest in the field of solar cells (Figure 6.3). CZ methods are used to fabricate crystalline solar cell (Li et al. 1992). In this technique, high-purity silicon is melted in a crucible. Certain doping is added to improve the conductivity of the semiconductor. Usually, phosphorous or boron is used as a dopant. A seed crystal is immersed into the molten silicon and pulled upward and revolves continuously by this way single silicon ingot is formed from the melting silicon. Single-crystal silicon has a uniform molecular structure.

6.2.2 Gallium Arsenide Solar Cell

GaAs is formed by a mixture of gallium and arsenic that have characteristics similar to silicon (Figure 6.4). The conversion efficiency of GaAs is higher than that of crystalline silicon solar cells. GaAs solar cells are thinner but more expensive than a crystalline solar cell. The GaAs solar cell accounts for about 30% efficiency (Aleksic et al. 2002; Yablonovitch et al. 2012)

6.2.3 Thin-Film Solar Cell

Thin-film solar cells are fabricated by depositing multiple thin film layers of semiconductor materials on a substrate, that is, glass, plastic foil, or metal. The thickness of the film is

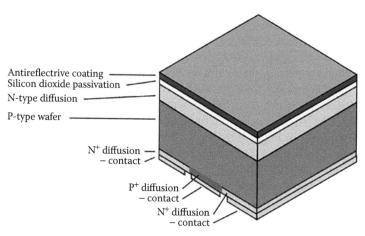

FIGURE 6.3
Single-crystalline silicon solar cell.

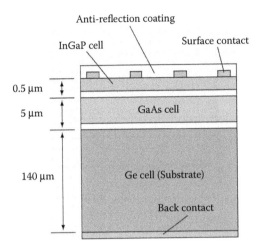

FIGURE 6.4
Gallium arsenide solar cell.

FIGURE 6.5
Thin film solar cell.

less than a micrometer compared to traditional solar cells. This results in more lightweight and flexible solar cells. However, the efficiency of thin-film solar cells is less compared to the crystalline solar cell. Figure 6.5 shows thin-film solar cell.

6.2.4 Amorphous Silicon Solar Cell

An amorphous silicon (a–Si) solar cell is fabricated by using a noncrystalline form of silicon, that is, a–Si solar cell as shown in Figure 6.6. These can be used in small-scale applications as the thickness is about 1 μm, which is less than the monocrystalline solar cell. A non-crystalline form of silicon has higher light absorptive capacity than a single silicon crystalline structure since the thin-film structure makes it more flexible and of less weight.

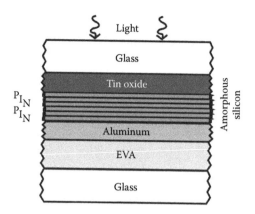

FIGURE 6.6
Amorphous solar cell.

6.2.5 Cadmium Telluride Solar Cell

A cadmium telluride (CdTe) solar cell is a compound made of cadmium and tellurium. Figure 6.7 shows different layers of CdTe. CdTe solar cells are grown on the substrate. Polyimide, metal foils, and glass are commonly used as the substrates. Their high efficiency, low cost, stability, and the potential for low-cost production make them suitable for large-area applications. The CdTe solar cell has high absorption coefficient and has shown laboratory efficiency as high as 16.5%. The commercial module efficiency is similar to an a–Si solar cell, that is, 7%–9%.

6.2.6 Organic Solar Cells

A CdTe solar cell is a compound made of cadmium and tellurium. CdTe solar cells are grown on a substrate as shown in Figure 6.8. Polyimide, metal foils, and glass are

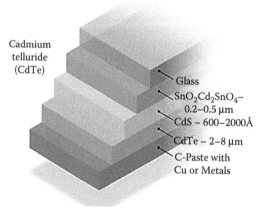

FIGURE 6.7
Cadmium telluride solar cell.

FIGURE 6.8
Organic solar cell.

commonly used as the substrates. Their high efficiency, low cost, stability, and the potential for low-cost production make them suitable for large-area applications. The CdTe solar cell has high absorption coefficient and has shown laboratory efficiency as high as 16.5% (Lynn 2011) (Fthenakis 2004). The commercial module efficiency is the same as an a–Si solar cell, that is, 7%–9% (Agrawal and Tiwari 2011).

6.2.7 Efficiency of Different Solar Cells

The National Renewable Energy Laboratory (NREL) maintains a plot of compiled values of highest confirmed conversion efficiencies for research cells, from 1976 to the present, for a range of photovoltaic (PV) technologies (Figure 6.9). Devices included in this plot have the current state-of-the-art efficiencies that are confirmed by independent, recognized test labs (e.g., NREL, AIST, Fraunhofer) and are reported on a standardized basis. The measurements for new entries must be with respect to Standard Test or Reporting Conditions (STC) as defined by the global reference spectrum for flat-plate devices and the direct reference spectrum for concentrator devices as listed in standards IEC 60904-3 edition 2 or ASTM G173. The reference temperature is 25°C and the area is the cell total area or the area defined by an aperture. Cell efficiency results are provided within different families of semiconductors: (1) multijunction cells, (2) single-junction gallium arsenide cells, (3) crystalline silicon cells, (4) thin-film technologies, and (5) emerging PV. Some 26 different subcategories are indicated by distinctively colored symbols. The most recent world record for each technology is highlighted along the right edge in a flag that contains the efficiency and the symbol of the technology. The company or group that fabricated the device for each recent record is bolded on the plot. The information plotted by NREL is provided in good faith.

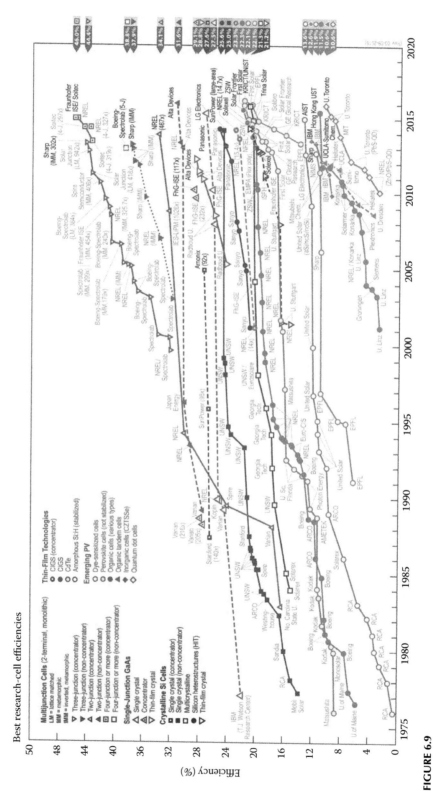

FIGURE 6.9
PV cells efficiencies. (From National Renewable Energy Laboratory, http://www.nrel.gov/ncpv/images/efficiency_chart.jpg, accessed March 16, 2016.)

6.3 Building Integrated Photovoltaic

The four main options for building integration of PV cells are on sloped roofs, flat roofs, facades, and shading systems. South-facing sloped roofs are usually best suited for PV installation because of the favorable angle with the sun. One option is to mount PV modules above the roofing system. Another option is PV modules that replace conventional building materials in parts of the building envelopes, such as the roofs or facades, known as buiding integrated photovoltaics (BIPVs). "BIPV is considered a functional part of the building structure, or they are architecturally integrated into the building's design" (Peng et al. 2011). The BIPV system serves as building envelope material and power generator simultaneously (Strong 2010). This can provide savings in materials and labor, reduces the electricity costs, and increases the importance of water-tightness and durability of the BIPV product.

6.3.1 Building Integrated Photovoltaic Products

There is a wide range of different BIPV products that can be categorized in different ways. In this work, the categorization is mainly based on how the manufacturer describes the product and the type of with which the product is combined. The product categories considered are foils, tiles, modules, and solar cell glazing products.

6.3.1.1 Building Integrated Photovoltaic Foil Products

The BIPV foil products are lightweight and flexible, which is ideal for easy installation and the weight constraints most roofs have. The PV cells are often made from thin-film cells to maintain the flexibility in the foil and the efficiency regarding high temperatures for use on nonventilated roof solutions.

6.3.1.2 Building Integrated Photovoltaic Tile Products

The BIPV tile products can cover the entire roof or just parts of the roof. They are normally arranged in modules with the appearance and properties of standard roof tiles and substitute a certain number of tiles. This is a good option for retrofitting of roofs. The cell type and tile shape vary. Some tile products resemble curved ceramic tiles and will not be area effective due to the curved surface area, but may be more aesthetically pleasing.

6.3.1.3 Building Integrated Photovoltaic Module Products

The BIPV module products presented are somewhat similar to conventional PV modules. The difference, however, is that they are made with weather skin solutions. Some of the products can replace different types of roofing, or they fit with a specific roof solution produced by its manufacturer, for example, Rheinzink's "Solar PV Click Roll Cap System" (Jelle et al. 2012a). These mounting systems increase the ease of installation. There are a lot of products on the market and some of them are promoted as BIPV products without functioning as weather skin. Other products are not very specific on how they are mounted, which leads to uncertainty whether they are BIPVs or BAPVs. Some of the products in this category are premade modules with insulation or other elements included in the body.

6.3.1.4 Solar Cell Glazing Products

Solar cell glazing products provide a great variety of options for windows, glassed or tiled facades, and roofs. Different colors and transparencies can make many different aesthetically pleasing results possible. The modules transmit daylight and serve as water and sun protection. "The technology involves spraying a coating of silicon nanoparticles onto the window, which works as solar cells" (Jelle et al. 2012b). The distance between the cells depends on wanted transparency level and the criteria for electricity production, but normally the distance is between 3 and 50 mm. The space in between cells transmits diffuse daylight. This way, both shading and natural lighting are provided while producing electricity.

6.3.1.5 Building Attached Photovoltaic Products

The BAPV products are, as mentioned earlier, added on rather than integrated into the roof or facade. These products are not focused on in this study, but it is still interesting to have a look at some of them.

6.4 Methods of Installation

BIPV systems provide many opportunities for innovative architectural design and can be aesthetically appealing. BIPVs can act as shading devices and also semi-transparent elements of fenestration (Jelle et al. 2012b). Amorphous silicon tiles can be used to make a BIPV roof look very much like a standard tiled roof (as shown in Figure 6.10), while on

FIGURE 6.10
Solar cell tiles. (From http://gajitz.com/the-roof-is-on-fire-solar-shingles-let-you-green-on-the-sly/.)

FIGURE 6.11
Semi-transparent modules. (From http://www.onyxsolar.com/photovoltaic-double-glazed-insulating-units.html.)

the other hand semi-transparent modules can be used in facades or glass ceilings to create different visual effects (as shown in Figure 6.11). Some architects enjoy presenting a BIPV roof as a roof giving a clear visual impression, while others want the BIPV roof to look as much as a standard roof as possible. Further information about building integration of solar energy systems in general and architectural integration of PV and BIPV in particular may be found in the studies by (Jelle et al. 2012a, b) respectively. The PVs can be two types as follows:

1. Architectural integration
2. Building integration

Integration of PV into buildings has the following advantages

- Part of the building is used for PV installation and so additional land is not required. This is particularly significant in the densely populated areas in the cities.
- The expense of the PV divider or rooftop can be counterbalanced contrasted with the expense of the building component it replaces.
- Power is generated on site and replaces electricity that would otherwise be purchased at commercial rates and avoids distribution losses.
- PV, if grid connected ensures the security of supply and avoiding the high cost of storage.
- Architecturally elegant, well-integrated systems will increase market acceptance.

6.4.1 Roof Integration of Photovoltaic

A PV system can be integrated into the roof in several ways. One choice is for the integrated system to be part of the external skin and therefore part of an impermeable layer in the construction. The other type is that the PV is glued onto insulation material. This kind of warm roof construction arrangement is very well suited to renovating large flat roofs. Using PV modules as a roof covering reduces the number of building materials needed, which is very favorable for a sustainable building and can help to reduce costs. There are also many products for small-scale use to suit the scale of the roof covering, for example, PV shingles and tiles. The roof integration of PV can be categorizing as follows:

- PVs on flat roof
- PVs on Inclined/pitched roof
- Roof with integrated PV tiles
- PV in saw-toothed north light roof/skylight
- PVs on Curved roof/wall
- PVs in Atrium/skylights

For PVs on the flat roof, the PVs are laid horizontally on the flat roofs, which are normally not visible on the ground and, hence, the significance of the aesthetic part of integration can be less. PV facilities on inclined or pitched roofs when facing in the right direction is suitable for good energy yield. Architectural aesthetics should be taken into consideration while integrating as these are visible parts of the building unlike the flat roofs (Figure 6.12). Roof with integrated PV tiles has been a normal practice to lie or integrated larger standard modules of PV on the roof for energy yield (Figure 6.13). However, when the integration is to be done on traditional tiled roofs, it may always not be possible to use large modules and the character of the roof may also be ruined.

Sawtooth roofs can be implemented as (semi)transparent or opaque. Glass saw-tooth roofs make optimal use of daylight, protect in contrast to direct sunlight, and thus reduce a building's cooling loads. The world's largest Integration of thin-film PV systems on the saw-toothed glass roof of Paul Lobe Haus, Berlin optimizes interior light conditions in addition to producing clean energy (Figure 6.14a) (CBD Energy). The flexibility with the integration of PV is further emphasized by the fact that PV modules can also be mounted on curved load-bearing surfaces. Arched surfaces and roofs are equally suited

(a)   (b)

FIGURE 6.12

PVs on flat roof: (a) integrated PV on the pitched roof of vacation house Bartholomä-Park in Germany and (b) integrated PV roofs in Schlierberg Solar Settlement in Freiburg. (From http://www.pvdatabase.org/; http://www.rolfdisch.de/.)

FIGURE 6.13
Integrated PV roof tiles on the roof. (From http://www.horizonrenewables.co.uk/; http://www.flickr.com/; http://www.archiexpo.com/.)

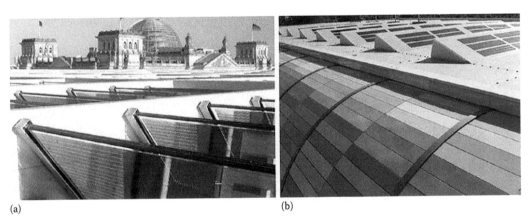

(a)   (b)

FIGURE 6.14
PV with saw-toothed glass roof: (a) integrated PV on the saw-toothed glass roof of Paul Lobe Haus, Berlin and (b) sawtooth roof with PV integration, DIY store, Hamburg. (From http://www.cbdenergy.com.au/; http://www.solarfassade.info/.)

for use with PV systems. This allows added freedom of design involving PV integration. The BP solar showcase in Birmingham, UK is a good example of PV integration on the curved roof (Figure 6.15). PVs can equally be integrated as multifunctional elements in transparent roof structures or atriums that allow controlled light into the interior. As semitransparent roof units, they can protect the building from heat, sunlight, glare, and the weather (Figure 6.16).

FIGURE 6.15
PV integration on the curved roof of BP solar showcase (Birmingham). (From http://www.solartechnologies.
co.uk/.)

(a)   (b)

FIGURE 6.16
PV with transparent roof structures: (a) atrium with PV modules, Ludesch/Vlbg., Austria and (b) PV on colored
skylights at Bejar Market, Salamnca, Spain. (From http://www.solarfassade.info/; http://www.onyxsolar.com/.)

6.4.2 Facade Integration of Phototvoltaic

It is a normal building practice that external walls of buildings are covered with insulation and protective cladding. This covering can be wood, metal sheets, panels, glass, or PV

modules. For luxury office buildings where the cladding is often expensive, integrating PV modules as cladding on opaque parts of the building are not more expensive than other commonly used materials like natural stones, granite or aluminum cladding. In the Solar XXI building (SHC 2012) in Portugal, vertical bands of PV panels are integrated into the south facade, with an alternative rhythm with the glazing, resulting in an elevation based on the concept of modularity and repetition.

Façade integration of PV module is installed on walls depending upon inclination.

- PV on inclined walls
- PVs as sunshades

6.5 Challenges of Building Integrated Photovoltaic

6.5.1 Price

BIPV systems generally carry a larger price tag than do flat panel systems, though the reasons for this are somewhat unclear, given the lack of BIPV market data available. The following list of factors can account for some of the price differentials:

- Customer perception that these products should cost more because of their specialty function and their willingness to pay premiums for that function.
- Supply chain issues for products and services (e.g., difficulties in establishing distribution channels and getting product to market).
- BIPV modules may include additional materials (e.g., adhesives and framing and flashing materials).
- Additional labor costs deriving from specialized architectural design, engineering design, and installation, according to a Greentech Media Report.
- It is important to note that BIPV prices are variable by the market and by the application (i.e., the structure-specific design of the module), and so pricing is something of a moving target.

Despite reportedly higher prices, BIPV systems may offer an offset value in the construction process through, among other things, the replacement of traditional building materials and the dispensation of rack-mounting hardware. A recent NREL report on BIPV in the residential sector cautions, however, that "past market experiences suggest that realizing these cost reductions can be very challenging." And without significant reductions in installed costs (~5%), BIPV's cost of energy comes up short of competitive with flat-panel PV.

6.5.2 Performance

There are some important performance variables to consider when calculating energy costs of a BIPV system. For starters, BIPV modules may experience higher operating temperatures because, unlike rack-mounted PV, they are flush with the building surface and do not permit airflow between the module and host structure. Higher temperatures may degrade the semiconducting material of the module, which could decrease the conversion efficiency more quickly and precipitate early failure. Some PV materials—for example, amorphous silicon, which has a flexible form factor and a potentially greater integration potential—are more susceptible to thermally accelerated degradation than others. Also, PV materials with greater integration potential, such as thin films and flexible PV technologies, generally have lower efficiencies, to begin with, and this may contribute to higher energy costs.

Finally, because BIPV modules typically contain less semiconducting material than traditional PV modules, a BIPV system will likely produce less electricity than a flat-panel system of the same size. And even though BIPV can increase the PV-suitable space of a building (i.e., more than just the roof is eligible for installation), the suboptimal angle of irradiation on these non-horizontal surfaces, combined with the obstructions posed by surrounding buildings, create diminishing returns on increased module deployment.

6.5.3 Codes and Standards

Because BIPV modules serve dual functions, they must hew to the codes and standards of two separate industries (PV and construction). Currently, PV modules (including BIPV) are subject to the qualification and design standards devised by the International Electrotechnical Commission and the Underwriters Laboratory. But BIPV may be required to meet additional criteria as a structural component, and this can act as a market handicap. For example, the International Code Council, whose pervasive International Building Codes have been adopted by all 50 states and Washington, DC, has established criteria for BIPV as a roofing material that dictates its performance on stability, wind resistance, durability, and fire safety.

Even something as simple as measurement standards could complicate BIPV deployment. The construction industry employs square meter units, which denotes area, and the PV industry uses watt units, which measure electrical output. If this incongruence remains unresolved, it could create some headaches for installers in the building trade.

For now, BIPV keeps awkward toeholds in both the PV and construction industries, without an integrated set of standards and codes to carve out the middle ground. The establishment of this middle ground through a clear set of guidelines and expectations for the manufacturing and construction process will serve as a growth platform for the BIPV industry.

6.5.4 Market Limitations

Unlike flat-panel PV, where module designs do not vary greatly from one application to another, BIPV manufacturers' products vary by façade type (e.g., roof shingles, windows, and awnings). These emphases on custom-design segments the BIPV market and, in turn, hobble the technology's path to scalability. The fact that BIPV does not compete in the utility-scale, ground-mount space (in other words, it is limited to residential and commercial building applications) further hinders its scalability. Without the kind of capital accumulation, economies of scale, and learning curve progress that comes from a manufacturing and deployment scale-up, BIPV may not realize the kinds of cost reductions that could facilitate its adoption.

6.6 Role of Building Integrated Photovoltaic for Developing Smart Cities

In January 2016, the Indian government announced a list of 20 cities to be developed into smart cities in India. Today's era comprises of everything that is smart—be it smartphones, smart people, or the concept of the smart city. Basically, a smart city is an advanced urban city/town that has well connected infrastructure and communications through data centers and automated networks. India is expecting to develop 98 such cities in all.[1] These cities will become a model township for a sustainable and ultramodern lifestyle.

Solar applications such as solar street lights, solar water heaters, rooftop solar, and so on, can go a long way in imparting a clean and green living style to these smart cities. Not only will these smart cities improve the conditions in India in terms of employment generation and an urban living style, it will also go a long way in promoting the use of renewable forms of energy and thus help the country fight the growing concerns of global warming and pollution.

It has already been mandated that *10% of the smart cities' energy requirement will come from solar energy* and at least *80% buildings should be energy efficient and green buildings*. With a plan to develop approximately 98 such cities as shown in Table 6.2, the rate of renewable energy usage will go up in the country.

TABLE 6.2

List of Smart Cities

1	Port Blair	26	Ranchi	51	Pune	76	Erode
2	Vishakhapatnam	27	Mangaluru	52	Imphal	77	Thoothukudi
3	Tirupati	28	Belagavi	53	Shillong	78	Chennai
4	Kakinada	29	Shivamogga	54	Aizawl	79	Greater Hyderabad
5	Pasighat	30	Hubballi-Dharwad	55	Kohima	80	Greater warangal
6	Guwahati	31	Tumakuru	56	Bhubaneswar	81	Agartala
7	Muzaffarpur	32	Davanegere	57	Rourkela	82	Moradabad
8	Bhagalpur	33	Kochi	58	Oulgaret	83	Aligarh
9	Biharsharif	34	Kavaratti	59	Ludhiana	84	Shaharanpur
10	Chandigarh	35	Bhopal	60	Jalandhar	85	Bareilly
11	Raipur	36	Indore	61	Amritsar	86	Jhansi
12	Bilaspur	37	Jabalpur	62	Jaipur	87	Kanpur
13	Diu,	38	Gwalior	63	Udaipur	88	Allahabad
14	Silvassa	39	Sagar	64	Kota	89	Lucknow
15	NDMC	40	Satna	65	Ajmer	90	Varanasi
16	Panaji	41	Ujjain	66	Namchi	91	Ghaziabad
17	Gandhinagar	42	Navi Mumbai	67	Tiruchirapalli	92	Agra
18	Ahmedabad	43	Nasik	68	Tirunelveli	93	Rampur
19	Surat	44	Thane	69	Dindigul	94	Dehradun
20	Vadodara	45	Greater Mumbai	70	Thanjavur	95	New Town Kolkata
21	Rajkot	46	Amaravati	71	Tiruppur	96	Bidhannagar
22	Dahod	47	Solapur	72	Salem	97	Durgapur
23	Karnal	48	Nagpur	73	Vellore	98	Haldia
24	Faridabad	49	Kalyan-Dombivalli	74	Coimbatore		
25	Dharamsala	50	Aurangabad	75	Madurai		

Note: http://timesofindia.indiatimes.com/india/Full-list-of-98-smart-cities/articleshow/48694723.cms.

6.7 Summary

The use of solar cells in the buildings is one of the most promising, reliable, and environmentally friendly forms of renewable energy technology and has the potential to contribute significantly to the energy and environmental system in this area. The present chapter represents the overview of BIPV panel in the present scenario. The BIPV panel plays an important role in the energy-efficient buildings (EEBs). It generates power and acts as insulators in the buildings thereby reducing the cooling loads in buildings. The reduction in operational energy was due to the declination in the heat gain and cooling load.

The BIPV has great potential for the power generation and reduces cooling load in buildings although it has some barriers in their installation, price, marketing, and performance. Above all, it is essential that there be close working relationships between architects, planners, and industry—through an exchange of information and training so that the full potential of BIPV may be exploited. With BIPV, solar systems are becoming a standard building component, just like glass panes or doors. This allows homeowners and architects to take energy consumption into account when designing a home, without compromising energy efficiency or aesthetics.

References

Absi Halabi M, Al-Qattan A, Al-Otaibi A. Application of solar energy in the oil industry—Current status and future prospects. *Renewable and Sustainable Energy Reviews* 2014;43:296–314.

Agrawal B, Tiwari GN. Building integrated photovoltaic thermal systems: For sustainable developments. *Royal Society of Chemistry*, 2011. doi:10.1039/9781849732000.

Aleksic J, Zielke P, Szymczyk JA. Temperature and flow visualization in a simulation of the czochralski process using temperature-sensitive liquid crystals. *Annals of the New York Academy of Sciences* 2002;972(1):158–163.

Chaudhari JR, Tandel PKD, Patel PV. Energy saving of green building using solar photovoltaic systems. *International Journal of Innovative Research in Science, Engineering and Technology* 2013;2(5):1407–1416.

EIA. *Annual Energy Outlook 2007 With Projections to 2030.* www.eia.doe.gov/ (Accessed: April 1, 2017).

Fthenakis VM. Life cycle impact analysis of cadmium in CdTe PV production. *Renewable and Sustainable Energy Reviews* 2004;8(4):303–334.

Jelle BP, Breivik C, Røkenes HD. Building integrated photovoltaic products: A state-of-the-art review and future research opportunities. *Solar Energy Materials and Solar Cells* 2012a;100:69–96.

Jelle BP, Hynd A, Gustavsen A, Arasteh D, Goudey H, Hart R. Fenestration of today and tomorrow: A state-of-the-art review and future research opportunities. *Solar Energy Materials and Solar Cells* 2012b;96(1):1–28.

Kats G, Capital E. *Green Building Costs and Financial Benefits.* Massachusetts Technology Collaborative, Boston, MA, 2003.

Li Z, Kraner HW, Verbitskaya E, Eremin V, Ivanov A, Rattaggi M et al. Investigation of the oxygen-vacancy (A-center) defect complex profile in neutron irradiated high resistivity silicon junction particle detectors., *IEEE Transactions on Nuclear Science* 1992;39(6):1730–1738.

Lynn PA. *Electricity from Sunlight: An Introduction to Photovoltaics.* John Wiley & Sons, Chichester, UK, 2011.

Peng C, Huang Y, Wu Z. Building-integrated photovoltaics (BIPV) in architectural design in China. *Energy and Buildings* 2011;43(12):3592–3598.

Strong S. Building Integrated Photovoltaics (BIPV), Whole Building Design Guide [Internet]. June, 2010. Available from: https://www.wbdg.org/resources/bipv.php (Accessed: April 1, 2017).

Yablonovitch E, Miller OD, Kurtz SR. The opto-electronic physics that broke the efficiency limit in solar cells. *Photovoltaic Specialists Conference (PVSC), 2012 38th IEEE.* IEEE, Austin, TX, 2012, pp. 1556–1559.

Endnote

1. http://smartcities.gov.in/

7

Energy Conservation Potential through Thermal Energy Storage Medium in Buildings

C. Veerakumar and A. Sreekumar

CONTENTS

7.1   Introduction 132
7.2   Energy Resources 132
7.3   Energy Demand and Consumption 132
7.4   Energy Efficiency and Energy Conservation 134
7.5   Thermal Energy Storage 134
    7.5.1   Sensible Heat Storage 135
    7.5.2   Latent Heat Storage 135
    7.5.3   Thermochemical Heat Storage 137
7.6   Applications of Thermal Energy Storage in Energy Conservation of Buildings 138
    7.6.1   Thermal Energy Storage for Heating in Buildings 138
        7.6.1.1   Domestic Hot Water 138
        7.6.1.2   Space Heating 139
    7.6.2   Thermal Energy Storage for Cooling in Buildings 140
        7.6.2.1   Building Materials 141
        7.6.2.2   Building Components 142
7.7   Green Building Technology 145
7.8   Case Studies 146
    7.8.1   Case Study 1 146
        7.8.1.1   Campus District Cooling System with Large Scale Thermal Energy Storage at Cairns Campus, James Cook University, QLD, Australia 146
    7.8.2   Case Study 2 147
        7.8.2.1   Floating Pavilion, Rotterdam, The Netherlands 147
        7.8.2.2   Potential Research Work at Solar Thermal Energy Lab, Centre for Green Energy Technology, Pondicherry University 148
Acknowledgment 149
References 149

7.1 Introduction

With steady increase in overall global energy demand, it is necessary to focus on wise use of fossil fuel-based energy resources. Shifting toward renewable sources will help in reducing the energy consumption up to a certain extent. Building sector remains as a major energy consumer across the globe. The increasing awareness on environmental impact of climate change, greenhouse gas (GHG) emission, and so on, have led to energy conservation by adapting energy-efficient techniques. The reason behind this concept is reduction in extensive utilization of fossil fuels which leads to carbon emission. In the scenario of increasing energy consumption and depletion of fossil fuel resources, the thirst for energy storage technologies is recently gaining momentum. This chapter is designed to explain the application of thermal energy storage (TES) for energy conservation in buildings. The demand side management can be accomplished by developing efficient TES systems. The overall energy storage capacity and operational performance of the TES system can be enhanced by integrating renewable energy systems in buildings. The need for heating and cooling of buildings plays a major part of overall energy consumption. The integration of TES system with conventional heating and cooling system of buildings helps to facilitate the development of energy-efficient and sustainable buildings.

7.2 Energy Resources

Energy resources refer to the reserves of energy that help human beings survive and sustain their environment. The system used for extracting energy from its resources is also classified as the energy resource. Our earth has a large number of energy resources such as solar, hydro, wind, biomass, geothermal, and so on. The techniques used for extracting energy from these resources should be done in an efficient manner. The extraction of energy from its resource is not only the major task; it should be transformed into a social need. For developed and developing countries, it is of upmost importance to reduce the gap between energy generation and energy consumption. The global energy potential is shown in Figure 7.1.

7.3 Energy Demand and Consumption

Energy production and consumption play an important role in energy conservation opportunities. Sector-wise total global energy consumption is shown in Figure 7.2. When discussing buildings, energy use generally refers to the energy consumption by a building in its life cycle when it is in operational phase. In a domestic residential building, the operational phase involves space heating or cooling, lighting, cooking, water heating, and operation of various energy consuming appliances. In developed countries, the major energy consumption is for space heating and cooling. The major source of energy for them is fossil fuels, which lead to high adverse environmental impact. Table 7.1 gives the estimated worldwide annual primary energy consumption by buildings [1]. Increasing awareness on climate change led many countries to emphasize promoting energy conservation and energy efficiency by developing innovative policies and popularizing eco-friendly energy devices.

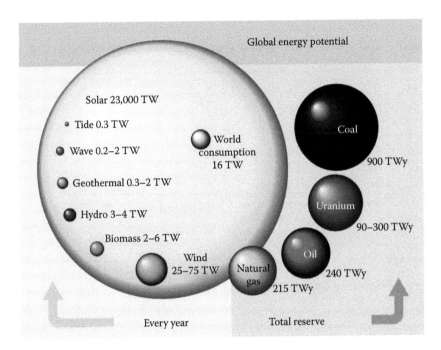

FIGURE 7.1
Global energy potential. (From Perez, R., and Perez, M., A fundamental look at energy reserves on the planet, 2009.)

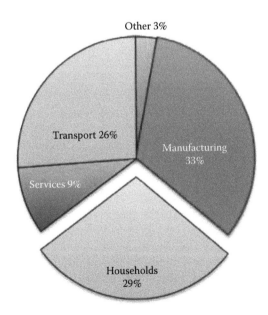

FIGURE 7.2
Total final energy consumption per sector. (From IEA, *Worldwide Trends in Energy Use and Efficiency*, Key Insights from IEA Indicator Analysis, Paris, France, 2008.)

TABLE 7.1

Estimated Annual Primary Energy Consumption Worldwide
by Buildings

Year	Energy Consumption (Quads year^{-1})
2004	72.2
2010	82.2
2015	90.7
2020	97.3
2025	103.3
2030	109.7

Source: Praditsmanont, A., and Chungpaibulpatana, S., *Energ. Build.*,
40, 1737–1746, 2008.

7.4 Energy Efficiency and Energy Conservation

The term energy efficiency refers to the ability to obtain the required outcome with less energy usage. Energy conservation refers to the reduced energy consumption depending on the type of application. The reduction in energy consumption should not influence the specific output. For energy conservation, the practices that are easy to implement with less investment and time are categorized as minor opportunities and those that are complex in nature with high investment and time are categorized as major opportunities. The need for energy efficiency measures are listed as follows:

- Deficiency of primary energy source
- GHG and carbon emission
- Climate change, global warming, ozone depletion
- Fuel price hike

The integration of renewable energy concepts will result in augmentation of energy conservation potential.

7.5 Thermal Energy Storage

TES is a method which stocks the thermal energy when it is available and it can be used at a later time. The main advantage of TES is to balance the time of mismatch between the energy generation and usage. The thermal energy generated through any energy conversion technique can be stored in a medium during low demand period. During peak demand conditions, the stored thermal energy can be discharged from the medium. TES works on a cyclic process, which involves three steps: (1) charging, (2) storing, and (3) discharging. During the charging process, the thermal energy is supplied to a medium and then it is stored for a certain time and then discharged when it is needed which accomplishes the complete cycle.

The thermal storage systems are generally of two types: (1) active and (2) passive systems. In active thermal storage systems, the heat transfer fluid will also serve as a storage medium. While in a passive system, the heat transfer fluid passes through a storage medium where the thermal energy is being stored.

The benefits of implementing TES system are listed below [2].

- Better economics: reducing capital and operational cost
- Better efficiency: achieving a more efficient use of energy
- Less pollution of the environment and less CO_2 emissions
- Better system performance and reliability

TES can be stored by means of three methods based on its characteristics and need:

1. Sensible heat storage
2. Latent heat storage
3. Thermochemical heat storage

7.5.1 Sensible Heat Storage

In sensible heat storage, the heat is stored by changing the temperature of a storage medium. The amount of thermal energy stored is directly proportional to the mass, specific heat capacity, and change in temperature of the storage medium. The storage medium can be solid, liquid, or gas. The amount of thermal energy stored as sensible heat is calculated by the following equation:

$$Q = \int_{T_i}^{T_f} mC_p \Delta T \tag{7.1}$$

where:
Q is the amount of sensible heat stored (kJ)
m is the mass of storage medium (kg)
C_p is the specific heat capacity of the storage medium (kJ kg^{-1} K^{-1})
ΔT is the temperature difference during the process (K)

Some of the common materials used as a sensible storage medium are presented in Table 7.2 [3]. It is necessary to be focused on several properties such as specific heat capacity, density, thermal conductivity, chemical stability, and cost while selecting materials for sensible heat storage.

7.5.2 Latent Heat Storage

In latent heat storage, the energy storage material undergoes phase transition. The phase transition may be either solid to liquid or liquid to solid. The materials used in latent heat storage are known as phase change materials. The energy storage occurs in the isothermal process. The amount of energy storage is directly proportional to the latent heat of fusion

TABLE 7.2

Common Materials Used as Sensible Storage Medium

Material	Density (kg m⁻³)	Specific Heat Capacity (kJ kg⁻¹ K⁻¹)	Thermal Conductivity (W m⁻¹ K⁻¹)
Aluminum	2700	0.945	238.4
Copper	8300	0.419	372
Iron	7850	0.465	59.3
Brick	1800	0.840	0.5
Concrete	2200	0.879	1.279
Granite	2750	0.892	2.9
Limestone	2500	0.740	2.2
Sandstone	2200	0.710	1.8
Soil (Gravelly)	2040	1.840	0.59
Water	988	4.182	0.6

Source: Hahne, E., *Storage of Sensible Heat*, vol. I. University of Stuttgart, ITW, Stuttgart, Germany.

of the material. The amount of energy stored in a latent heat storage system is given by the following equation:

$$Q = \int_{T_i}^{T_m} mC_p\Delta T + mH + \int_{T_m}^{T_f} mC_p\Delta T \qquad (7.2)$$

where:

Q is the amount of latent heat stored (kJ)
m is the mass of storage medium (kg)
C_p is the specific heat capacity of the storage medium (kJ kg⁻¹ K⁻¹)
ΔT is the temperature difference (K)
H is the latent heat of fusion of the material (kJ kg⁻¹)

Some of the typical materials used in latent heat TES are given in Table 7.3.

TABLE 7.3

Typical Materials Used in Latent Heat TES System

Material	Melting Temperature (°C)	Latent Heat of Fusion (kJ kg⁻¹)
Water–salt solution	−100–0	200–300
Water	0	330
Clathrates	−50–0	200–300
Paraffins	−20–100	150–250
Salt hydrates	−20–80	200–600
Sugar alcohols	20–450	200–450
Nitrates	120–300	200–700
Hydroxides	150–400	500–700
Chlorides	350–750	550–800
Carbonates	400–800	600–1000
Fluorides	700–900	>1000

Source: Mehling, H., and Cabeza, L. F., *Heat and Cold Storage with PCM: An Up to Date Introduction into Basics and Applications*, Springer, Heidelberg, Germany, 2008.

7.5.3 Thermochemical Heat Storage

Thermochemical energy storage system depends on the energy absorbed and released during a chemical reaction by breaking and reforming molecular bonds. The chemical reaction should be completely reversible. The amount of energy stored is based on the nature of reacting materials. The amount of energy stored in a thermochemical energy storage system is given by the following equation:

$$Q = m\Delta h_r \tag{7.3}$$

where:

Q is the amount of energy stored by thermochemical reaction (kJ)

m is the mass of the reacting material (kg)

Δh_r is the endothermic heat of reaction (kJ kg^{-1})

The comparison of different types of TES is presented in Table 7.4 [4].

TABLE 7.4

Comparison of Different Types of Thermal Energy Storage

	Sensible	Latent	Thermochemical
Storage medium	Water, gravel, pebble, soil, and so on	Organics, inorganics	Metal chlorides, metal hydrides, metal oxides, and so on
Type	Water based system (Water tank, Aquifer) Rock- or ground-based system	Active storage Passive storage	Thermal sorption (Adsorption, Absorption) Chemical reaction (Normally for high-temperature storage)
Advantage	Environmentally friendly cheap material, relatively simple system, easy to control, reliable	Higher energy density than sensible heat storage Provide thermal energy at constant temperature	Highest energy density, compact system Negligible heat losses
Disadvantage	Low-energy density, huge volumes required because of low specific heat capacity Self-discharge and heat losses problem, high cost of site construction, geological requirements	Lack of thermal stability Crystallization, Corrosion, high cost of storage material	Poor heat and mass transfer property under high density condition Uncertain cyclability High cost of storage material
Present status	Large-scale demonstration plants	Material characterization, laboratory-scale prototypes	Material characterization, laboratory-scale prototypes
Future work	Optimization of control policy to advance the solar fraction and reduce the power consumption, optimization of storage temperature to reduce heat losses, simulation of ground-/soil-based system with the consideration of affecting factors (e.g. underground water flow)	Screening for better suited PCM materials with higher heat of fusion Optimal study on store process and concept Further thermodynamic and kinetic study, noble reaction cycle	Optimization of the particle size and reaction bed structure to get constant heat output Optimization of temperature level during charging/discharging process Screening for more suitable and economical materials. Further thermodynamic and kinetic study, noble reaction cycle

Source: Xu, J. et al., *Sol. Energ.*, 103, 610–638, 2014.

7.6 Applications of Thermal Energy Storage in Energy Conservation of Buildings

7.6.1 Thermal Energy Storage for Heating in Buildings

In a building, the TES for heating can be used for domestic hot water and space heating. There are several requirements for the implementation of heat storage. The general problem is that the consumer needs thermal energy when it is not available or not cost-effective. So, it is necessary to store the surplus heat when it is easily available. While developing a TES system for building, the amount of heat stored and storage duration has to be taken into account. In order to have low payback period, the TES system should possess high utilization rate, high energy storage density, high thermal output efficiency, and environmental compatibility.

7.6.1.1 Domestic Hot Water

Domestic hot water needs are responsible for 17%–34% of overall household energy demand [5]. The TES by hot water can be stored by means of direct renewable heating systems such as solar thermal collectors. The latent heat storage system (LHTES) can be used along with the hot water storage to lengthen the time of storage and increase the overall efficiency. While incorporating latent heat storage in domestic hot water application, special encapsulation is required for holding the phase change materials. The overall energy storage capacity can be increased by LHTES. The schematic of simple solar water heater with TES is given in Figure 7.3.

Seddegh et al. presented a review on solar domestic hot water system using latent heat storage medium [6]. This review compares the conventional water heating system and solar water heating system with Phase Change Materials (PCM) TES and also discussed the need for research in TES integration techniques. Frazzica et al. experimentally tested a hybrid sensible-LHTES for domestic hot water application [7]. The combined system uses organic and salt hydrate mixture and the result shows 13% increased energy storage density.

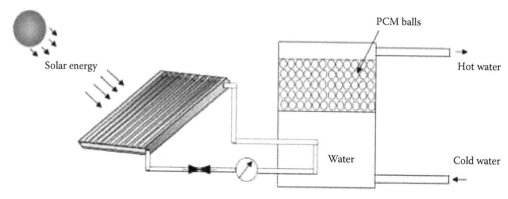

FIGURE 7.3
Domestic hot water system with PCM thermal energy storage.

7.6.1.2 Space Heating

The application of space heating was in practice prior to space cooling. But recently, space cooling receives more focus because of its need. However, the requirement of space heating is considerably increasing, and research is progressing to improve the technology.

7.6.1.2.1 Space Heating System with Hot Water

In this type of space heating system, water is used as heat transfer fluid. The heat storage medium stores heat in the form of sensible or latent heat. In case of sensible heat storage, the water acts as heat transfer fluid as well as sensible storage medium. In latent heat storage, the water acts as heat transfer fluid and the storage medium will be any phase change material. The hot water will flow through the space to be heated and will recirculate within the system. Streicher discussed solar thermal technologies for domestic hot water preparation and space heating [8]. He also describes components of the solar thermal collectors, its hydraulics and stagnation behavior. The future research and development demand for the systems are also presented. Analysis of the performance of a tankless water heating combo system for space heating was done by Der et al. [9]. Tankless water heating system for combined domestic water heating and space heating are the recent advancement in water heating and control technology. Elmegaar et al. presented a study on integration of space heating and hot water supply in low temperature district heating [10]. This study also involves reduction in heat loss during the supply to many consumers in a district heating system. This will enhance the efficiency of the combined heating system. Figures 7.4 and 7.5 shows the technique of hot water-based space heating system.

FIGURE 7.4
Warmboard radiant heat floor. (From http://www.builditsolar.com.)

FIGURE 7.5
Innovative radiant floor heating system. (From http://arcsolar.com.)

7.6.1.2.2 *Space Heating System with Hot Air*

Space heating system using hot air has been in practice since ancient times. During that period, wood and other fuels were used to produce hot air. But in recent years, renewable technologies are used to heat the air in order to reduce the energy consumption. In the case of heating the air with solar collectors, the heating can be combined with the supply of fresh air to the collector and the mixture of fresh and warm air is finally supplied to the building.

Tyfour et al. designed and tested a ready-to-use standalone hot air space heating system [11]. The developed system is simple, expandable, and can be operated parallelly with other conventional space heating systems and also shows higher efficiency. The problem in using air as heat transfer medium is its low heat capacity. The use of PCM seems to be a promising solution due to its high energy storage density. Researchers are working on an efficient way to integrate thermal storage with the air heating system. Belmonte et al. presented a study on integration of PCM fluidized bed energy storage in air-based solar systems for buildings [12]. The usage of PCM fluidized bed offers an interesting alternative to the usual energy storage system and fluidized bed enables faster charging and discharging. This will enable the operation of solar air heating system with less solar radiation. The schematic of solar air heating system with fluidized bed PCM TES is shown in Figure 7.6.

7.6.2 Thermal Energy Storage for Cooling in Buildings

Cooling is one of the major energy consuming processes. Reducing the cooling load in the building leads to energy conservation. Providing thermal comfort to the occupant of the building is essential for the successful operation of the building. If the temperature increases above the comfort range, it is necessary to bring down it to acceptable level.

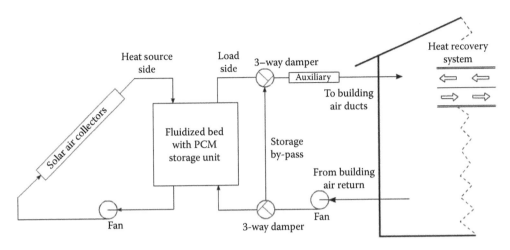

FIGURE 7.6
Schematic of solar heating system with fluidized bed PCM storage unit.

In sensible storage method, the cooling is provided by means of chilled water technique. The cooled water is stored in a tank when power is available in surplus (off-peak period) or when power from renewable source is available and it can be used for space cooling by circulating it. But this type of system is bulk because of low energy storage capacity. By employing PCM for energy storage, the system size will get reduced because of its high energy density and isothermal output. In high thermal mass buildings, free cooling/night ventilation can be used to trap the cold available in nighttime ambient air, and it can be used for cooling air in the daytime. The PCM can be incorporated into the building in two ways: (1) building materials and (2) building components.

7.6.2.1 Building Materials

The phase change materials can be incorporated into the building materials to reduce the temperature fluctuations and to cut down the peak energy demand. There are several constraints in using PCM in building materials such as PCM leakage, evaporation, and so on. Encapsulation of the PCM is a possible solution. The techniques for using PCM with building materials are discussed in this section.

7.6.2.1.1 Gypsum Board with Phase Change Materials

For the construction of lightweight building, gypsum boards are used. The PCM can be incorporated into the gypsum board and has the potential to reduce the building energy demand. This technique of incorporating PCM in gypsum boards has been in practice for more than two decades. It was first investigated by Shapiro et al. [13]. The commercially available KNAUF PCM smartboard is shown in the Figure 7.7.

7.6.2.1.2 Micro/Nanoencapsulated Plasters and Concrete

Another method for integrating PCM into building materials is using plaster and concrete which contains micro/nanoencapsulated PCM. The company Maxit and FHG–ISE in a project, developed a plaster composed of 20 wt.% of micronal PCM with a heat storage capacity of 18 kJ kg^{-1} in the temperature range of 23°C–26°C. There are many literatures available on the use of encapsulated PCM materials with concrete which is further used

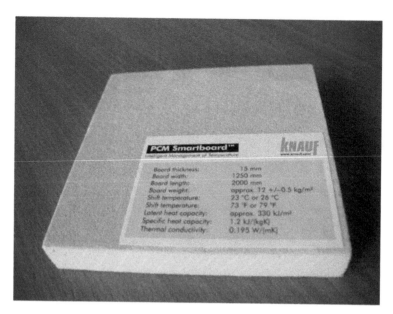

FIGURE 7.7
KNAUF PCM smartboard.

in buildings. The recent researches are also focused on the development of innovative techniques for improving the thermal efficiency of the building with advanced building materials. Figures 7.8 and 7.9 shows the plaster and concrete with microencapsulated PCM [14,15].

7.6.2.2 Building Components

The building components and structures can be equipped with phase change materials for controlling the indoor temperature. Based on the method of installation, there are different types of PCM integrated building components. These components can be easily installed just by screws or nails and can be removed easily.

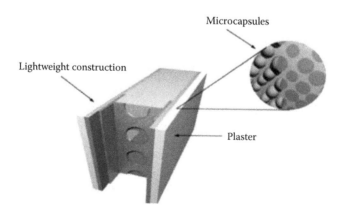

FIGURE 7.8
Plaster with microencapsulated PCM. (From Cui, M.Y. and Riffat, S., *Appl. Mech., Mat.*, 71–78, 1958–1962, 2011.)

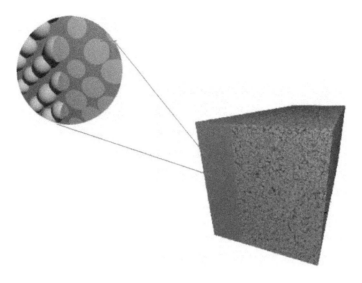

FIGURE 7.9
Concrete with microencapsulated PCM. (From F. I. I. Material, New building material with microencapsulated phase change materials.)

7.6.2.2.1 Phase Change Materials-Integrated Ceiling and Wall

The building components equipped with macroencapsulated PCM can be suspended in the ceilings or can be fixed in the walls. This way of PCM integration is efficient in passive cooling of buildings. The free cooling/night ventilation system is a best example for PCM integration in ceiling. The encapsulated PCM is kept in a channel in the ceiling. At night, the cold air passes through the channel and is discharged outside. The cold is accumulated in the PCM. In daytime, the warm air from the room is passed through the channel. The air is then cooled and supplied to the room. Figure 7.10 shows the general concept of cooling with PCM integrated into the ceiling. The PCM integrated into the ceiling board is shown in Figure 7.11.

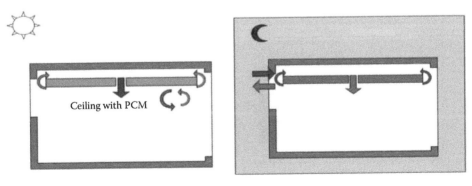

FIGURE 7.10
General concept of cooling with PCM integrated into ceiling.

FIGURE 7.11
PCM-integrated into ceiling. (From Souayfane, F. et al., *Energ. Build.*, 129, 396–431, 2016.)

7.6.2.2.2 *Phase Change Material-Integrated Floor Cooling*

The concept of PCM-integrated floor cooling is same as integration of PCM in ceiling and wall. The free cooling/night ventilation can also be implemented by this concept. The PCM is placed under the floorboards. Like PCM-integrated gypsum boards, PCM floor tiles are also available in market to maintain the thermal comfort. Instead of using nighttime cold air, artificial cooling can also be used at off-peak time. Additionally, some amount of fresh air is supplied along with the recirculated air from the room. Figure 7.12 presents the concept of PCM-integrated floor cooling [16].

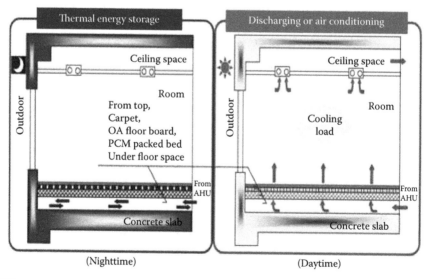

FIGURE 7.12
PCM-integrated floor cooling. (From Souayfane, F. et al., *Energ. Build.*, 129, 396–431, 2016.)

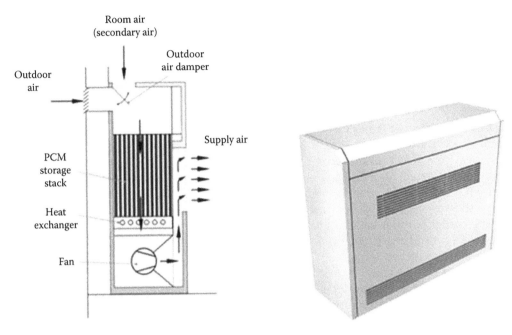

FIGURE 7.13
Decentralized PCM cooling system with heat exchanger. (From http://www.pcm-ral.org.)

7.6.2.2.3 Decentralized Phase Change Material Cooling System

Decentralized PCM cooling system is a concept, which combines cooling and ventilation. The energy storage unit is kept outside the building structure instead of integrated into the building. The PCM is kept inside the heat exchanger or any macroencapsulated structure; it is connected to the building space through ducts. Another setup of this independent system is the PCM kept inside the ventilation ducts with special arrangements. A decentralized PCM cooling system with heat exchanger is shown in Figure 7.13.

7.7 Green Building Technology

Green building refers to the building structures, which follow environmentally responsible practices and resource-efficient techniques during its life cycle. The ideal green building would allow preservation of natural environment around its location. New technologies are being developed to accomplish the objectives of green building. TES is one of such practices, which helps in improving the energy efficiency of the building structure. There are several other techniques such as net zero energy buildings that promote energy conservation in building structures. Use of minimum energy, advanced techniques for space heating and cooling, energy-efficient lighting system, and self-sufficient energy production by renewable technologies are some of the practices in zero energy buildings.

7.8 Case Studies

7.8.1 Case Study 1

7.8.1.1 Campus District Cooling System with Large Scale Thermal Energy Storage at Cairns Campus, James Cook University, QLD, Australia

The Cairns campus of James Cook University (JCU) is situated in the coastal tropics with a total air-conditioned floor area of 25,000 m². The peak summer air-conditioning loads are high with a year-round requirement of cooling; the annual energy usage of the air-conditioning system represents a significant part of the university's operating cost. In future, JCU plans to expand the campus with ten new buildings that would effectively double the air-conditioned floor area (54,000 m²). As it was not possible to upgrade the existing chiller plant, it was decided to construct a new campus district cooling system including a central energy plant to house high efficiency chillers and cooling towers and an adjacent TES system tank.

The system was designed by leading experts in central energy plants with TES utilizing stratified chilled water storage. It includes a 9ML tank positioned in a higher elevation than the buildings in the surrounding greenscape. The Central Energy Plant is the centralized plant for the district cooling. It contains the chillers, cooling towers, pumps, and Thermal Energy (chilled water) Storage tank. It offers the benefits of high efficiency, a centralized plant, reduced maintenance, ease of expansion, and technological upgrades as technology advances. The project was completed in the year 2012 and shown in the Figure 7.14.

TES makes use of periods of the day or night when the site demand for cooling is less than the average demand. During these times the central chilled water plant cools return water (15°C) back to chilled water (6°C). During times when the site demand exceeds the

FIGURE 7.14
Campus district cooling system with chilled water storage tank. (From https://www.jcu.edu.au.)

average demand (typically in the afternoon), the chilled water is drawn from the storage tank. From here, the precooled water is then reticulated throughout the campus and delivered to air conditioning units within each building. The installation of air-conditioning systems within the buildings themselves remains essentially the same as any conventional chilled water system, except that the chiller plant takes the form of one efficient centralized plant rather than numerous different cooling plants. The central energy plant can be up to 2.5 times more efficient than the aged smaller chiller plant that was previously used. The central energy plant operates at a system coefficient of performance (COP) between 5.5 and 7.5 compared with a conventional air-cooled package plant at 2.8–3.1. A COP of 5 simply means every 1 kW of electrical power is converted into 5 kW of cooling. JCU's central chiller plant consumes 50% less energy compared with an air-cooled package plant. In addition to the energy savings, the reduction in Site Power Demand provided further operating cost savings of 40% over traditional systems.

7.8.2 Case Study 2

7.8.2.1 Floating Pavilion, Rotterdam, The Netherlands

The city of Rotterdam has planned to develop a sustainable floating city. They are implementing advanced techniques and it has been demonstrated by the floating pavilion in Rijnhaven. The pavilion houses an information and exhibition centre that features life on the water. The complex consists of three interconnected half spheres, each 12 m high and 18.5, 20, and 24 m in diameter, respectively, as shown in the Figure 7.15. The trio has a combined floor area of 46 × 24 m. The three domes share one buoyancy chamber; the second

FIGURE 7.15
Floating Pavilion, Rotterdam, The Netherlands. (From http://www.sustainableinsteel.eu/p/558/pcm_projects. html.)

buoyancy chamber serves as a meeting plaza. The floating bodies are made up of five layers of expanded polystyrene (EPS) from 20 to 75 cm thick. In the thickest layer, there is a grid of concrete beams with a concrete floor on top and concrete slabs on all sides, which serves as a hard shell to protect the island from the influence of wave action. To a large extent, the pavilion can meet its own energy needs. Division into different "climate zones" enables the energy to be used where needed at any time. The energy for heating and cooling comes from solar collectors and the surface water. In the auditorium—the smallest dome—PCMs have been used, due to the rapidly changing heat load caused by changing occupancies. The function of the PCMs is to cushion the temperature fluctuations in the rooms. Demand for cooling during the day is met by recirculation of air round the PCM and by indirect evaporative cooling in combination with sorption and solar collectors. In the evening an air-conditioning unit ensures that the PCM is "reloaded."

7.8.2.2 Potential Research Work at Solar Thermal Energy Lab, Centre for Green Energy Technology, Pondicherry University

Solar thermal research lab (STEL) at Centre for Green Energy Technology (CGET), Pondicherry University is involved in potential research work related to building energy conservation through latent heat TES. At present, the research group is working on novel eutectic phase change materials for the application of indoor thermal comfort and identified some potential material for air-conditioning application. The differential scanning calorimetric curve of the identified material is shown in Figure 7.16. The thermophysical analysis shows that the prepared eutectic PCM is well suited for building a cooling application. The research group is also working on solar thermal collectors for air heating and water heating applications integrated with latent heat TES system for extended and efficient operation. Design and development of novel phase change material-based TES system for energy efficiency and building thermal comfort is also in progress.

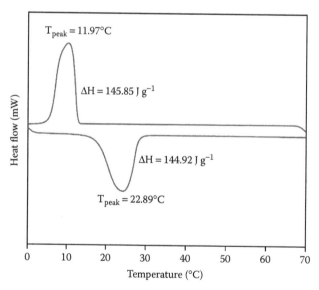

FIGURE 7.16
DSC curve of the eutectic PCM.

Acknowledgment

The authors acknowledge Science and Engineering research board, DST, Govt. of India for the financial support and CIF, Pondicherry University, STIC, CUSAT for providing characterization facility.

References

1. A. Praditsmanont and S. Chungpaibulpatana, Performance analysis of the building envelope: A case study of the Main Hall, Shinawatra University, *Energ. Build.*, 40(9): 1737–1746, 2008.
2. L. F. Cabeza, *Advances in Thermal Energy Storage Systems: Methods and Applications*. Elsevier, Waltham, MA, 2014.
3. E. Hahne, *Storage of Sensible Heat*, vol. I. University of Stuttgart, ITW, Stuttgart, Germany.
4. J. Xu, R. Z. Wang, and Y. Li, A review of available technologies for seasonal thermal energy storage, *Sol. Energ.*, 103: 610–638, 2014.
5. P. Armstrong, D. Ager, I. Thompson, and M. McCulloch, Improving the energy storage capability of hot water tanks through wall material specification, *Energy*, 78: 128–140, 2014.
6. S. Seddegh, X. Wang, A. D. Henderson, and Z. Xing, Solar domestic hot water systems using latent heat energy storage medium: A review, *Renew. Sustain. Energ. Rev.*, 49: 517–533, 2015.
7. A. Frazzica, M. Manzan, A. Sapienza, A. Freni, G. Toniato, and G. Restuccia, Experimental testing of a hybrid sensible-latent heat storage system for domestic hot water applications, *Appl. Energ.*, 183: 1157–1167, 2016.
8. W. Streicher, 2—Solar thermal technologies for domestic hot water preparation and space heating, in *Renewable Heating and Cooling*, 2016, pp. 9–39.
9. J. P. Der, L. W. Kostiuk, and A. G. McDonald, Analysis of the performance of a tankless water heating combo system: Space heating only mode, *Energ. Build.*, 137: 1–12, 2017.
10. B. Elmegaard, T. S. Ommen, M. Markussen, and J. Iversen, Integration of space heating and hot water supply in low temperature district heating, *Energ. Build.*, 124: 255–264, 2016.
11. W. R. Tyfour, G. Tashtoush, and A. Al-Khayyat, Design and testing of a ready-to-use stand-alone hot air space heating system, *Energ. Proc.*, 74: 1228–1238, 2015.
12. J. F. Belmonte, M. A. Izquierdo-Barrientos, A. E. Molina, and J. A. Almendros-Ibáñez, Air-based solar systems for building heating with PCM fluidized bed energy storage, *Energ. Build.*, 130: 150–165, 2016.
13. M. M. Shapiro, D. Feldman, D. Hawesm, and D. Banu, PCM thermal storage in drywall using organic phase change material, *Passiv. Sol. J.*, 4(4): 419–438, 1987.
14. M. Y. Cui and S. Riffat, Review on phase change materials for building applications, *Appl. Mech. Mat.*, 71–78(25): 1958–1962, 2011.
15. F. I. I. Material, New building material with micro-encapsulated phase change materials.
16. F. Souayfane, F. Fardoun, and P. H. Biwole, Phase change materials (PCM) for cooling applications in buildings: A review, *Energ. Build.*, 129: 396–431, 2016.
17. H. Mehling and L. F. Cabeza, *Heat and Cold Storage with PCM: An Up to Date Introduction into Basics and Applications*. Springer, Heidelberg, Germany, 2008.
18. R. Perez and M. Perez, A fundamental look on energy reserves at the planet, 2009.
19. IEA, *Worldwide Trends in Energy Use and Efficiency*. Key Insights from IEA Indicator Analysis, Paris, France, 2008.

8

Silica Aerogel Blankets as Superinsulating Material for Developing Energy Efficient Buildings

Kevin Nocentini, Pascal Biwole, and Patrick Achard

CONTENTS

8.1 Introduction..151
8.2 Aerogel Blanket..152
    8.2.1 Method 152
8.3 Characterization of the Needle Glass Fibers Aerogel Blanket...............153
    8.3.1 Texture...153
        8.3.1.1 Microstructure...153
        8.3.1.2 Specific Surface, Porosity, and Density 155
    8.3.2 Chemical Characteristics 155
    8.3.3 Hydric Properties..156
        8.3.3.1 Water Uptake 156
        8.3.3.2 Contact Angle 157
        8.3.3.3 Water Vapor Transmission...158
    8.3.4 Thermal Properties..159
        8.3.4.1 Thermal Conductivity 159
8.4 Specific Heat Capacity...160
8.5 Conclusion 163
References...163

8.1 Introduction

The current global energy situation is characterized by a need to reduce our energy consumption, to delay the fossil fuel shortage, and to slow down greenhouse gas (GHG) emissions.

Recently, there has been a tendency of the national governments to release more energy-saving policies, especially in the energy-intensive sectors. According to the Directive 2012/27/EU of the European Parliament and the Council on efficiency (2012) [1], the building sector is responsible for 40% of the primary energy consumption in Europe. Increasing thermal insulation of the building envelope remains the most efficient way to reduce this consumption [2,3]. Thermal insulation reduces the heat exchange from inside to outside during heating and cooling time and has to guarantee thermal comfort for the occupants. Conventional insulation materials for buildings are mineral or organic wools and polymer foams, with thermal conductivities between 30 and 60 mW.m^{-1}.K^{-1} [4]. To improve the overall thermal insulation, one can either increase its thickness or decrease its thermal conductivity. Indoor insulation thickness is typically limited to a few centimeters, especially in urban areas where the living space is scarce and expensive. The second solution has been studied for years, and we have seen the emergence of a new kind of insulation material called "superinsulating" materials, with a thermal conductivity lower than that of still air. (25 mW.m^{-1}.K^{-1}). Vacuum insulation panels [5] and ambient pressure aerogel blankets [6] are the two principal types of superinsulating materials and some commercial products already exist [7–11]. However, their manufacturing cost remains high, so they are not used as generic thermal insulations. The latest research efforts are focused on the reduction of time and cost in the manufacturing process of such superinsulating materials. In the future, the superinsulation global market should increase threefold, and the prices of the materials should decrease [12]. In addition, the use of superinsulation in niche sectors as space or cryogenic installation could be developed [13,14].

8.2 Aerogel Blanket

Silica aerogels are highly porous and open-cell thermal insulating materials based on the aggregate of silica nanoparticles (2–10 nm), made by the sol–gel process and dried by CO_2 supercritical technique [15]. The resulting transparent material has a high porosity (85%–99%), a low density (30–150 kg.m^{-3}), a low sound velocity (50–300 m.s^{-1}), and a very low thermal conductivity (0.01–0.02 W.m^{-1}.K^{-1}). These properties offer a great potential as insulating materials in the building sector and could reduce by 75% the heat losses through walls. However, some disadvantages prevent a proper use of silica aerogel as superinsulating materials. The silica skeleton is brittle because of the small contact surface between each nanoparticle. The transparency of the silica aerogel between the wavelength 2–8 μm is responsible for the increase of the heat flux at higher temperatures. Finally, the supercritical drying step is considered as a major drawback for large-scale industrialization because it is a long discontinuous process that can only dry batches of material.

Regarding the latter problems, a likely solution is an impregnation of a fibrous batting (usually polymer or mineral fibers) by the sol with the addition of a hydrophobic agent. The addition of fibers and the hydrophobic agent allows a drying either supercritical or at ambient pressure. The resulting composite material is called aerogel blanket and offers a macroscopic flexibility while keeping a very low thermal conductivity. At present, the main physical drawback of superinsulating aerogel blankets is still their dust release behavior. The dust release is not dangerous for the health, but it may contaminate surrounding equipment when the material is cut or handled.

8.2.1 Method

For the present study, the aerogel blankets are made by Enersens Company in the framework of the HOMESKIN European project [16]. Initially, a needle glass fibers network (NGF) is impregnated by prehydrolyzed TEOS (Tetraethoxysilane) precursors. Next, a silica organogel is obtained by the in situ gelation of the precursor. Then, the organogel is made hydrophobic by silylation with HMDSO (Hexamethyldisiloxane) in an acid medium to prevent any modifications of the structure in case of capillary condensation, which would lead to the degradation of mechanical and thermal properties.

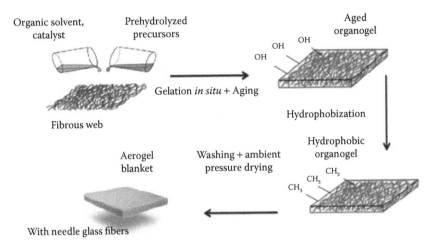

FIGURE 8.1
Elaboration of the aerogel blankets.

TABLE 8.1

Basic Properties of the Aerogel Blankets

	Native Silica Aerogel	Needle Glass Fibers Wool	NGF Aerogel Blanket
Length (side) (mm)	Granular	150	150
Thickness (mm)	X	6	8
Volume fraction of fibers	X	4%	4%
Thermal conductivity (mW.m^{-1}.K^{-1})	14–15	31	15–17
Density (g.cm^{-3})	100–110	100	150
Flexural strength (kPa)	<20	X	206.6

The silylation reaction substitutes hydrolysable groups such as silanol and ethoxy groups by nonhydrolyzable groups and permits the silica gel to return to its initial dimensions after the relaxation of capillary stress (springback effect [17,18]). Finally, the composite obtained containing the fibrous network and the hydrophobic silica gel is dried at ambient pressure for 1 hour. The improvement of the silica gel by the addition of fibers as well as the hydrophobic agent allows us to get a dry material, with good mechanical and thermal properties without using supercritical drying. Figure 8.1 shows the elaboration of the discussed aerogel blankets, and Table 8.1 their basic properties.

8.3 Characterization of the Needle Glass Fibers Aerogel Blanket

8.3.1 Texture

8.3.1.1 Microstructure

The microstructure was investigated using scanning electron microscopy which provides qualitative information about the arrangement of fibers in the aerogel matrix, pores, fibers, and cracks dimension. The images in Figures 8.2 through 8.4 were taken with a SEM Philips XL30 for small magnifications (fibers scale), and with a SEM FEG Zeiss SUPRA 40 for higher magnifications (pores scale). For high magnifications, the materials need to be coated with platinum to eliminate charging effect that may cause distorted images. The fibers used are typically a few millimeters in length and 5–25 μm in diameter and are principally not contacting after casting the sol in the wool batting.

Aerogel blankets are composed of a fibrous phase, generally less than 5% vol to limit the solid conduction through fibers. Fibers are presented as woven or unwoven mat or wool, or even bulk fibers, and put perpendicular to the heat flux. Some fibers can also be woven in the transverse direction to enhance mechanical properties and prevent delamination issues. Aerogel blanket is also composed of aerogel phase (>85% vol) responsible for the low thermal conductivity of the composite. The aerogel phase is made of amorphous silica nanoparticles, micropores (<2 nm), mesopores (2–50 nm), and macropores (50 nm–few microns). At least 70% of pores are mesoporous and confine the air in a rarefaction regime.

Therefore, unlike native silica aerogel, aerogel blanket has 10%–15% of cracks which appear between fibers and the aerogel phase during the drying process leading to a slight increase in thermal conductivity.

FIGURE 8.2
SEM image of a needle glass fibers wool.

FIGURE 8.3
SEM images of an aerogel blanket.

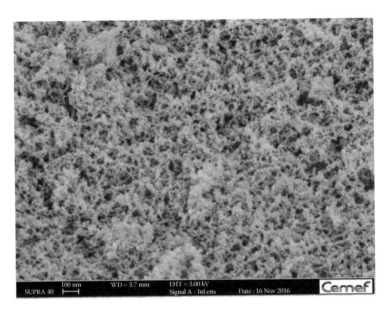

FIGURE 8.4
SEM image of the silica aerogel phase.

8.3.1.2 Specific Surface, Porosity, and Density

The native aerogel (i.e., silica phase without fibers) has a specific surface of 820 $m^2.g^{-1}$ calculated using Brunauer-Emmett-Teller method (BET). Its bulk density was measured to be 110 $kg.m^{-3}$. Assuming that the skeletal density is 2000 $kg.m^{-3}$, we deduce that the porosity of the aerogel is 95%.

Aerogel blankets have a BET specific surface of 300 $m^2.g^{-1}$ due to the presence of fibers and cracks. BET specific surface is commonly related to the fiber weight fraction with a linear dependence.

The aerogel blanket bulk density is measured to be 110–200 $kg.m^{-3}$ depending on the amount of fiber added, and the volume fractions of phases are respectively: 5% for the silica skeleton, about 90% for the air, and less than 5% for the fibers.

8.3.2 Chemical Characteristics

After drying, untreated silica aerogels contain residual Si-OH groups that give them a hydrophilic behavior. For building thermal insulation applications, it is necessary to have materials that do not absorb water vapor. Commonly, the addition of a hydrophobization agent before the gelation process replaces residual hydrophilic groups Si-OH by hydrophobic groups Si-CH$_3$. Fourier transformed infrared spectroscopy analyzes chemical bonds of the material and can validate the graft of the Si–CH$_3$ groups. The infrared spectra were obtained with TENSOR 27 spectrometer in Attenuated Total Reflectance (ATR) mode using germanium crystal. The analyses are purely qualitative, and no quantitative information was investigated from the peak areas. Before the analysis, the samples were dried in a climatic chamber at 70°C for 2 hours. The drying is done to limit the adsorption peaks due to water molecules in the material. Analyses were performed over the range 4000–600 cm^{-1}, which corresponds to the range of middle infrared energy. Figure 8.5 shows the spectra of the silica aerogel with and without hydrophobization agent.

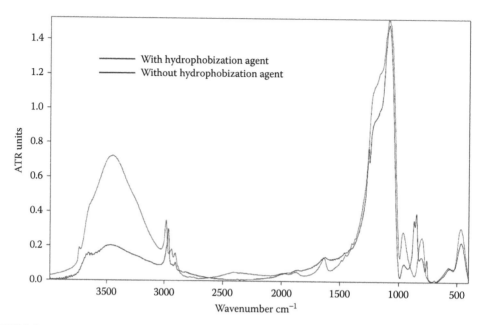

FIGURE 8.5
FTIR spectra of silica aerogel phase.

The spectra reveal a broad band between 3600 and 3000 cm^{-1} that corresponds to the overlapping of the O–H stretching band of hydrogen-bonded water molecules H_2O and Si–OH stretching of surface silanols hydrogen-bonded to molecular water. The intensity of this broad band is related to the amount of water absorbed in the sample. Here, the addition of the hydrophobization agent sharply decreases the intensity of this peak. It is an indirect proof of the agent effectiveness. The apparition of a peak at 760 cm^{-1}, a double peak in the 820–860 cm^{-1} range and another peak at 1260 cm^{-1} are respectively due to the rocking vibration of –CH_3, the deformation vibration of Si–C of the Si–CH_3 bonds, and the stretching vibration of C–H of the Si–CH_3 bonds. These peaks confirm the successful reaction between the hydrophobization agent and the silica aerogel.

8.3.3 Hydric Properties

8.3.3.1 Water Uptake

Aerogel blanket is a porous material, that is, it is made of a solid part and air inside open pores. Air and humidity flow in the porous network and can damage the material. Humidity is the primary factor limiting the lifespan of insulation. Humidity in the aerogel blanket may cause mold, cracks, expansion/shrinkage cycles, and degradation of thermal properties. In order to understand how the insulation behaves with humidity, a quantitative tool is the sorption/desorption graph that shows the hygroscopic storage properties of a porous material. It describes the water quantity adsorbed by the insulation (water uptake) versus the relative humidity. The water uptake u is determined as follows:

$$u = \frac{\text{mass}_{\text{humid}} - \text{mass}_{\text{dry}}}{\text{mass}_{\text{dry}}}$$

where:
   mass$_{humid}$ is the mass of the sample at a specified relative humidity
   mass$_{dry}$ is the mass of the sample once dried

Samples of aerogel blankets are dried at a relative humidity of 10% and a temperature of 70°C. We consider that the dry mass is reached when the mass does not vary more than 0.1% for three days. Then, the samples are placed in successively increasing humidity atmospheres with constant temperature (23°C), and weighted for each atmosphere during the equilibrium phase (constant mass). The same method is used with decreasing humidity atmospheres (desorption curve). The Figure 8.6 shows sorption/desorption graph plotted between 10% and 98% of relative humidity for the blanket aerogel and the fibrous wool before impregnation.

For both materials, Figure 8.6 reveals a hysteresis between the sorption and desorption curve. This type of sorption isotherm is characteristic of porous materials according to the IUPAC classification [19]. The fibrous wool has an excellent hydrophobic behavior; it does not influence much the aerogel blanket water uptake. Residual Si–OH bonds of the aerogel blanket increase a bit its water uptake because they can undergo strong hydrogen bonding with water. However, the aerogel blanket water uptake remains low with 0.02 kg/kg for 98% of relative humidity. For comparison, silica aerogels are considered as hydrophobic for a water uptake lower than 0.06 kg/kg [20–22].

8.3.3.2 Contact Angle

Another technique for assessing the degree of hydrophobicity is a measurement of the contact angle that a water droplet makes with the blanket surface. The droplet shape analyses were performed with a goniometer KRUSS DSA 10. The fibrous wool was first characterized before impregnation by the sol, and then the composite material was characterized.

Figure 8.7 confirms the surface hydrophobicity of the wool and of the aerogel blanket. The contact angle is measured to be higher than 90°, representative of a hydrophobic surface.

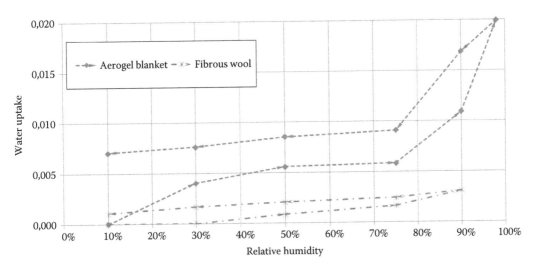

FIGURE 8.6
Sorption and desorption curve of the aerogel blanket.

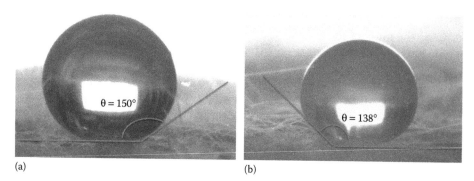

(a)   (b)

FIGURE 8.7
Water droplet on NGF aerogel blanket (a) and NGF wool (b).

8.3.3.3 *Water Vapor Transmission*

Measuring surface hydrophobicity is insufficient to prove that the materials are water-repellent. Aerogel blankets are highly porous materials, and water vapor can pass through the pores. It is relevant to assess both the hydrophobicity of the interior of the aerogels and the extent to which water vapor can permeate aerogels that have hydrophobic surfaces. Usually, a vapor barrier is set into the wall to stop the vapor, but each layer of the wall participates to the reduction of the vapor flux.

The water vapor diffusion resistance factor μ was chosen as the main parameter for this study. It indicates the relative magnitude of the water vapor resistance of the product and that of an equally thick layer of stationary air at the same temperature.

Test samples were sealed to the open side of a test dish containing an aqueous saturated salt solution (humid condition) or a desiccant (dry condition). The assembly was then placed in a test atmosphere whose temperature and humidity were respectively set to 23°C and 50%. Because of the difference between the partial water vapor pressures in the test assembly and the test atmosphere, water vapor flowed through the tested sample (Figure 8.8). Periodic weighings of the assembly were conducted to determine the rate of water vapor transmission when the steady state was reached.

Once the steady state is reached, the water vapor diffusion resistance factor μ is calculated as follows:

$$\mu = \frac{\delta_{air}}{\delta} = \frac{\delta_{air}}{W.d} = \frac{\delta_{air}.Z}{d} = \delta_{air}.\frac{A.\Delta p}{G.d}$$

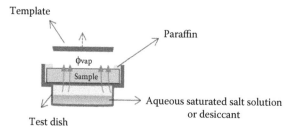

FIGURE 8.8
Principle of the water vapor transmission experiment.

where

δ_{air} is the water permeability of the air in the test condition ($\delta_{air} = 0.72$ mg.h^{-1}.m^{-1}. Pa^{-1}) for the experiment

δ is the water permeability of the test sample

W is the water vapor permeance

Z is the water vapor resistance

d is the thickness of the test sample

A is the exposed area of the test sample

Δp is the water vapor pressure difference. For our set of test conditions, $\Delta p = 1350$ Pa

G is the mass change of the specimen in milligrams per hour after reaching the steady state

Another common parameter used is the water vapor diffusion equivalent air layer thickness Sd calculates as follows: Sd = μ.d

Table 8.2 summarizes the vapor transmission results of the tested aerogel blankets in dry and wet condition. The μ value is acceptable for a thermal insulating material. Typical mineral wools have $\mu = 1$–3, and organic insulations (polyurethane, expanded polystyrene, extruded polystyrene) have $\mu = 60$–150.

8.3.4 Thermal Properties

8.3.4.1 Thermal Conductivity

The most interesting property of aerogel blanket is its low thermal conductivity $\lambda = 0.015$–0.017 W.m^{-1}.K^{-1} at ambient temperature, pressure, and relative humidity. The addition of fibers to the native aerogel increases only slightly the thermal conductivity, whatever the fibers are. Fibers do not directly influence the heat transfer since they are present in low quantities and they are principally not contacting. The phonon path through fibers is almost infinite. Fibers also play the role of opacifier. Unlike the aerogel phase, fibers scatter the infrared radiation, so they have a higher extinction coefficient, especially in the 2–8 μm wavelengths range, where silica aerogel phase is translucent to radiation. The opaque property of aerogel blanket allows using the thermal insulation at higher temperatures.

Therefore, fibers are responsible for cracks inside the aerogel blanket. Inside cracks, the air is not confined as it is in mesopores, thus the heat transfer increases. Figure 8.9 shows the thermal conductivity of our lab made aerogel blankets dried either in supercritical or ambient drying. Spaceloft® aerogel blanket from Aspen company was also tested for comparison. We submitted our samples to a hydric cycle to check the thermal conductivity behavior regarding the relative humidity. The hydric cycle is described as follows: (1) Relative humidity = 50%, (2) relative humidity = 90%, and (3) relative humidity = 10%. For all the experiment, the temperature is set to 23°C.

TABLE 8.2

Water Vapor Resistance Properties of NGF Aerogel Blanket

Test Condition	d (mm)	G (mg.h^{-1})	W (mg.m^{-2}.h^{-1}.Pa^{-1})	Z (m^2.h. Pa.mg^{-1})	Δ (mg.h^{-1}.Pa^{-1})	μ	Sd (m)
Dry	19.0	70.5	4.6	0.22	0.09	8.2	0.16
Wet	13.1	151.4	9.9	0.10	0.13	5.5	0.07

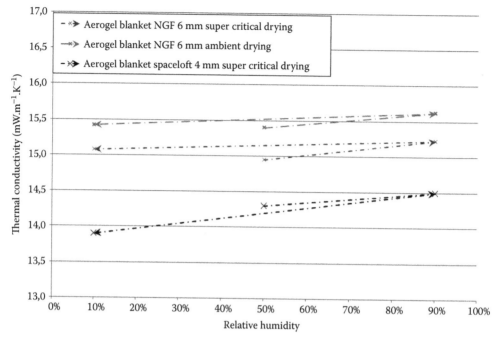

FIGURE 8.9
Thermal conductivity of aerogel blankets.

Figure 8.9 shows that the humidity does not affect much the aerogel blanket thermal conductivity. A hydrophilic aerogel would have doubled its thermal conductivity at 90% of relative humidity and would not have been considered as a superinsulating material anymore.

Ambient drying seems to increase the effective thermal conductivity of NGF aerogel blanket. Indeed, the supercritical drying avoids capillary stresses inside the pores, but the ambient drying does not. Even though the addition of the hydrophobization agent and the fibers prevent the structure from collapsing, the stresses in the smallest pores are important and can lead to the breakdown of some of them.

Considering the manufacturing cost saving by using ambient drying, the thermal conductivity increase is satisfactory.

The aerogel blanket Spaceloft® dried in supercritical remains better than the NGF sample. It may be explained by the type of fibers and the volume of fibers used as well as the application of an additional carbon opacifier at the blanket surface.

Additionally, for low-density fibrous material, thinner samples are known to have a lower effective thermal conductivity. This consequence is called "thickness effect" and is discussed in [23,24].

8.4 Specific Heat Capacity

The specific heat capacity of insulating materials is an essential parameter for the study of the dynamic heat transfer through building's walls. Specific heat capacity is related to the thermal mass of walls (thermal energy storage capability) and significantly influences their time lag and decrement factor [25].

The specific heat capacity of aerogel blankets was measured with a differential scanning calorimeter (DSC 8500) over the temperature range of 0°C–40°C.

The specific heat capacity of native silica aerogel and needle glass fibers was measured with such a device.

The heat capacity of the blankets was deduced according to the mixture law:

$$C_{p \text{ blanket}} = f_m C_{p \text{ fibers}} + (1 - f_m) C_{p \text{ aerogel}}$$

where:
C_p is the specific heat capacity
f_m the mass fraction of fibers

Figure 8.10 shows the specific heat capacity of the NGF aerogel blanket over the range 0°C–40°C. The specific heat measured is very close to the one of amorphous silica with a linear temperature-dependence for the studied temperature range. These values are within the typical range for building insulating materials 0.7–1.4 J g^{-1}C$^{\circ-1}$.

8.4.1 Building Application of Aerogel Blankets

At present day, aerogel blankets as superinsulating materials are found in various sectors such as off-shore oil and gas, aeronautic and aerospace, high temperature and cryogenic applications. Only two American companies share the aerogel blanket market for over the past ten past years. ASPEN manufactures Spaceloft® blanket ($\lambda = 15$ mW.m^{-1}.K^{-1}), which is a reference product today for building superinsulation. The company also sells aerogel blankets for high-temperature applications with Pyrogel ® ($\lambda = 21$ mW.m^{-1}.K^{-1}) and low-temperature applications with Cryogel® ($\lambda = 17$ mW.m^{-1}.K^{-1}).Cabot is its only competitor and sells aerogel blankets Thermalwrap® for ambient thermal insulation. Their products are said to be dust-free but a bit less efficient ($\lambda = 21$ mW.m^{-1}.K^{-1}). More recently, Enersens is known as the first French company working in the superinsulation market with its product Skogar® ($\lambda = 15$ mW.m^{-1}.K^{-1}). The materials distinguish itself thanks to an innovative ambient drying process set by Enersens and would be able to compete with the previously mentioned blankets.

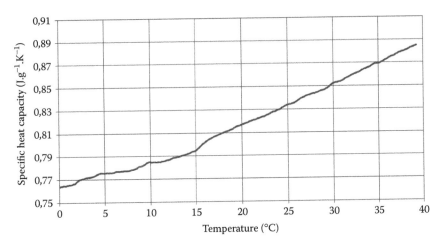

FIGURE 8.10
Specific heat of the NGF aerogel blanket.

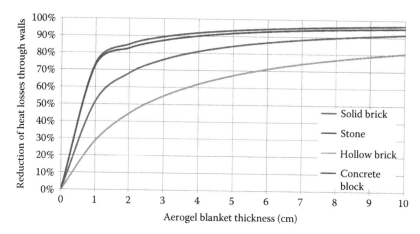

FIGURE 8.11

Reduction of heat losses through walls due to the retrofitting by aerogel blanket for different structural materials.

The most promising application for aerogel blankets seems to be thermal insulation for buildings. In Europe, in 2012, the building sector accounted for about 40% of total final energy consumption with 70% due to the space heating. It appears that about 65% of accommodations don't have any insulation and account for 75% of the energy consumption. Figure 8.11 shows the effect of 2 centimeters of aerogel blanket interior retrofitting for houses built before 1974.

The first centimeters of insulation are the most important and can save 50%–90% of heat losses through the building's walls. However, for now, the most economical way to reduce the heat losses through walls is to use thicker insulation at a low cost. Aerogel blanket provides better results but at a significantly higher price. Its actual price is about 2000 €/m³ but should decreases to 1500 €/m³ by 2020. This will be reached thanks to economies of scale, more productive plants located closer to the markets, the integration of the precursor manufacturing on site and optimized distribution networks.

On the other hand, aerogel blankets can offer the same insulation as conventional insulating materials (mineral wool or cellular foam) with a reduced thickness.

The purpose is not only to save the living space and decrease the wall thickness, but the extra surface means immediate margin on the global cost of construction or renovation for the building companies but also meaningful additional use and rental value for inhabitants and end users.

Table 8.3 shows a comparison between a glass wool interior insulation renovation and an aerogel blanket insulation renovation of a 60 m² flat, real estate value of which is 3000 €/m² and rental value is 10 €/m²/month.

TABLE 8.3

Real Estate Cost Saving Using Aerogel Blanket

60 m² Insulated Wall Surface Over a 25 m Perimeter	Cost of the Insulation Solution	Saved Surface (m²)	Save Real Estate Value	Saved Rental Value Over 10 years
Glass wool	2300 €	0	0	0
Aerogel blanket	6600 €	1.5	4500 €	>3000 €

According to the building code in France, the thermal resistance of walls needs to be 2.85, which implies 40 mm of aerogel blanket versus 100 mm of glass wool ($\lambda = 35$ mW.m^{-1}.K^{-1}).

The calculation includes in both cases 30 €/m^2 for worksite costs, setup, and insulation system modules (labor, plasterboard, framework, screws, and plain finish painting). The cost of the glass wool used is 80€/m^3. The loss of rental value has been amortized over an 18-year period, which is shorter than the lifetime of the insulation system.

The use of aerogel blanket is adequate for applications that need a space saving or that are space limited.

The main potential applications are as follows:

- Renovation of historical buildings
- Flat-roof balcony constructions
- Aesthetic architecture with slim exterior insulation, lightweight elements
- Inner city thin constructions

Most of new constructed buildings and renovation will still use conventional insulation materials because the spare space saved by slimmer insulation only counterbalances the added cost in high-prized location. That is why aerogel blankets always have difficulties finding their way into the building insulation market. Nonetheless, this trend is currently evolving with more and more new aerogel-based insulation products and demonstration projects emerging these days.

8.5 Conclusion

During the past 40 years, there has been no substantial change in thermal insulation materials employed in the building sector. Aerogel blankets could be a worthy successor to conventional insulation in the future. Characterization results show a flexible, easy-to-handle material with a thermal conductivity two times better than the actual best insulation products. Aerogel blankets are now hydrophobic and can be dried in ambient conditions and still have the same properties of blankets made by the supercritical route. They can be used for the retrofitting of insulated houses and reduce their consumption actively.

Nevertheless, considering the high cost of manufacturing, they are limited to few applications in niche sectors. The principal arguments for the use of aerogel insulation systems are space-saving, reducing operating cost, longevity, and chemical resistance. In addition, the global volume market of superinsulation has a large growth potential with an enlargement of 50% per year. Lower prices will lead to equilibrium between aerogel blanket and traditional insulation regarding the cost/benefit ratio.

References

1. Directive of European parliament and of the council amending Directive 2012/27/EU on energy efficiency, 2012.
2. G. Verbeeck and H. Hens, Energy savings in retrofitted dwellings: Economically viable? *Energ. Build.*, 37(7): 747–754, 2005.

3. Energy Efficiency Watch, Final Report, Improving and implementing national energy efficiency strategies in the EU framework, 2013.
4. D. M. S. Al-Homoud, Performance characteristics and practical applications of common building thermal insulation materials, *Build. Environ.*, 40(3): 353–366, 2005.
5. S. E. Kalnaes and B. P. Jelle, Vacuum insulation panel products: A state-of-the-art review and future research pathways, *Appl. Energ.*, 116: 355–375, 2014.
6. R. Baetens, B. P. Jelle, and A. Gustavsen, Aerogel insulation for building applications: A state-of-the-art review, *Energ. Build.*, 43(4): 761–769, 2011.
7. Enersens Granules et poudres, available at: http://enersens.fr/fr/granules-et-poudres/ (Accessed: October 2016).
8. OKAGEL Insulating Glass with Aerogel, available at: http://www.okalux.de/en/solutions/solution-gallery/okagel-insulating-glass-with-aerogel/ (Accessed: October 19, 2016).
9. All Insulation Products from Aspen Aerogels, available at: http://www.aerogel.com/products-and-solutions/all-insulation-products/ (Accessed: October 19, 2016).
10. Product Lines|Airloy Ultramaterials: Strong Aerogels from Aerogel Technologies, available at: http://www.airloy.com/category/productlines/ (Accessed: October 19, 2016).
11. Cabot Aerogel Blanket, *Cabot Corporation,* available at: http://www.cabotcorp.com/solutions/products-plus/aerogel/blanket (Accessed: October 19, 2016).
12. M. Koebel, A. Rigacci, and P. Achard, Aerogel-based thermal superinsulation: An overview, *J. Sol. Gel Sci. Technol.*, 63(3): 315–339, 2012.
13. S. M. Jones, Aerogel: Space exploration applications, *J. Sol. Gel Sci. Technol.*, 40(2–3): 351–357, 2006.
14. B. E. Coffman et al., Aerogel blanket insulation materials for cryogenic applications, 2010, 913–20, https://doi.org/10.1063/1.3422458.
15. A. C. Pierre and A. Rigacci, SiO_2 aerogels, in *Aerogels Handbook*, Springer, London, UK, 2011, pp. 21–45.
16. HomeSkin|HomeSkin—Thinner Insulation Systems, available at: http://homeskin.net/ (Accessed: October 2016).
17. D. M. Smith, R. Deshpande, and C. J. Brinke, Preparation of low-density aerogels at ambient pressure, in *MRS Proceedings*, vol. 271, 1992, p. 567.
18. A. Bisson, E. Rodier, A. Rigacci, D. Lecomte, and P. Achard, Study of evaporative drying of treated silica gels, *J. Non-Cryst. Solids*, 350: 230–237, 2004.
19. K. S. Sing, Reporting physisorption data for gas/solid systems with special reference to the determination of surface area and porosity (Recommendations 1984), *Pure Appl. Chem.*, 57(4): 603–619, 1985.
20. A. V. Rao and M. M. Kulkarni, Effect of glycerol additive on physical properties of hydrophobic silica aerogels, *Mater. Chem. Phys.*, 77(3): 819–825, 2003.
21. P. Wagh and S. Ingale, Comparison of some physico-chemical properties of hydrophilic and hydrophobic silica aerogels, *Ceram. Int.*, 28(1): 43–50, 2002.
22. T. M. Tillotson, K. G. Foster, and J. G. Reynolds, Fluorine-induced hydrophobicity in silica aerogels, *J. Non-Cryst. Solids*, 350: 202–208, 2004.
23. D. L. McElroy and R. P. Tye, *Thermal Insulation Performance: Symposium.* ASTM International, 1980.
24. C. J. Shirtliffe, Effect of thickness on the thermal properties of thick specimens of low-density thermal insulation, in *Thermal Insulation Performance.* ASTM International, 1980.
25. S. A. Al-Sanea, M. F. Zedan, and S. N. Al-Hussain, Effect of thermal mass on performance of insulated building walls and the concept of energy savings potential, *Appl. Energy*, 89(1): 430–442, 2012.

9

Heating Ventilation and Air-Conditioning Systems for Energy-Efficient Buildings

Karunesh Kant, Amritanshu Shukla, and Atul Sharma

CONTENTS

9.1 Introduction 165
9.2 Types of Heating, Ventilation, and Air-Conditioning and Energy Consumption 167
    9.2.1 Types of Heating, Ventilation, and Air-Conditioning Systems 167
        9.2.1.1 Heating and Air-Conditioning Split System 167
        9.2.1.2 Hybrid Heat Split System 168
        9.2.1.3 Duct-Free Split Heating and Air-Conditioning System 168
        9.2.1.4 Packaged Heating and Air-Conditioning System 168
    9.2.2 Energy Consumptions in Heating, Ventilation, and Air-Conditioning 169
9.3 Technology Overview 169
9.4 Energy-Saving Strategies for Heating, Ventilation, and Air-Conditioning 170
    9.4.1 Evaporative Cooling Systems 170
    9.4.2 Evaporative-Cooled Air-Conditioning System 171
    9.4.3 Ground-Coupled Heating, Ventilation, and Air-Conditioning Systems........... 172
    9.4.4 Thermal Storage Systems 173
    9.4.5 Heat Recovery Systems 174
    9.4.6 Effect of Building Behavior 174
9.5 Role of Heating, Ventilation, and Air-Conditioning in Energy-Efficient Buildings 177
9.6 Conclusions 178
References 178

9.1 Introduction

The need for energy demand has increased with the increase in the living standard. This requires researchers and engineers to focus on energy efficiency in buildings to reduce the consumption of fossil fuel and reduce the global warming. Buildings account for about 45% of the global energy consumption and contribute over 30% of the CO_2 emissions [1]. Figure 9.1 represents sector wise global energy consumptions. From Figure 9.1, it is clear that the maximum energy is consumed by building sector. Heating, ventilation, and air-conditioning (HVAC) systems, which play a significant role in confirming thermal comfort in the buildings, are amongst the major energy consumers in buildings. Nearly 50% of the

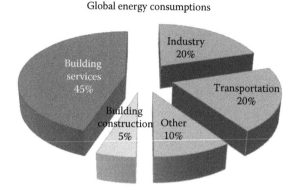

Global energy consumptions

FIGURE 9.1
Sector-wise global energy consumption.

energy demand is used to support indoor thermal comfort situations in commercial structures [2]. Additionally, as most people spend more than 90% of their time inside the buildings [3], the growth of energy-efficient HVAC systems that do not depend on fossil fuels will play a strategic role in dropping energy ingesting. In India, air-conditioning systems account for around 32% of electricity consumption of a building [4]. It was projected that world energy consumption will grow by 56% between 2010 and 2040, from 524 quadrillion British thermal units (Btu) to 820 quadrillions Btu [5]. However, approximately 80% of the energy usage still comes from fossil fuels [6,7].

HVAC systems control the indoor temperature, moisture content in the air and quality of air to the buildings to a set of selected conditions. A typical air HVAC system is shown in Figure 9.2. To accomplish this, the systems require transfer of the heat and moistness into and out of the air in addition to controlling the air pollutants level, by either directly removing them or by diluting them to tolerable levels. The heating systems increase the temperature in a space to compensate for heat losses concerning the interior space and outside. The ventilation systems supply air to space and extract contaminated air from it. The cooling of air available in the space is required to get the temperature down where heat gains have risen due to the presence of people, equipment, or the sun and are affecting human discomfort. HVAC systems differ extensively in terms of size and the functions they perform. Some systems are large and central to the building services; these were probably designed when the building was initially commissioned and use ventilation to deliver heating and cooling. Other systems may provide heating through boilers and radiators, with some limited ventilation to provide fresh air or cooling to confident portions of the building such as conference rooms.

In some cases, individual comfort cooling units have been added to a building to overcome a specific overheating problem that had not been thought of at the time of the original design. By considering HVAC systems as a separate component rather than as an intermingling system, it would be informal to supervise the main area of energy wastage—that one element might influence one another, for example, it would be uneconomical to upsurge heating inside a building while the cooling system is combat to decrease temperatures. Consequently, it is worthwhile to look at how the components of an HVAC system interrelate with each other and fine tune every portion to save energy and money. Inappropriate design and indecorous installation of the HVAC system have an adverse effect on thermal comfort inside the buildings and on energy bills. Inappropriate design

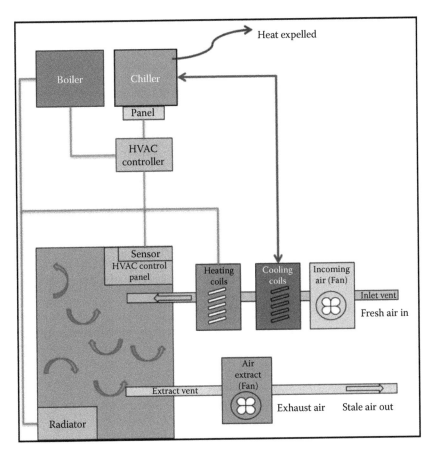

FIGURE 9.2
Heating ventilation and air-conditioning systems.

and installation of an HVAC system can intensely reduce the quality of air in a space. Poorly designed and poorly installed air circulation ducts can make unsafe conditions that may diminish the human comfort and degrade indoor air quality or even threaten the health of the homeowners.

9.2 Types of Heating, Ventilation, and Air-Conditioning and Energy Consumption

9.2.1 Types of Heating, Ventilation, and Air-Conditioning Systems

9.2.1.1 Heating and Air-Conditioning Split System

Split systems are the most classic of the heating and air-conditioning systems. These are the traditional types of HVAC system where you have components of the whole system that are both inside and outside the building. HVAC split systems will typically have

- An air-conditioner that cools the refrigerant
- Furnaces and a fan or evaporator coil to convert the refrigerant and circulate the air
- Ducts that carry air all through the building
- A control panel/thermostat to manage the system
- The occasional optional accessories for quality indoor air such as air cleaners, purifiers, humidifiers, UV lamps, and so on

9.2.1.2 Hybrid Heat Split System

The hybrid heat split system is an advanced version of the classic HVAC split system that has an improved energy efficacy. When included in these types of HVAC systems, a heat pump will allow the option of having an electrically fueled HVAC up and above the typical gas furnaces. An ideal hybrid heat split system that is cost effective will have

- A heat pump that heats or cools the refrigerant
- Furnaces plus the evaporator coil for conversion of the refrigerant and circulation of air
- The ducts to channel the air around the building
- An interface for adjusting and controlling the system
- Optional accessories for more quality indoor air

9.2.1.3 Duct-Free Split Heating and Air-Conditioning System

A duct-free HVAC provides good installations for places and areas where the conventional systems with ducts can't go. These systems are also ideally great compliments to existing ducted types of HVAC systems. Duct-free systems will have the following:

- The heat pump or an air-conditioner to heat\cool the refrigerant
- A fan coil that is compact
- Wires and tubing for the refrigerant, connecting the outdoor unit to the fan coil
- The thermostat or control panel
- Optional accessories to clean the air and make it more pleasant before its distribution through the house

9.2.1.4 Packaged Heating and Air-Conditioning System

A packaged HVAC system is the solution to those homes and offices without adequate spaces for all the separate multiple components of the split systems. Packaged heating and air-conditioning systems will sort out confined spaces that range from entire homes to the one-roomed units, all in one package. Packaged HVAC systems will contain the following:

- The air-conditioner/heat pump together with the evaporator/fan coil in one unit
- Thermostat/control interface for a complete control of the system

- Optional air quality improvements, including the air purifiers, cleaners, ventilators or UV lamps, which are geared toward making the air extra clean before it circulates your home or office.

9.2.2 Energy Consumptions in Heating, Ventilation, and Air-Conditioning

HVAC can account for the majority of money spent by an organization on energy. Even small adjustments to these systems can significantly improve the working environment and at the same time, save money.

There are five important factors that determine the energy use of an HVAC system:

- The design, layout, and operation of the building—this affects how the external environment impacts on internal temperatures and humidity
- The required indoor temperature and air quality—more extreme temperatures, greater precision, and more refined air quality consume more energy
- The heat generated internally by lighting, equipment, and people—all of these have an impact on how warm your building is
- The design and efficiency of the HVAC plant—provides heat, cooling, and moisture control exactly where it is needed in the building
- The operating times of the HVAC equipment and ability of the controls—these limit operation exactly when it is needed

9.3 Technology Overview

HVAC systems vary widely in terms of the individual components that make them up and how they are set up within a building. Most systems contain some common basic components as follows:

- *The boiler* produces hot water (or sometimes steam) for distributing to the working space. This is done either by heating coils that heat air as part of the ventilation system or through hot water pipes to radiators.
- *Cooling equipment* chills water for pumping to cooling coils. Treated air is then blown over the chilled water coils into the space to be cooled through the ventilation system. As part of the refrigeration cycle in the chiller, heat must also be rejected from the system via a cooling tower or condenser.
- *Pumps* are used throughout the system to circulate the chilled and hot water to the required areas throughout the building.
- Stale air is extracted, usually using a fan, via separate ducts and expelled outside.
- Controls are used to make components work together efficiently. They turn equipment on or off and adjust chillers and boilers, air and water flow rates, and temperatures and pressures. A controller incorporating one or more temperature sensors inside the workspace sends a signal to the heating or cooling coils to activate.

- If there is a demand for heating or cooling, then the controls may also send a signal to the chiller and boiler to operate as required. There are often other control panels on the chiller or boiler too, allowing users to have greater control.

9.4 Energy-Saving Strategies for Heating, Ventilation, and Air-Conditioning

With the growing dependency on HVAC systems in the commercial, the residential, and industrial sector there has been a huge upsurge in energy consumption, predominantly in the summer months. It is essential to develop energy-efficient HVAC systems to protect users from surging power costs and to protect the environment from the effects of greenhouse gas emissions caused by the use of energy-inefficient electrical appliances. With the technological advancement in the field of science and technology, there are several approaches that can be used to achieve higher energy-efficient HVAC systems. In order to increase the energy efficiency of these systems, though, a clear understanding of building comfort conditions is necessary. The thermal comfort is all about human satisfaction within their thermal environment. The design and calculation of air-conditioning systems to control the thermal environment in a way that also achieves an acceptable standard of air quality inside a building should comply with the ASHRAE standard 55-2004 [8]. In recent past, diverse control and optimization approaches have been used to reduce the energy consumption rates of HVAC systems [9]. Figure 9.3 represents the approaches used to attain superior HVAC energy efficiency.

9.4.1 Evaporative Cooling Systems

The evaporative cooling methods have been extensively used for over a century [10]. The direct evaporative cooling (DEC) systems have small setup and operating costs and have been confirmed to considerably improve a building's cooling and ventilation capability with marginal energy use. The ozone-destroying elements chlorofluorocarbons and hydrochlorofluorocarbons can be avoided by using water as the working fluid. The evaporative cooling systems can offer thermal comfort via the transformation of sensible heat to latent heat; however, the lowest temperature of DEC systems can reach the wet-bulb temperature of the external air. Jiang et al. [10] estimated the perspective of dropping the yearly energy ingesting of a central cooling system by joining with regenerative evaporative cooling technique to achieve the energy conversation objective. Delfani et al. [11] studied the effect of conjoining the indirect evaporative cooling system with a packed unit air-conditioner on electrical energy consumption. In this configuration, the air is precooled by using the indirect evaporative cooler and then passes through the packaged cooling unit as demonstrated in Figure 9.4. The obtained results showed that the indirect evaporative chiller can decrease cooling load up to 75%, which causes a 55% decrease in power ingesting of the packaged air-conditioning system.

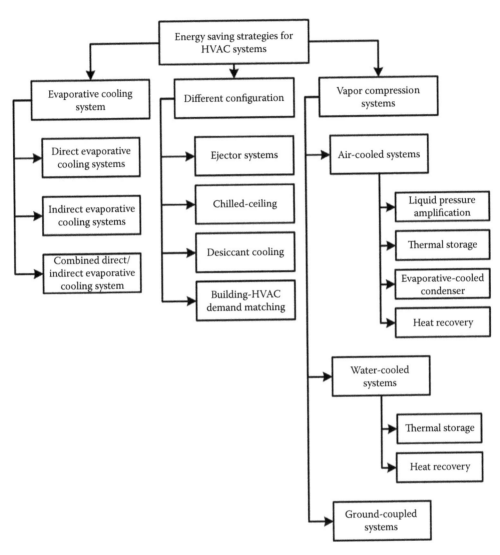

FIGURE 9.3
HVAC energy-saving strategies discussed in this study. (From Vakiloroaya, V. et al., *Energ. Conver. Manag.*, 77, 738–754, 2014.)

9.4.2 Evaporative-Cooled Air-Conditioning System

Recent research reveals that air-conditioning systems grounded on mechanical vapor compression consume substantial amounts of electricity. Consequently, enhancing the coefficient of performance (COP) of evaporatively cooled air-conditioning methods with air-cooled condensers is an interesting problem. Precooled air goes to the condenser coil, and the condenser is capable to reject higher heat. Consequently, cooling capacity upsurges though energy demand and usage decreases. As condensing temperatures are dropped, the head pressure decreases to allow the compressor to run less regularly, resulting in an energy consumption reduction. The typical design of these systems necessitates a frame to be constructed and occupied by evaporative media pads which are connected to the

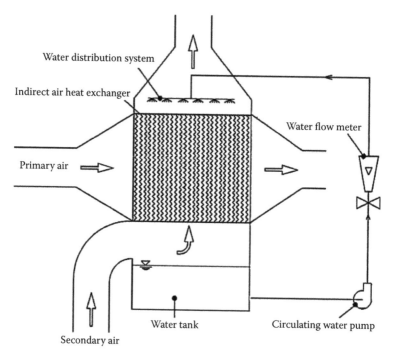

FIGURE 9.4
The schematic diagram of the IEC system.

air-cooled condenser. A water circulation system, consisting of a small pump, a tank and pipes, is added. The water then is injected on the top of the media pad. Hot ambient air passes the wet pad and then the condenser to improve the system performance. As the hot air from ambient is drawn through the media, the water engrosses the heat and evaporates, dropping the temperature of the ambient air and making a cooler operating environment for the air-cooled condenser that allows the condenser to reject additional heat into the atmosphere. The compression ratio is then reduced, resulting in reduced energy usage when the compressor is run. In a comparable design, the mist is scattered directly into the ambient air in advance passing the air-cooled condenser.

9.4.3 Ground-Coupled Heating, Ventilation, and Air-Conditioning Systems

The ground-coupled HVAC technology depends on the fact that, at depth, the earth/ground has a comparatively persistent temperature that is colder than the air temperature in summer and warmer than the air temperature in winter. In this system, under cooling mode operation, heat is discharged to a ground loop that provides a lower temperature heat sink than ambient outdoor air temperature. During winter heating operations, heat is extracted from a source that is at a higher temperature than ambient outdoor air. This system has been used on a residential and commercial scale since the 1920s [12]. The studies on vertical closed-loop ground-coupled heat pumps give a yearly reduction of 30%–70% in electrical energy ingesting for heating and cooling when compared to air-to-air heat pump systems in a southern climate [13]. As stated in one study [14], the COP of a ground source heat pump (GSHP) was greater than that of the air-source heat pump (ASHP) by 74%, due to lower condensing temperatures in the GSHP system. A different study [15] compared the GSHP and ASHP for an archives building; results showed that while the initial cost for GSHP is more than ASHP, the operating cost of the GSHP can be decreased by 55.8% with a pay-back time of about 2 years. On the other hand, GSHPs capture only a small percentage of the heating and cooling market due to the high cost of mounting the ground heat exchanger, which can upsurge system budgets by 20%–30% [16] and the preliminary capital cost by 30%–50% when compared to air source units [17].

9.4.4 Thermal Storage Systems

Thermal energy storage (TES) systems shift the energy usage of the HVAC systems from peak to off-peak periods to avoid peak demand charges. The TES systems are also capable of rate variance concerning energy supply and demand to save energy consumption [18]. In this type of system, energy for cooling is stowed at lower temperatures generally under 20°C for cooling, although energy for heating is kept at temperatures generally above 20°C [19]. Compared to conventional HVAC systems, TES systems provide numerous benefits for heating and cooling systems, such as energy and capital cost savings, system operation enhancements, prolonging system capacity, and reducing equipment size lessening, resulting in an expertise that is extensively castoff. Yau and Rismanchi stated that in the early 1990s, around 1500–2000 units of TES systems were engaged in the United States for office, school, and hospital buildings [20]. Cooling thermal storage can be categorized according to the thermal medium as obtainable by Al-Abidi et al. [21] and presented in Figure 9.5. Ice and chilled-water storing systems

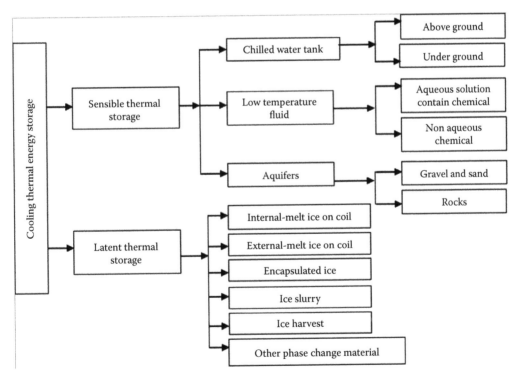

FIGURE 9.5
Classification of cooling thermal energy storage. (From Al-Abidi, A.A. et al., *Renew. Sustain. Energ. Rev.*, 16, 5802–5819, 2012.)

are two most common TES systems. There are numerous types of ice storing methods. An ice harvester system uses an open insulated storage tank and a vertical plate surface that is located above the tank. During the charging period, water flows on the outside surface of the evaporator and forms ice sheets. Ice slurry is another type of the ice storage system in which a glycol–water solution passes through pipes submerged in an evaporating refrigerant to form the ice. The produced ice elements are then dropped into the storage container.

9.4.5 Heat Recovery Systems

Heat recovery techniques can be used to improve energy that might otherwise be fruitless. The motive of heat recovery is to decrease the operating cost of an HVAC system by transferring heat between two fluids such as exhaust air and fresh air. According to *ASHRAE Handbook of HVAC Systems and Equipment* (2012) [22], there are three types of heat recovery systems, that is, (1) comfort-to-comfort, (2) process-to-comfort, and (3) process-to-process. Comfort-to-comfort systems use exhaust air that is captured and use again the waste heat energy to pre-condition the fresh air coming into the HVAC system. This type of heat recovery system can be used as the sensible heat recovery mode and in total heat transfer mode. Typically, rotary wheel heat exchangers are used for comfort-to-comfort heat recovery systems [22]. Process-to-comfort and process-to-process systems accomplish sensible heat recovery. The different types of heat recovery systems, which are cast off to recover energy between deliveries and exhaust air flows, are composed of the fixed plate, rotary wheel, heat pipe, and run-around coil [22]. Heat and moisture recovery can save nearly 70%–90% of the energy that is cast off for cooling and dehumidifying the fresh air [22], but it should be noted that the investment and running costs are the major expenses associated with heat recovery systems [23].

9.4.6 Effect of Building Behavior

The energy consumption of an HVAC system depends not only on its performance and operational parameters but also on the characteristics of the heating and cooling demand and the thermodynamic behavior of the building. The actual load of the HVAC systems is less than it is designed for most operating periods due to building behavior. Therefore, the most important factors that contribute to HVAC energy usage reduction in a given building is proper control of the heating and cooling demand [24]. Integrated control of building cooling load components, such as solar radiation, lighting and fresh air, can result in significant energy savings in a building's cooling plant. It is estimated that around 70% of energy savings are possible through the use of better design technologies to coordinate the building demand with its HVAC system capacity [25]. The comparison of various HVAC energy-saving strategies is tabulated in Table 9.1.

TABLE 9.1

Comparison of Different HVAC Energy-Saving Strategies

System Type	Capital Cost	Projected Improvement	Merits	Demerits
Direct evaporative cooling (DEC) systems	Very cheap among all air-conditioning systems	Their efficiency can be considerably enhanced using several methods	Reduction of pollution emissions/life cycle cost effectiveness/reduction of peak demand	Cannot effectively work when ambient relative humidity if higher than 40%
Indirect evaporative cooling (IEC) systems	The running and capital cost is higher than DEC systems but lower than vapor compression air-conditioning systems	Still, there are several opportunities to improve their performance	Their air quality is considerably higher than DEC, more energy-efficient compared with vapor compression refrigeration systems	Their installation and operation are more complex than DEC systems
Evaporative-cooled air-conditioning systems	Water usage cost is increased while electricity usage cost is reduced. In total, this method adds less cost than other discussed methods to the air-conditioning system	Control methods can reduce the water usage of the evaporative-cooled condenser	Can significantly reduce the energy consumption of the air-conditioning system in peak demand conditions	Their energy-saving potential is limited to a period that ambient temperature is high
Liquid pressure amplification (LPA) systems	More expensive than conventional vapor compression refrigeration systems but cheaper than thermal storage, heat recovery, ground-coupled, and desiccant cooling systems	This method can be applied in more air-cooled vapor compression air-conditioning systems in operation today worldwide	Results in significant energy savings for vapor compression refrigeration systems while provides reasonable time frame payback of the additional costs	Their energy-saving potential is limited to a period that ambient temperature falls to low temperature
Thermal storage systems	Expensive in both capital and running cost	Robust control method can enhance the efficiency of this system	They significantly reduce the electric energy cost/ Required smaller ducting system than conventional air systems	Their coefficient of performance is less than conventional vapor compression air-conditioners

(Continued)

TABLE 9.1 (*Continued*)

Comparison of Different HVAC Energy-Saving Strategies

System Type	Capital Cost	Projected Improvement	Merits	Demerits
Heat recovery systems	Expensive in both capital and running cost	Their dimension can be reduced by increasing the efficiency of heat recovery devices	Highly energy efficient in temperate climates	Larger than conventional air-handling units
Ground-coupled systems	Expensive in both capital and running cost	Investigations should be done to optimize the installation method and cost	Compared with standard vapor compression systems, this system creates less noise and reduces greenhouse gas emissions	Requirements of deep below the earth's surface/Very high upfront costs
Chilled-ceiling systems	Reasonable timeframe payback of the additional costs	More study is required to reduce the risk of condensation using new materials	They require cooled water instead of chilled water, which leads to less refrigeration	Are unable to moderate indoor humidity/risk of condensation at cold surface
Desiccant cooling systems	Expensive in both capital and running cost	This system can gradually attain wider market penetration	When used in conjunction with conventional air-conditioning systems, humidity control is improved	Some desiccants are corrosive/response time is relatively large/crystallization may be a problem
Ejector cooling systems	Reasonable time frame payback of the additional costs	Studies should be carried out to enhance the coefficient of performance of this system	More installation, maintenance and construction simplicity than conventional vapor compression refrigeration systems	This system needs a heat source with the temperature more than 80°C/compared with conventional vapor compression systems, the ejector cooling systems have lower COP
Variable refrigerant flow (VRF) systems	Expensive in both capital and running cost	Several efforts are currently carrying out to enhance the performance of this system	Efficient in part load conditions	Required extra control systems/cannot provide full control of humidity

9.5 Role of Heating, Ventilation, and Air-Conditioning in Energy-Efficient Buildings

The HVAC systems play an important role in building energy efficiency. The efficient HVAC systems provide better comfort conditions and reduced energy consumption in the buildings. Most of the case studies emphasized either simply setting a higher summer set point temperature (SST) or implementing a wider/varying range of indoor design temperature for a different time of the day and different outdoor conditions. Two major types of control techniques have been proposed for the heating and cooling systems. The first type involves diverse thermostat strategies such as changes of the setback period, SST and set-back temperature [26]. Attempts have also been made to correlate cooling energy use with corresponding thermostat operation mode to have a better understanding of the trade-off between energy consumption and thermal comfort [27,28]. The second type deals with the dynamic control of the SST based on adaptive comfort models [29,30]. Table 9.2 shows a summary of some of the case studies involving adaptive comfort models and/or raising the SST [31–38]. Substantial energy savings could be achieved for both office and residential buildings, from 6% reduction in HVAC electricity consumption in Australian office buildings by raising 1°C in the SST [37] to 33.6% reduction in total energy cost in the hot desert area in Riyadh [36]. Apart from energy-saving potential, raising the SST could also substantially reduce the peak electricity demand as demonstrated by the work on residential buildings in Las Vegas [38]. This could have significant energy policy implications as it helps to alleviate and/or delay the need for new power plants to meet the expected increase in power demand due to economic and population growth.

TABLE 9.2

Summary of Energy Savings in Cooled Buildings

City (Climate) (Year)	Building	Measure	Cooling Energy Savings
Hong Kong SAR (subtropical) (1992)	Office	Raise SST from 21.5°C to 25.5°C (SST = summer set point temperature)	Cooling energy reduced by 29%
Montreal (humid continental) (1992)	Office	Raise SST from 24.6°C to 25.2°C (during 09:00–15:00) and up to 27°C (during 15:00–18:00)	Chilled water consumption reduced by 34%–40% and energy budget for HVAC by 11%
Singapore (tropical) (1995)	Office	Raise SST from 23°C to 26°C	Cooling energy reduced by 13%
Islamabad (humid subtropical) and Karachi (arid) (1996)	Office	Change the 26°C SST to a variable indoor design temperature ($T_c = 17 + 0.38 \times T_o$; T_c = comfort temperature, T_o = mean monthly outdoor temperature)	Potential energy savings of 20%–25%

(Continued)

TABLE 9.2 (*Continued*)

Summary of Energy Savings in Cooled Buildings

City (Climate) (Year)	Building	Measure	Cooling Energy Savings
Hong Kong SAR (subtropical) (2003)	Office	Change SST from 24°C (average) to adaptive comfort temperature ($T_c = 18.303 + 0.158T$)	Energy consumption by cooling coil reduced by 7%
Riyadh (hot desert) (2008)	No specific building type	Change yearly-fixed thermostat setting (21°C–24.1°C) to optimize monthly fixed settings (20.1°C–26.2°C)	Energy cost reduced by 26.8%–33.6%
Melbourne (oceanic), Sydney (temperate) and Brisbane (humid subtropical) (2011)	Office	Static (raise SST 1°C higher) and dynamic (adjust SST in direct response to variations in ambient conditions)	HVAC electricity consumption reduced by 6% (static) and 6.3% (dynamic)
Las Vegas (subtropical desert) (2012)	Home	Raise SST from 23.9°C to 26.1°C (during 16:00–19:00)	Peak electrical energy demand reduced by 69%

Source: Yang, L. et al., *Appl. Energ.*, 115, 164–173, 2014.

9.6 Conclusions

This chapter represents an overview of HVAC; introduces the main energy-saving opportunities for buildings; and demonstrates how simple actions save energy, cut costs, and increase profit margins. Reducing energy use saves money, enhances corporate reputation, and helps everyone in the fight against climate change. Conventional HVAC systems rely heavily on energy generated from fossil fuels, which are being rapidly depleted. This together with a growing demand for cost-effective infrastructure and appliances has necessitated new installations and major retrofits in occupied buildings to achieve energy efficiency and environmental sustainability.

References

1. L. Yang, H. Yan, J.C. Lam, Thermal comfort and building energy consumption implications-a review, *Applied Energy* 115 (2014) 164–173.
2. N. Enteria, K. Mizutani, The role of the thermally activated desiccant cooling technologies in the issue of energy and environment, *Renewable and Sustainable Energy Reviews* 15 (2011) 2095–2122. doi:10.1016/j.rser.2011.01.013.
3. R. Qi, L. Lu, H. Yang, Investigation on air-conditioning load profile and energy consumption of desiccant cooling system for commercial buildings in Hong Kong, *Energy and Buildings* 49 (2012) 509–518. doi:10.1016/j.enbuild.2012.02.051.

4. V. Vakiloroaya, B. Samali, A. Fakhar, K. Pishghadam, A review of different strategies for HVAC energy saving, *Energy Conversion and Management* 77 (2014) 738–754. doi:10.1016/j.enconman.2013.10.023.

5. EIA, International Energy Outlook 2016, 2016. doi:www.eia.gov/forecasts/ieo/pdf/0484(2016) (Accessed: April 1, 2017).pdf.

6. Q. Ma, R.Z. Wang, Y.J. Dai, X.Q. Zhai, Performance analysis on a hybrid air-conditioning system of a green building, *Energy and Buildings* 38 (2006) 447–453. doi:10.1016/j.enbuild.2005.08.004.

7. U. Desideri, S. Proietti, P. Sdringola, Solar-powered cooling systems: Technical and economic analysis on industrial refrigeration and air-conditioning applications, *Applied Energy* 86 (2009) 1376–1386. doi:10.1016/j.apenergy.2009.01.011.

8. W.A. Dunn, G.S. Brager, K.A. Brown, D.R. Clark, J.J. Deringer, J.J. Hogeling et al., *Ashrae Standard Thermal Environmental Conditions for Human Occupancy*, American Society of Heating, Refrigerating and Air-Conditioning Engineers, Inc., 2004.

9. Z. Ma, S. Wang, X. Xu, F. Xiao, A supervisory control strategy for building cooling water systems for practical and real time applications, *Energy Conversion and Management* 49 (2008) 2324–2336. doi:10.1016/j.enconman.2008.01.019.

10. Y. Jiang, X. Xie, Theoretical and testing performance of an innovative indirect evaporative chiller, *Solar Energy* 84 (2010) 2041–2055. doi:10.1016/j.solener.2010.09.012.

11. S. Delfani, J. Esmaeelian, H. Pasdarshahri, M. Karami, Energy saving potential of an indirect evaporative cooler as a pre-cooling unit for mechanical cooling systems in Iran, *Energy and Buildings* 42 (2010) 2169–2176. doi:10.1016/j.enbuild.2010.07.009.

12. A. Mustafa Omer, Ground-source heat pumps systems and applications, *Renewable and Sustainable Energy Reviews* 12 (2008) 344–371. doi:10.1016/j.rser.2006.10.003.

13. O. Zogou, A. Stamatelos, Effect of climatic conditions on the design optimization of heat pump systems for space heating and cooling, *Energy Conversion and Management* 39 (1998) 609–622.

14. Y. Hwang, J.K. Lee, Y.M. Jeong, K.M. Koo, D.H. Lee, I.K. Kim et al., Cooling performance of a vertical ground-coupled heat pump system installed in a school building, *Renewable Energy* 34 (2009) 578–582. doi:10.1016/j.renene.2008.05.042.

15. X.Q. Zhai, Y. Yang, Experience on the application of a ground source heat pump system in an archives building, *Energy and Buildings* 43 (2011) 3263–3270. doi:10.1016/j.enbuild.2011.08.029.

16. S. Hackel, A. Pertzborn, Effective design and operation of hybrid ground-source heat pumps: Three case studies, *Energy and Buildings* 43 (2011) 3497–3504. doi:10.1016/j.enbuild.2011.09.014.

17. A. Hepbasli, Experimental study of a closed loop vertical ground source heat pump system, *Fuel and Energy Abstracts* 44 (2003) 254. doi:10.1016/S0140-6701(03)82141-0.

18. C. Li, J. Zhao, Experimental study on indoor air temperature distribution of gravity air-conditioning for cooling, *Energy Procedia* 17 (2012) 961–967. doi:10.1016/j.egypro.2012.02.194.

19. M.M. Rahman, M.G. Rasul, M.M.K. Khan, Feasibility of thermal energy storage systems in an institutional building in subtropical climates in Australia, *Applied Thermal Engineering* 31 (2011) 2943–2950. doi:10.1016/j.applthermaleng.2011.05.025.

20. Y.H. Yau, B. Rismanchi, A review on cool thermal storage technologies and operating strategies, *Renewable and Sustainable Energy Reviews* 16 (2012) 787–797. doi:10.1016/j.rser.2011.09.004.

21. A.A. Al-Abidi, S.B. Mat, K. Sopian, M.Y. Sulaiman, C.H. Lim, T. Abdulrahman, Review of thermal energy storage for air conditioning systems, *Renewable and Sustainable Energy Reviews* 16 (2012) 5802–5819.

22. *ASHRAE Handbook, Heating, Ventilating, and Air-Conditioning Systems and Equipment*, American Society of Heating, Refrigerating and Air-Conditioning Engineers, Inc., 2012.

23. K. Zhong, Y. Kang, Applicability of air-to-air heat recovery ventilators in China, *Applied Thermal Engineering* 29 (2009) 830–840. doi:10.1016/j.applthermaleng.2008.04.003.

24. X. Jin, Z. Du, X. Xiao, Energy evaluation of optimal control strategies for central VWV chiller systems, *Applied Thermal Engineering* 27 (2007) 934–941. doi:10.1016/j.applthermaleng.2006.08.015.

25. I. Korolija, L. Marjanovic-Halburd, Y. Zhang, V.I. Hanby, Influence of building parameters and HVAC systems coupling on building energy performance, *Energy and Buildings* 43 (2011) 1247–1253. doi:10.1016/j.enbuild.2011.01.003.

26. J.W. Moon, S.H. Han, Thermostat strategies impact on energy consumption in residential buildings, *Energy and Buildings* 43 (2011) 338–346. doi:10.1016/j.enbuild.2010.09.024.

27. R. Karunakaran, S. Iniyan, R. Goic, Energy efficient fuzzy based combined variable refrigerant volume and variable air volume air conditioning system for buildings, *Applied Energy* 87 (2010) 1158–1175. doi:10.1016/j.apenergy.2009.08.013.

28. C. Tzivanidis, K.A. Antonopoulos, F. Gioti, Numerical simulation of cooling energy consumption in connection with thermostat operation mode and comfort requirements for the Athens buildings, *Applied Energy* 88 (2011) 2871–2884. doi:10.1016/j.apenergy.2011.01.050.

29. L. Peeters, R. de Dear, J. Hensen, W. D'haeseleer, Thermal comfort in residential buildings: Comfort values and scales for building energy simulation, *Applied Energy* 86 (2009) 772–780. doi:10.1016/j.apenergy.2008.07.011.

30. M.K. Singh, S. Mahapatra, S.K. Atreya, Adaptive thermal comfort model for different climatic zones of North-East India, *Applied Energy* 88 (2011) 2420–2428. doi:10.1016/j.apenergy.2011.01.019.

31. T.T. Chow, J.C. Lam, Thermal comfort and energy conservation in commercial buildings in Hong Kong, *Architectural Science Review* 35 (1992) 67–72. doi:10.1080/00038628.1992.9696715.

32. R. Zmeureanu, A. Doramajian, Thermally acceptable temperature drifts can reduce the energy consumption for cooling in office buildings, *Building and Environment* 27 (1992) 469–481. doi:10.1016/0360-1323(92)90045-Q.

33. S.C. Sekhar, Higher space temperatures and better thermal comfort—A tropical analysis, *Energy and Buildings* 23 (1995) 63–70. doi:10.1016/0378-7788(95)00932-N.

34. F. Nicol, S. Roaf, Pioneering new indoor temperature standards: The Pakistan project, *Energy and Buildings* 23 (1996) 169–174. doi:10.1016/0378-7788(95)00941-8.

35. K.W.H. Mui, W.T.D. Chan, Adaptive comfort temperature model of air-conditioned building in Hong Kong, *Building and Environment* 38 (2003) 837–852. doi:10.1016/S0360-1323(03)00020-9.

36. S.A. Al-Sanea, M.F. Zedan, Optimized monthly-fixed thermostat-setting scheme for maximum energy-savings and thermal comfort in air-conditioned spaces, *Applied Energy* 85 (2008) 326–346. doi:10.1016/j.apenergy.2007.06.019.

37. A.C. Roussac, J. Steinfeld, R. de Dear, A preliminary evaluation of two strategies for raising indoor air temperature setpoints in office buildings, *Architectural Science Review* 54 (2011) 148–156. doi:10.1080/00038628.2011.582390.

38. S.B. Sadineni, R.F. Boehm, Measurements and simulations for peak electrical load reduction in cooling dominated climate, *Energy* 37 (2012) 689–697. doi:10.1016/j.energy.2011.10.026.

10

Improving Energy Efficiency in Buildings: Challenges and Opportunities in the European Context

Delia D'Agostino

CONTENTS

10.1 Energy Consumption in Europe.. 181
    10.1.1 Overall Trends in Buildings 183
    10.1.2 The Residential Sector 185
    10.1.3 The Non-Residential Sector 186
10.2 European Policies for Energy Efficiency.. 187
10.3 Efficient Buildings... 188
    10.3.1 High Performing Building Categories 188
    10.3.2 Nearly Zero Energy Buildings 189
10.4 Cost-Optimality 191
10.5 Improving Energy Efficiency in Existing Buildings.............................. 193
    10.5.1 Barriers and Challenges towards Efficient Buildings........ 196
    10.5.2 Policies, Measures, and Best Practices in Member States... 198
10.6 Role of Innovative Technology in Efficient Buildings 202
    10.6.1 Overview of Building Envelope, Systems, and Renewables
         Technologies.. 202
10.7 Conclusion... 204
References... 206

10.1 Energy Consumption in Europe

Energy consumption in buildings is a huge concern in Europe as buildings are estimated to account for approximately 40% of primary energy and 36% of greenhouse emissions [1]. In some Member States this share even exceeds 45% making the building sector the largest end-use sector in the Europe. The residential sector consumes more than a quarter of total energy and accounts for two-thirds of building's consumption.

The European Climate and Energy package established the renowned "20–20–20" targets according to which a 20% increase of energy from renewables, a 20% decrease of greenhouse gases (GHGs) emissions, and a 20% reduction of primary energy consumption must be reached in buildings by 2020. These targets have been updated by the 2030 Climate and Energy framework that fixes GHG emissions reduction at 40% from 1990 levels, the share for renewable energy at 27%, and the improvement in energy efficiency at 27%. Figure 10.1 shows primary energy consumption at European level and energy efficiency targets for 2020.

Energy savings represent the effect of the reduction in unit consumption at a level of up to 30 subsectors or end-use, as derived from an indicator that measures the energy

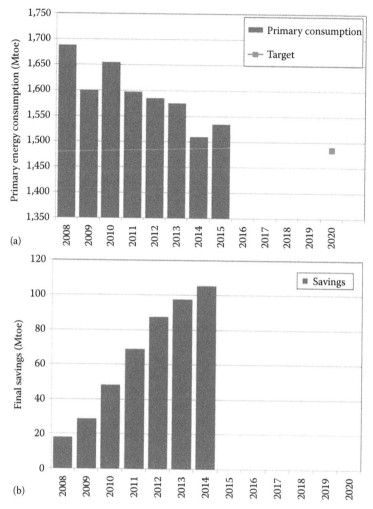

FIGURE 10.1
(a) Primary energy consumption at European level and the 2020 target; (b) final savings at European level. (From Odyssee-Mure database, available at: http://www.odyssee-mure.eu/; Eurostat, European System of Accounts—ESA 1995, Office for Official Publications of the European Communities, Luxembourg, Europe, 1996.) [2,3].

efficiency progress by sector (ODEX). Several benefits are linked to energy efficiency improvement, among them: energy security, job creation, fuel poverty alleviation, health, and indoor comfort [4].

With these aims, new policies have introduced technical and regulatory measures to promote a more rational use of energy over the last decade. These will be described in the next section. Their implementation generated an increase of savings in buildings since 2000 at European level (Figure 10.1). However, the economic crisis has slowed down the energy savings and the pace of energy efficiency improvements decreased from an average rate of 1.9% per year from 2000 to 2008, to 1.5% per year between 2008 and 2013. Although European Directives and national energy efficiency policies and measures are a major driver for energy efficiency in buildings, further attention to targeted initiatives is needed in several areas, such as building renovation, energy certificates, and public procurement.

10.1.1 Overall Trends in Buildings

A comparison of energy consumption for different end-uses in Europe and other countries is given in Figure 10.2.

It is evident how energy consumption is high in Europe in comparison to other countries. Energy is mainly consumed by heating, cooling, hot water, cooking, and appliances. In the U.S., primary energy consumption in residential buildings represents 54% of the overall consumption in the building sector and 22% of total primary energy consumption in the U.S. Commercial buildings consume about 46% of building energy consumption, with a growth rate exceeding 6% and 55% of this consumption in electricity resources.

In Australia, the energy consumption of the residential sector has been assessed at 9.60 Mtoe in the past decade and it is projected to reach 11.15 Mtoe by 2020. In Europe, the predominant energy end-use is space heating, which is responsible for about 70% of dwelling consumption. Fossil fuels represent the 37% of space heating consumption in the residential sector in 2014. Heating consumes a great part of the total energy consumption also in Australia (62%) and the United States (43%).

The most commonly used fuel in Europe is gas except for countries to the north and west (Austria, Belgium, Germany, France, Finland, Ireland, Luxembourg, Netherlands, Norway, Sweden, and UK) that use more oil. Central and Eastern European countries (Bulgaria, Czech Republic, Estonia, Latvia, Hungary, Slovakia, and Slovenia) consume the highest amount of coal in the residential sector. Renewable sources, such as solar, geothermal, biomass and waste, are the 21% of the total final consumption in Central and Eastern countries, 12% in South (Cyprus, Greece, Italy, Malta, Portugal, Spain, and Malta) and 9% in north and west countries.

The total stock in European Member States is reported in Figure 10.3 together with the number of new dwellings divided per typology.

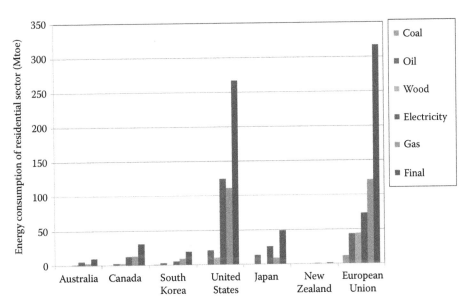

FIGURE 10.2
Coil, oil, wood, electricity, gas, and final consumption for the residential sector in different nations (year 2012). (From Odyssee-Mure database, available at: http://www.odyssee-mure.eu/.)

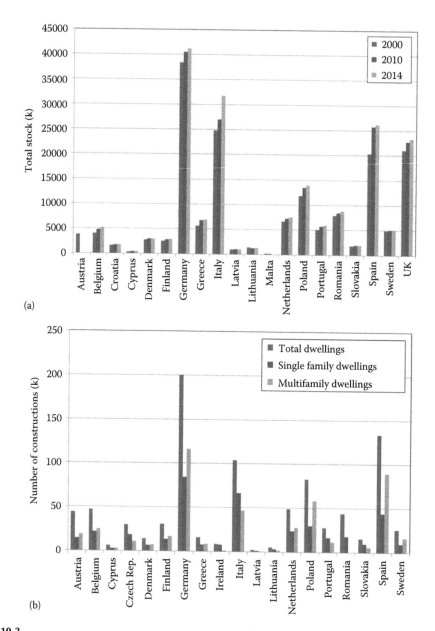

FIGURE 10.3

(a) Total stock in European Member States and (b) new dwellings constructions in 2012. (From Odyssee-Mure database, available at: http://www.odyssee-mure.eu/.)

In Europe there are about 25 billion m² of building useful floor space. Half of the total estimated floor space is located in the North and West countries while about 36% and 14% are located in South and Central and East countries. The stock performance depends on several factors such as the efficiency of heating system and envelope, climatic conditions, occupant behavior and social aspects. Data on energy consumption of the existing stock by age show that the largest energy-saving potential is associated with the older building stock characterized by a lack of building envelope insulation.

10.1.2 The Residential Sector

The annual new construction growth rate is assessed at around 1% in the European residential sector. The decrease in the rate of new constructions in the last decade is mainly due to financial crisis of the construction sector.

Figure 10.4 reports electricity, gas, oil, and final energy consumption in the residential sector in different European Member States.

Germany, France, Italy, Poland, and Spain are among the countries with higher energy consumption. The final residential energy consumption per dwelling has been decreasing from 2005 to 2014 in European Member States. In 2008, the residential consumption per dwelling was 1.58 Mtoe while it was is 1.21 Mtoe per dwelling in 2014.

European energy imports reached 53.5% in 2014. In particular, natural gas is the dominant energy carrier for heating, and its dependency rate has been assessed at 67.4%. The imported gas mainly comes from Russia and Norway, while Eastern and Baltic countries are highly exposed to disruption of Russian gas supplies.

Figure 10.5 shows how household energy consumption per dwelling has been decreasing regularly in most countries since 2000.

Energy efficiency in the residential sector has improved by 1.7% per year in Europe since 2000. This is mainly due to energy efficiency improvement in space heating that is assessed around 2.3% per year at European level. A low construction rate has limited the impact of new houses since 2009 and, as a result, the share of space heating in total household consumption is declining.

Electricity consumption per household has been diminishing in most countries since 2008 thanks to the diffusion of efficient appliances (e.g., energy labels A+ to A++). Small electrical appliances have been growing rapidly, so that they are a high share of the total appliances consumption. Thanks to labeling and eco-design directives, large appliances are more and more efficient, with efficiency gains around 35% for cold appliances, such

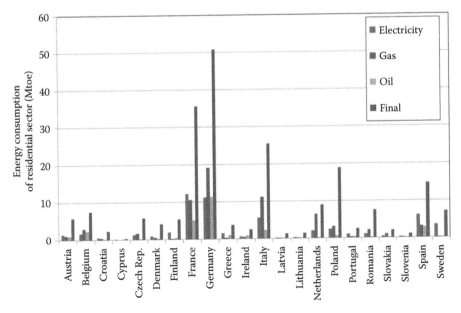

FIGURE 10.4
Electricity, gas, oil, and final energy consumption in the residential sector in Europe (year 2014). (From Odyssee-Mure database, available at: http://www.odyssee-mure.eu/.)

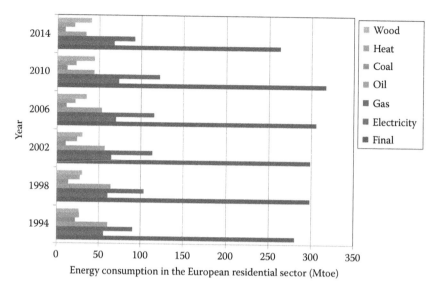

FIGURE 10.5
Evolution of energy consumption in the European residential sector from 1994 to 2014.

as refrigerators and freezers, washing machines and dishwashers. Furthermore, lighting consumption has been decreasing in half at European level since 2000 thanks to the diffusion of Compact Fluorescent Lights (CFLs) and Light Emitting Diode (LEDs). Without energy efficiency improvements, the energy consumption of households would have been 60 Mtoe higher in 2012 considering the increasing number of dwellings and appliances.

10.1.3 The Non-Residential Sector

Non-residential buildings account on average for 25% energy consumption of the total European building stock, representing a heterogeneous sector compared with the residential.

Figure 10.6 reports the share of residential and non-residential buildings in European final energy consumption.

This sector holds a big percentage of total final energy consumption, especially in relation to commercial and hospital buildings [5]. The average specific energy consumption in the non-residential sector is on average 280 kWh/m² (covering all end-uses), which is at least 40% greater than the equivalent value for the residential sector.

In the non-residential sector (e.g., offices, shops, schools, and hospitals) consumption related to lighting, ventilation, heating, cooling, refrigeration, IT equipment, and appliances vary greatly from one category to another [6]. Over the past decade, electricity consumption has increased at the European level between 2008 and 2012 mainly due to a growing number of new appliances, IT devices, new telecommunication types, as well as air-conditioning [7].

Non-residential buildings require on average 55% more electrical energy than residential buildings, 286 kWh/m² compared to 185 kWh/m²; in 2012 the average annual specific consumption per m² was assessed around 210 kWh/m². This specific consumption varies significantly among EU countries: it is 80% lower in Bulgaria and Spain than in Finland. Such differences are partly explained by climatic conditions, country energy dependence,

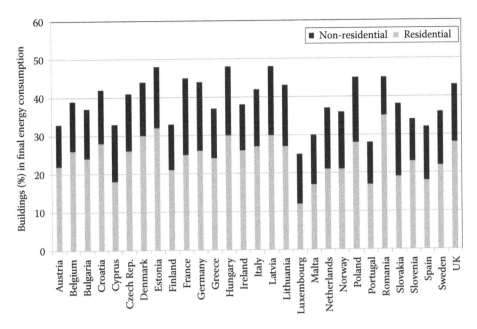

FIGURE 10.6
Share of buildings in final energy consumption (2012). (From Eurostat, European System of Accounts—ESA 1995, Office for Official Publications of the European Communities, Luxembourg, Europe, 1996.)

and imports [8]. Even if the diffusion of Information and Communication Technology (ICT) and air-conditioning increased comfort and productivity, energy savings have contributed to reduce growth in electricity consumption [9].

10.2 European Policies for Energy Efficiency

The European Union set up a policy framework focused on reducing energy consumption in obtaining important savings from buildings.

To this end a major step forward is represented by the Energy Performance of Buildings Directive recast (EPBD recast, Directive 2010/31/EC) that establishes the implementation of nearly zero energy buildings (NZEBs) as the building target from 2018 onwards [10,11]. In more details, Member States have to ensure that new buildings occupied by public authorities and properties are NZEBs by December 31, 2018 and that all new buildings are NZEBs by December 31, 2020. Another important provision of the EPBD is the introduction of the concept of cost-optimality. Delegated Regulation No 244/2012 supplementing the EPBD recast describes a comparative methodology framework to derive cost-optimal levels of minimum energy performance requirements for buildings and building elements [12,13].

The Energy Efficiency Directive (EED) (EU, 2012/27/EU) and the Renewable Energy Directive (RED) (EU, 2009/28/EU) set out a package of measures to create the conditions for improvements in the energy performance of Europe's building stock [14,15].

The EED deals with more efficient energy use throughout the energy chain, from its production to its final consumption. Member States have to submit National Energy Efficiency

Action Plans (NEEAPs) and establish a long-term strategy beyond 2020 for mobilizing investment in renovation of residential and commercial buildings. It states that the public sector should purchase energy efficiency in buildings, products, and services with a view to improving the building stock energy performance [16]. According to the Directive, energy distributors or retail energy sales companies must achieve 1.5% energy savings per year through the implementation of energy efficiency measures. Among the introduced measures, there are legal obligations to establish energy-saving schemes, energy efficiency national incentives, and energy services and audits.

The RED Directive establishes strategies for promoting and producing energy from renewables. The Directive specifies national renewable energy targets for each country, taking into account their overall potential for renewable production. This aspect is crucial in NZEBs as these buildings must combine high efficiency technologies with renewable production.

In relation to policies on products, the Energy Labeling Directive (2010/30/EU) [17] focuses on labeling and standard products on energy consumption, while the Eco-design Directive (2009/125/EU) [19] creates the criteria for eco-design requirements identifying specific product categories. The indication given in both directives are important as high efficiency appliances, products, and systems have an impact on NZEBs energy performance.

10.3 Efficient Buildings

In recent years, the topic of efficient buildings has been widely analyzed and discussed, but it is still subject to discussion at international level on the technical meaning of different terms such as NZEBs and others that have been launched. Among these terms, there are near zero, zero, net, and plus energy buildings that are discussed in this section. Such building categories have been placed on the market to underline their quality as environmentally friendly buildings [19].

10.3.1 High Performing Building Categories

U.S. DOE defined ZEB as the following: "An energy-efficient building where, on a source energy basis, the actual annual delivered energy is less than or equal to the on-site renewable exported energy [20]." The source energy is the site building energy plus the energy consumed in the extraction, processing and transport of primary fuels, such as coal, oil, and natural gas; energy losses in thermal combustion in power generation plants; and energy losses in transmission and distribution to the building site. The building energy is the energy consumed at the building site as measured at the site boundary. This includes heating, cooling, ventilation, domestic hot water, indoor and outdoor lighting, plug loads, process energy, elevators and conveying systems, and intrabuilding transportation systems.

ZEB refers to a Zero Energy Building and Zero Emission Building [21]. The first indicates the energy consumed by a building in its day-to-day operation, the second the carbon emissions released into the environment as a result of its operation.

In general terms, a ZEB can be described as a building with greatly reduced energy needs and/or carbon emissions, achieved through efficiency gains, such as the balance of energy needs supplied by renewable energy [22].

Another category differentiates between Autonomous ZEB and Net ZEB [23]. An autonomous ZEB does not require connection to the grid. Stand-alone buildings can supply their own energy needs, being able to store energy for nighttime or wintertime use. A Net ZEB is a yearly energy neutral building that delivers as much energy to the supply grids as it draws back. It does not need fossil fuel for any energy use although it can be supplied by the grid. An Energy Plus Building (+ZEB) produces more energy from renewable energy sources (RES) than it imports over a year. However, even if a common customer sense of "positive" means greater than zero, it should be clear that a positive building cannot create energy, violating the energy conservation principle. Buildings using less delivered energy than exported on-site renewable energy would mean that its source energy is negative.

Four other different concepts around zero-energy buildings have been listed depending on boundaries and the metrics [24]. A Net Zero Site Energy building is a building that produces at its location at least the amount of energy that it uses. A hierarchy of renewable supply options has been proposed. This encourages the reduction of site energy use through low-energy technologies and the use of renewables available within the building footprint or at the site.

Four types of ZEBs can be distinguished in reference to the energy demand and the installed renewables [21]. A PV–ZEB is a building with a relatively low electricity demand and has a photovoltaic (PV) system, while a Wind-ZEB has a relatively low electricity demand and a small on-site wind turbine. A PV-Solar thermal-heat pump ZEB is characterized by a low heat and electricity demand as well as by a PV installation in combination with solar thermal collectors, heat pumps, and heat storage. A Wind-Solar thermal-heat pump ZEB has a low heat and electricity demand and a wind turbine in combination with a solar thermal collector, a heat pump, and heat storage.

10.3.2 Nearly Zero Energy Buildings

The EPBD recast defines NZEBs in Article 9 as the following: "Nearly zero energy building means a building that has a very high energy performance […]. The nearly zero or very low amount of energy required should be covered to a very significant extent by energy from renewable sources, including energy from renewable sources produced on-site or nearby."

The NZEB topic has gained a growing attention in the last decade, but the achievement of an agreed and approved NZEBs definition is not yet reached across Europe [24].

The progress towards the establishment of NZEBs definitions has been evaluated based on different sources of information, such as National Plans, templates, the Commission report of 2013 and its update of October 2014, information from the EPBD Concerted Action (CA), Energy Efficiency Action Plans (NEEAP), and National Codes [25]. Among the aspects to be defined there are the following: building category, typology, physical boundary, type and period of balance, included energy uses, RES, metric, normalization, and conversion factors. The analysis assessed progress in many Member States compared with the first attempts to establish NZEBs definitions [26]. About the energy calculation, the following are the most discussed topics:

- The main included energy uses are: heating, domestic hot water (DHW), ventilation, and cooling. Auxiliary energy and lighting are taken into account in almost all EU Member States. Several Member States also include appliances and central services.

- The most common choice regarding the energy balance calculation is the difference between the primary energy demand and the energy generated, over a period one year, and considering annual constant weightings/factors (e.g., primary energy factors)

- Single building or building unit is the most frequent indicated physical boundary for the calculation, but the overall impression is that the differences among building unit/site/zone/part need to be better addressed.

- As regards the normalization factors, conditioned area is the most agreed on choice in Member States. Although other options, such as net floor area and treated floor, are also selected.

- The most common considered RES option is the on-site generation, but many countries also consider external generation and nearby generation (but probably not always with the same meaning).

- Almost all Member States prefer the application of low-energy building technologies and available RES. The most used technologies are PV, solar thermal, air- and ground-source heat pumps, geothermal, passive solar, passive cooling, wind power, biomass, biofuel, micro-CHP, and heat recovery.

The current situation towards the establishment of applied national NZEBs definitions in European Countries has improved in comparison with the 2013 Commission progress report [27]. In the last year, many NZEBs definitions have been implemented at national level. Consolidated and systematic information has been submitted through the templates, and Member States benefited from more guidance and clarifications.

Table 10.1 reports a qualitative evaluation of the current status of NZEBs development in Member States and compliance with the EPBD requirements. It includes the main aspects discussed in this report, such as the NZEBs applied definitions, the inclusion of RES in the NZEBs concept, intermediate targets (qualitative and quantitative), as well as measures to promote NZEBs renovation [28].

Different system boundaries and energy uses are the cause of high variations within the described definitions. The level of energy efficiency, the inclusion of lighting and appliances, as well as the recommended renewables to be implemented vary from country to country.

In particular, the requirements provided by Member States in terms of primary energy show a significant variability and reflect different national and regional calculation methodologies and energy flows. National energy policies have evolved with new legislation and methodologies introduced with technical regulatory measures to improve the energy efficiency of buildings and RES generation.

The proportion of renewable energy production in Cyprus has been defined as a percentage: 25%. Other countries have indicated more ambitious values, such as Germany (60%), and Denmark (56%).

The reduction of energy demand through energy-efficient measures and the utilization of RES to supply the remaining demand have reached common agreement towards the implementation of the NZEBs concept across Europe.

In relation to intermediate targets for improving the energy performance of new buildings by 2016, most Member States, presented only qualitative targets (e.g. strengthening building regulations, obtaining energy performance certificates by a certain year). The targets appear extremely variable, and the quantitative targets (about the number or share of NZEBs) are almost never defined (e.g., 60,000 new NZEBs dwellings by 2015 in the Netherlands). However, in many MS there are no targets, or they are not consistent with long-term climate mitigation targets. Targets are necessary to evaluate the impact of a policy instrument. Interim targets are necessary to design suitable instruments and to monitor their attainment. For example, in Germany there are targets for the reduction of the energy demand of the building stock: by 2050 a reduction of primary energy consumption by 80% shall be achieved.

TABLE 10.1

Evaluation of the NZEBs Development in Europe (✓=satisfactory development, /= partial development, and X = not defined/unclear)

Country	NZEB Definition	RES Included in the NZEB Concept	Qualitative and Quantitative Intermediate Targets	Measures Promoting Deep or NZEB Renovation
Austria	✓	✓	/	/
Belgium—Brussels	✓	/	/	✓
Belgium—Flanders	✓	/	/	✓
Belgium—Wallonia	✓	/	/	✓
Bulgaria	/	/	/	/
Cyprus	✓	✓	X	/
Czech Republic	✓	✓	/	✓
Germany	/	/	/	✓
Denmark	✓	✓	/	✓
Estonia	✓	✓	X	/
Greece	X	X	X	/
Spain	X	X	X	/
Finland	/	X	X	✓
France	✓	✓	✓	✓
Croatia	✓	/	/	/
Hungary	/	✓	X	/
Ireland	✓	✓	/	✓
Italy	✓		/	/
Latvia	✓	✓	X	/
Lithuania	✓	✓	/	/
Luxembourg	✓	✓	/	✓
Malta	✓	/	/	✓
Netherlands	✓	✓	✓	✓
Poland	✓	/	/	✓
Portugal	/	X	/	X
Romania	✓	✓	/	✓
Slovenia	✓	✓	✓	/
Slovakia	✓	✓	/	/
Sweden	/	X	X	✓
United Kingdom	/	X	/	✓

Source: D'Agostino, D. et al., Synthesis Report on the National Plans for NZEBs, EUR 27804 EN. http://publications.jrc.ec.europa.eu/repository/bitstream/JRC97408/reqno_jrc97408_online%20nzeb%20report(1).pdf.

10.4 Cost-Optimality

As defined by the EPBD recast [11], Member States have also to account for cost-optimality. This concept requires the evaluation of minimum energy performance requirements leading to the lowest building costs. The Directive introduces a methodology to set benchmark requirements for national standards [12,13]. The cost-optimal level is defined as "the energy performance level which leads to the lowest cost during the estimated economic lifecycle."

According to Article 5, energy-related investment costs, maintenance, operating costs and, where applicable, disposal and replacements costs, have to be considered in the analysis.

Delegated Regulation No. 244/2012 and its Guidelines describe the methodology to be followed to derive cost-effectiveness from a technical and economic perspective. The methodological framework comprises both new and existing buildings undergoing major and nonmajor renovation of their structural and technical components.

The methodology involves the definition of a reference building and the application of energy efficiency measures to reduce primary energy consumption and address the choice of the most economically advantageous solutions. According to the methodology, construction alternatives have to be included and compared in terms of costs [29] and energy performance [30] among the available solutions. The cost-optimal level is the minimum level of ambition for the energy performance. In relation to new buildings, it will be determined by the best technology that is available and well introduced on the market at that time, considering financial aspects, legal and political considerations at national level. The cost-optimal configuration presents the lowest costs maintaining a high performance. It can be identified in the lower part of the curve that reports global costs ($€/m^2$) versus energy consumption (kWh/m^2y) (Figure 10.7).

Different studies have shown that several parameters can alter the shape of the curve, among them geometrical building features, technical systems, energy prices, discount rates, and costs [31,32]. A sensitivity analysis can be performed to understand and reduce this variability within calculations [33,33]. In the implementation of a cost-optimal approach, Member States have to decide on many important aspects: the choice of reference buildings, the selection of packages of energy-efficient measures, reference construction and maintenance costs of building elements, reference lifetime of building elements, discount rates, and energy prices [35].

A heterogeneous situation characterizes European countries as each building and climate types present a different cost-optimal level of energy-efficient measures [36]. A broad overview on the implementation of the cost-optimal methodology in Member States has been published by Buildings Performance Institute Europe (BPIE) [37]. The study reports calculation examples for Austria, Germany, and Poland as well as the impact of discount rates, simulation variants, costs, and energy prices on the final results.

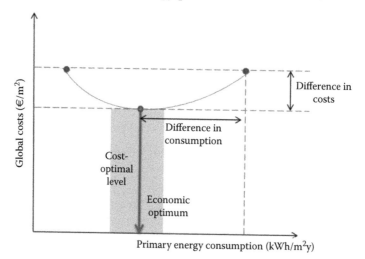

FIGURE 10.7

The cost-optimal solution modified from [31].

TABLE 10.2

NZEBs Level of Performance (kWh/m²y) per Building Type According to the European Climate

Building type	Climate			
	Mediterranean Zone 1: Catania (others: Athens, Larnaca, Luga, Seville, Palermo)	Oceanic Zone 4: Paris (others: Amsterdam, Berlin, Brussels, Copenhagen, Dublin, London, Macon, Nancy, Prague, Warszawa)	Continental Zone 3: Budapest (others: Bratislava, Ljubljana, Milan, Vienna)	Nordic Zone 5: Stockholm (others: Helsinki, Riga, Stockholm, Gdansk, Tovarene)
	Level of Performance (kWh/m²y)			
Office buildings				
Net primary energy	20–30	40–55	40–55	55–70
Primary energy use	80–90	85–100	85–100	85–100
On-site RES sources	60	45	45	30
New single family house				
Net primary energy	0–15	15–30	20–40	40–65
Primary energy use	50–65	50–65	50–70	65–90
On-site RES sources	50	35	30	25

The European Commission Recommendation on Guidelines for the promotion of NZEBs [38] states that there cannot be a single level of ambition for NZEBs across Europe. Flexibility is needed to account for the impact of climatic conditions on heating and cooling needs and on the cost-effectiveness of packages of energy efficiency and RESs measures.

At EU level, the establishment of numeric benchmarks for NZEBs primary energy use indicators is most useful when the values to be compared with these benchmarks result from transparent calculation methodologies. Standards are currently under finalization to allow for transparent comparison of national and regional calculation methodologies. Benchmarks are usually provided in terms of energy needs that are the starting point for the calculation of primary energy. A very low level of energy need for heating and cooling is a vital precondition for nearly zero primary energy buildings. Very low energy needs are also a precondition to achieve a significant share of energy from RES and nearly zero primary energy.

The values suggested for NZEBs projecting the 2020 prices and technologies, benchmarks for the energy performance are in the ranges reported in Table 10.2 for the different climatic zones.

10.5 Improving Energy Efficiency in Existing Buildings

The European existing building stock is old and renovated at a slow pace. Besides efforts to design new buildings with low energy demand and RES availability, it is essential to tackle the high energy consumption in existing buildings, characterized by an average age of about 55 years with many buildings having been in use for hundreds of years. While more stringent building codes and policies caused this value to decrease slightly in residential buildings since 2007, the final energy consumption in non-residential buildings remained quite stable over the past decade [6]. This is also due to the increasing cooling needs, leading to a 6% rise of primary energy use per m².

More than 40% of the European residential buildings have been constructed before the 1960s when energy building regulations were very limited. Countries with the largest components of older buildings include Czech Republic, Denmark, France, Sweden, and the United Kingdom.

It is undeniable that the economic crisis since 2007 has also influenced the energy trends of recent years. On the one hand, it led to a reduction in energy consumption due to an increase in the poverty levels of households, and on the other hand, it curbed building renovation activity [39]. The recent evaluation of the Europe 2020 strategy reveals that, because of the economic crisis, the number of people at risk of poverty increased from 80 million prior to the crisis to 124 million in 2012.

Reaching the NZEBs target in new buildings appears to be feasible according to design studies [40] and studies on energy performance optimization have been performed especially in new buildings [41,42]. However, the challenge of achieving energy-efficiency targets in Europe remains for the existing built environment. The renovation rate of existing buildings is currently low due to the economic crisis which started in 2007. In 2011, the renovation rate of the European building stock has been assessed at between 0.5% and 2.5% per year [43]. Buildings dating between 1945 and 1980 have the largest energy demand. Moreover, the existing stock is characterized by a high heterogeneity in terms of uses, climatic areas, construction traditions, and different system technologies [44].

Different levels of renovation can be distinguished depending on the type of intervention and savings obtained. Renovation can involve the installation of RES as well as the replacement or upgrade of all building elements to reduce energy consumption towards zero levels [45]. The refurbishment of a building façade (i.e. walls and windows) provides a different energy saving level compared to the retrofit of the overall building envelope and systems (HVAC, lighting, etc.).

The interest on the topic at EU level is highlighted by several projects aimed at the renovation of building stock in Member States (Table 10.3).

Among the European projects on the demonstration of energy efficiency implementation through retrofitting of buildings, there are the following:

- *Effesus*: Energy efficiency of European historic urban districts. The main goal is to develop and demonstrate a methodology for assessing and selecting energy efficiency interventions, based on existing and new technologies that are compatible with heritage values.

- *Resseepe*: Retrofitting solutions and services for the enhancement of energy efficiency in public education. The core idea is to technically advance, adapt, demonstrate, and assess a number of innovative retrofit technologies achieving reductions of 50% of energy consumption.

- *PassREg*: Passive house regions with renewable energies. It aimed to trigger the implementation of NZEBs using Passive House supplied as much as possible by renewable energies as the foundation.

- *Beem-up*: Building energy efficiency for massive market uptake. The goal is to demonstrate the economic, technical, and social feasibility of retrofitting to reduce energy consumption of 75% in heating energy demand in hundreds of dwellings located in Sweden, the Netherlands, and France.

TABLE 10.3

Selection of EU Projects for Building Stock Renovation

Project	Description
RetroKit: Toolboxes for systemic retrofitting	RetroKit demonstrates through three pilot projects how to increase the EU retrofitting rate and contribute to EU energy reduction commitments. The buildings are made of multifunctional, modular, low cost, and easy to install prefabricated modules. These innovative systems target existing multifamily residential buildings which represent more than 50% of the EU building stock and between 65% and 80% of their energy consumption. The project brings the aspect of multifunctional façade and roof elements into the retrofit sector. Special integrated solutions are analyzed for heating, ventilation, cooling, electricity, and ICT in a flexible way.
Cetieb: Cost-effective tools for better indoor environment in retrofitted energy efficient buildings	The project develops cost-effective and innovative solutions for better monitoring indoor environment quality and investigates active and passive systems. The focus lies on developing cost-effective solutions to ensure a wide application of the resulting systems. These solutions allow the monitoring of a large variety of indoor environmental factors, active systems to control natural ventilation improving indoor air quality and optimizing air flow in buildings. New structured surfaces based on titanium dioxide contribute to a cleaner and healthier environment by safely removing air pollutants and pathogenic microorganisms from air and building surfaces.
Episcope: Energy performance indicator tracking schemes for the optimization of the refurbishment process in European housing stocks	The goal of the project is to make the energy refurbishment processes in the European housing sector more transparent and effective. The objective is to implement pilot actions on different scales and to align and compare them using a common methodology. The conceptual framework is based on national residential building typologies developed during the project TABULA. The classification schemes for national building stocks will be extended to further countries. An upgrade of the Web Tool will also reflect the national interpretations of new buildings and Nearly Zero Energy Buildings (NZEBs). The main project activity is to track the energy refurbishment progress making comparisons among countries.
Request2: Action removing barriers to low carbon retrofit by improving access to data and insight into the benefits to key market actors	This project runs across nine European countries. Its focus is on how data from Energy Performance Certificates (EPCs) can be used to promote home energy efficiency. Governments, organizations, private companies, and individual households have a different role to play in making Europe's homes energy efficient. Each of these groups can benefit from energy saving - how much can different types of homes save by installing different energy efficiency measures? Information collected in the preparation of EPCs are a rich source of data and the project aims to make this data widely available.
Ecodistrict: Integrated decision support tool for retrofit and renewal towards sustainable districts	The project aims at developing an integrated decision-support tool that facilitates decision making on the retrofitting and renewal of existing districts and their buildings. It connects the main decision makers in urban district transformation programs, acting from different perspectives, with different time scales, to reach a coordinated approach that joins building retrofitting with district renovation. It provides trustworthy insights into (1) retrofitting and renewal projects, (2) the associated costs and benefits over the life cycle of the buildings, and (3) the impacts on resource efficiency, social aspects, indoor and outdoor quality of buildings and districts, and other environmental concerns.

- *E2ReBuild*: Industrialized, energy-efficient retrofitting of residential buildings in cold climates. It investigates, promotes, and demonstrates advanced, cost-effective, energy-efficient retrofit strategies.
- *School of the future*: The goal is zero emission with high-performance indoor environment. A project to design, demonstrate, evaluate, and communicate are shining examples of how to reach the future high-performance building level.

These projects demonstrate a huge interest in energy efficiency and a new holistic approach which involve envelope, service systems, integration of renewables, and building management systems. Several remarkable innovative technologies and materials are integrated in the retrofitting process including energy storage systems, nanotechnologies and smart materials, ICT, and intelligent building controls. This creates an added value for existing buildings and encourages energy-efficient behaviors and end-users to establish sustainable renovation solutions that reduce energy use.

10.5.1 Barriers and Challenges towards Efficient Buildings

The EED [14] calls Member States to evaluate and take appropriate measures to remove regulatory and non-regulatory barriers to energy efficiency (Article 19(1)(a)). These originate from the contrast between the possibilities of reduction in energy consumption and the real allocated investments. Barriers can be found in common worldwide or depend on the characteristics of a country.

An overview of the main barriers and challenges in relation to NZEBs appear common among the countries [46]. These are reported in Tables 10.4 and 10.5 in relation to the decision-making process and retrofit.

In the building sector, barriers are primarily political, but also technical, financial, and related to a lack of information and awareness of key actors and stakeholders. Energy efficiency policies can generate other barriers such as some invisible extra costs such as maintenance and transport, especially in developing countries.

A disconnection can be identified between developing innovative technologies from the building industry and the lack of uptake due to budget constraints. Awareness of how

TABLE 10.4

Barriers towards Efficient Buildings and NZEBs at EU Level

Technical	Existing building structure and technical system limit the choice of technical solutions that can be used, but where technical solutions can be found, they are often costly and not financially viable.
Financial	High equipment costs, limited access to investments, and the non-adequacy of financial models of micro-credit institutes with energy services and products.
Political and institutional	Short-term vision of efficiency implementation due to political instability. Non-rigorous administrative planning and little synergy between public and private authorities associated with the lack or absence of subsidies to building owners.
Social	Lack of knowledge, proper behaviors, lifestyle, culture, and interest for energy efficiency among residents and building owners, often due to lack of awareness combined with challenges with architectural and cultural values.
Environmental/health	Criteria for materials and waste, mix between comfort and efficiency.
Organizational/legal	The ownership structure and need for consensus among several home-owners can hinder renovations.

TABLE 10.5

Challenges towards Retrofit at EU Level

Technical	Existing building structure and technical systems limit the choice of technical solutions possible for renovations.
Financial	Building owners are unlikely to make a return on investment.
Social	The need for communication and information early in the renovation process to increase acceptance among residents.
Environmental/health	The risk of moisture must be taken into consideration when making a building more airtight.
Organizational/legal	The need for an extensive communication between involved organizations and actors early in the process.

users consume energy in residential buildings should be increased. Furthermore, it is widely recognized that energy targets are challenging for cultural and historic buildings.

In relation to NZEBs renovation, existing building structures set limits to what extent the existing technical solutions can be implemented. This limitation is more relevant where the architectural value of the building needs to be conserved, making the retrofit processes more challenging. Furthermore, existing technical solutions are perceived as expensive adding to the main financial challenge of having high investment in renovation projects. A return of the investment appears often as difficult apart from considering savings through the life cycle of the building; in this case the initial investment costs are lower than those of the overall operational costs. The payback period for renovation may take between 15 and 30 years, and often residents do not benefit from this period. Moreover, a landlord cannot, or does not want to raise rents and becoming uncompetitive in the market as the difference between non-efficient and efficient buildings is not considered by the tenants.

It is also common that a lack of knowledge regarding efficiency is spread among professionals and residents. Communication of best practices is important to increase knowledge among professionals and general public on energy-efficient renovation and technical solutions. A follow-up is important to ensure that residents use buildings properly. Communicating with residents and end users has been identified as necessary. End user behavior after a completed renovation is also a challenge in the retrofit process.

In relation to financial barriers, public authorities have a leading role in setting up financing schemes for national or local contexts. The level of ambition of financial programs rises in order to have greater impact and unlock further private investment for energy efficiency. Legislation and financial incentives also have a strong influence in developing NZEBs projects.

The cooperation between institutions and individuals is essential for the implementation of energy efficiency policies. Communication and information between involved actors and organizations of the renovation project, as well as with the residents, are among the factors that can provide a successful efficiency renovation. Involving the media in energy and environmental issues can raise customers' awareness.

To overcome financial barriers, market-based regulatory instruments like Energy Performance Contracting (EPC) can reduce transaction costs as well as researching financial support establishing partnerships with international bodies and institutions.

Spreading local energy audit programmes in public buildings can also help remove barriers as well as a global diffusion of new technologies using renewable resources. This is also important to fill the technological gap and ensure the effectiveness of energy efficiency measures.

TABLE 10.6

Estimated Benefits of NZEBs Implementation in Some Member States

	Poland	Romania	Bulgaria
CO_2 savings (million t)	31	68	4.7–5.3
Energy savings (TWh)	92	40	15.3–17
Additional investments (million Euro)	240–365	82–130	38–69
New full-time jobs	4100–6200	1390–2203	649–1180
Minimum requirements in 2015/2016			
Primary energy (kWh/m²y)	70	100	60–70
Renewable share (%)	>20	>20	>20
CO_2 emissions (KgCO₂/m²/y)	<10	<10	<8
Minimum requirements in 2020			
Primary energy (kWh/m²y)	30–50	30–50	30–50
Renewable share (%)	>40	>40	>40
CO_2 emissions (KgCO₂/m²/y)	<3–6	<3–7	<3–5

Starting from current construction practices, existing policy framework, and economic conditions, simulations have been carried out on energy performance and economic implications in NZEBs reference buildings. The estimated macro-economic benefits of implementing NZEBs between 2020 and 2050 are remarkable as detailed in Table 10.6.

As shown in Table 10.6, CO_2 savings are estimated as following: around 5 million tons in Bulgaria, 31 million tons in Poland, and 68 million tons in Romania. Energy savings are estimated around 16 TWh in Bulgaria, 92 TWh in Poland, and 40 TWh in Romania. New full-time jobs will be created: between 649 and 1180 in Bulgaria, between 4100 and 6200 in Poland, between 1390 and 2203 in Romania. Additional investments are expected between 38 and 69 million Euros in Bulgaria, between 240 and 365 million Euros in Poland, between 82 and 130 million Euros in Romania. Minimum primary energy requirements are foreseen between 70 kWh/m²y (Bulgaria and Poland) and 100 kWh/m²y (Romania) in 2015, but they will become between 30 kWh/m²y and 50 kWh/m²y in 2020. The percentage of renewable share will pass from 20% in 2015 to 40% in 2020. CO_2 emissions will pass from 8-10 KgCO₂/m²/y to 3-7 KgCO₂/m²/y in 2020.

10.5.2 Policies, Measures, and Best Practices in Member States

Successful policy measures towards the effective support of building renovation are summarized in Table 10.7 as extracted by the ODYSSEE–MURE database, which includes around 420 energy efficiency policy measures, including their impact, and about 225 measures explicitly related to the renovation of the residential and non-residential existing building stocks. The most recent ones, adopted in the last 10 years have been selected excluding the legislative-normative and focusing on those with a medium or high impact.

Front runner measures are lessons learned that can support the implementation of efficiency through both successful models and best practice examples which can be used as a basis for new policies. Number and types of policy measures vary significantly across European Member States. In both residential and tertiary sectors, financial measures aiming at investments in energy renovation, including renewables, are relevant. Other measures include mandatory energy audits, energy certificate,s and technology procurement as well as income tax reductions (in residential).

TABLE 10.7

Ongoing and Proposed Policy Measures on Building Renovation with Medium or High Impact

Member States	Sector	Measure Title	Starting Year
Austria	Residential	*klima*: Aktiv building—new standards for efficient buildings	2005
Bulgaria	Residential	*EU-related*: Energy Performance of Buildings EPBD (Directive 2010/31/EU)—National Program for Renovation of Residential Buildings in the Republic of Bulgaria, 2006–2020	2007
	Residential	Support for energy audits in multifamily buildings at guaranteed implementation of the recommended measures	2012
	Residential	Support for energy efficiency in multifamily buildings	2012
	Residential	Energy renovation of Bulgarian residential buildings	2012
	Tertiary	National strategy for financing the building insulation for energy efficiency 2006–2020—services	2006
Croatia	Residential	Plan for energy renovation of residential buildings	2011
	Residential	Integral multidwelling unit renovation incentives	2014
	Residential	Energy renovation of public sector buildings program	2014
	Tertiary	Energy reconstruction of commercial non-residential buildings	2011
	Tertiary	Energy renovation of commercial non-residential buildings	2012
Estonia	Residential	Building design and construction supervision support for apartment associations for making preparations for major renovation	2010
	Residential	Support scheme for reconstruction of apartment buildings	2010
	Residential	The program of renovation loan for apartment buildings (under the Operational Program for the Development of the Living Environment)	2009
	Tertiary	A program for reconstruction of public sector buildings	
Denmark	Residential	Strategy for energy renovation	2014
France	Residential	Zero-rated eco-loan	2009
	Residential	Social housing eco-loan	2009
	Residential and Tertiary	Energy Savings Certificates	2006
	Tertiary	"Moderning building and cities" program	2008
Germany	Residential	Tax incentives for Energy renovations	2015
	Residential	KfW Program "Energy-efficient refurbishment" (former CO_2 Building Rehabilitation Programme)	2009
Hungary	Residential	*"Our home" renovation subprogram*: Mitigation of heat demand of residential buildings (family homes and multioccupied residential buildings) with individual or central heating	2008
	Residential	*Liveable panel dwellings renovation sub-program*: Mitigation of district heating demand in residential buildings built by industrialized technologies	2008
Italy	Residential	Fiscal incentives for energy savings in the household sector	2008
Finland	Residential	Window energy rating system	2006
	Residential	Coordinated energy advice to the consumers	2010
	Tertiary	Renovation of state property stock	2009

(Continued)

TABLE 10.7 (*Continued*)

Ongoing and Proposed Policy Measures on Building Renovation with Medium or High Impact

Member States	Sector	Measure Title	Starting Year
Latvia	Residential	Grant scheme for renovation of residential buildings (2013–2016)	2013
	Residential	*EU-related*: Energy Performance of Buildings EPBD (Directive 2010/31/EU)—Energy Audits and Energy Certification of Residential Buildings	2009
	Residential	*Increasing heat energy efficiency in multiapartment buildings*: EU programming period of 2007–2013	2009
	Residential	*Increasing energy efficiency in multiapartment buildings*: EU programming period of 2014–2020	2015
	Tertiary	*Increasing energy efficiency in state public buildings*: EU Programming Period of 2014–2020	2015
	Tertiary	*Increasing energy efficiency in municipal buildings*: EU Programming Period of 2014–2020	2015
Lithuania	Residential	"Visagino Enervizija," Visaginas town Program for energy efficiency improvement in multiapartment buildings	2011
	Residential	Program for the renovation/upgrading of multiapartment buildings	2005
	Tertiary	EU Structural Funds 2007–2013	2007
	Tertiary	Renovation of State Institutions	2014
Netherlands	Residential	Reduced VAT rate on labor costs for insulation and glass and for maintenance and renovation of residential buildings (Verlaagd BTW tarief)	2009
Slovenia	Residential	Financial incentives for energy-efficient renovation and sustainable construction of residential buildings	2008
	Tertiary	Financial incentives for energy-efficient renovation and sustainable construction of buildings in the public sector	2008
Spain	Residential and Tertiary	State plan 2013–2016 for Rental Housing, Housing Rehabilitation, and Urban Regeneration and Renewal	2013
	Tertiary	*Action plan 2008–2012*: Energy Saving and Efficiency Plans in Public Administrations	2008

Source: Modified from D'Agostino, D. et al., Synthesis Report on the National Plans for NZEBs, EUR 27804 EN. http://publications.jrc.ec.europa.eu/repository/bitstream/JRC97408/reqno_jrc97408_online%20 nzeb%20report(1).pdf and Odyssee-Mure database, available at: http://www.odyssee-mure.eu/.

Table 10.8 reports current best practice in some Member States: Denmark, Germany France, the Netherlands, Sweden, and UK. Data are taken from the GBPN (Global Buildings Performance Network) policy tool for renovation which captures the performance of and enables their comparison [47]. The analysis supports the development of policy packages that drive the existing building stock towards deep renovation considering current best practice elements of policy packages for the residential building stock. Each policy package has been selected based on two main criteria: policies demonstration including elements that cover energy renovations; and a reduction of residential energy consumption.

The most diffused measures are financial (about 43%) and regulatory (about 25%). Financial measures mostly include co-financing programs offering grants and subsidies to building owners. Subsidy programs usually target residential sector and successful examples can be found in many countries.

TABLE 10.8

Best Practice Measures in a Selection of European Countries

Measure Type	Measure Description	Country
Financial Instruments Incentive Schemes	The KfWs Energy Efficient Construction and Rehabilitation (EECR) program for residential properties provides subsidized lending for the renovation of existing building stock. The funding is set according to the level of energy efficiency achieved, KfW can finance up to 100% of the loan. As required by the Government's Energy Concept 2050, from 2009 to 2012, funding of 500 million Euros was provided for the renovation of existing buildings.	Germany
	The Green Deal works hand-in-hand with the UK's Energy Company Obligation and aims to improve the energy efficiency of most of the 26 million homes in the United Kingdom. Individuals pay a part of the cost of improving their homes by taking a loan that is paid back via the savings they make on their fuel bills. The scheme, established by the Coalition Government through the Energy Bill, is designed to run between 2013 and 2027.	UK
Financial Instrument Taxation Mechanism	The renovation fund contains 1.5 billion DKK of subsidizes for private building projects. An income tax deduction has been reintroduced in 2013, which is estimated to amount to 1.5 billion DKK annually. The scheme will be unchanged and allows tax deductions for home renovation costs up to DKK 15,000 per person annually for renovation services.	Denmark
	The "Haushaltsnahen Dienstleistungen" tax incentive allows for 20% of the labor costs (up to 6000 Euros, tax relief on up to 1200 Euros) of home renovations associated with reducing the energy demand.	Germany
	Energy Tax is a tax on energy consumption that improve the cost-effectiveness of measures aimed at energy saving and renewable energy. The Energy Tax increases energy prices for small-scale consumers, such as households (up to 5000 m³ gas and 10000 kWh) to promote energy efficiency.	Netherlands
Economic Instrument Utility-Funded Energy Efficiency Program	Since 1996 there has been a legal obligation placed on the utilities and the latest 2009. Each utility is allowed to decide on the way in which it will accumulate savings. Each utility must present documentation to prove that they have realized their targets.	Denmark
	The Energy Company Obligation (ECO) was introduced in 2013 requiring the six biggest energy suppliers in the United Kingdom to deliver energy efficiency measures to domestic energy users. The financial support is provided when the energy companies and individuals communicate to identify and apply suitable saving measures. Up to 100% of the cost will be provided by the energy companies.	UK
Economic Instrument Market Instruments for Energy Efficient Renovation	The ESCO market is perceived to have grown due to a policy framework and to demonstration projects. France's National Energy Agency has been able to create a market for energy renovations in the public sector. The ESCOs market is under further development through a program called "Marché Public" that offers smaller amounts (< €5 million) for project financing.	France
Information and Capacity Building Training and Education Campaigns	The German Energy Agency is a center of expertise for energy efficiency, RES, and intelligent energy systems. It organizes campaigns, distributes information, supports the building sector to work in line with current standards and regulations and develops standards and labels for efficiency.	Germany
	The Swedish Energy Agency supports local authorities by training them to provide energy efficiency measures. A number of policy packages are available for the residential including information tools and economic incentives.	Sweden

Source: Modified from D'Agostino, D. et al., Synthesis Report on the National Plans for NZEBs, EUR 27804 EN. http://publications.jrc.ec.europa.eu/repository/bitstream/JRC97408/reqno_jrc97408_online%20nzeb%20 report(1).pdf and GBPN, Reducing energy demand in existing buildings: Learning from best practice renovation policies, July 2014, available at: http://www.gbpn.org/sites/default/files/08.%20Renovation%20 Tool%20Report.pdf.

Tax relief on energy efficiency upgrades for residential buildings is also adopted. In Italy, tax credit scheme for building renovations is a flag-ship measure for households. Loan programs are offered by a few countries, with the most successful program being KfW scheme in Germany.

Given the huge potential investments that exist at European level for renovation, new business models that stimulate energy renovation market and engagement of private capitals are necessary to boost large-scale energy renovation of buildings in the EU. Some innovative financing concepts should be further developed and utilized, including EPC, on-bill financing schemes and revolving funds with the use of European structural and investment funds.

10.6 Role of Innovative Technology in Efficient Buildings

Energy savings can be achieved through the adoption of different efficient technological measures. As previously pointed out, construction alternatives have to be considered when buildings are designed to search for cost-effective solutions according to the EPBD recast. These alternatives mainly include envelope, energy sources, installations, and systems. There is a wide range of new technologies that are becoming an integral part of buildings.

In the view of a new building concept, technology innovation is proposing new solutions: the envelope represents a key element able to conduct dynamically the interaction between indoors and outdoors, renewable energy production has to compensate energy consumptions, ICT can provide a smarter use of energy, insulation can guarantee less thermal losses, and systems are becoming more and more efficient.

The solutions enable more dynamic and interactive buildings. In efficient buildings and NZBEs, technologies are used in conjunction with optimum design techniques that minimize summer heat gains and winter heat losses, the use of passive heating and cooling techniques, a rational use of daylight to reduce lighting and renewables.

Industry is more and more focused on dynamic construction techniques, using automation and robotics, and more sustainable materials, buildings components and systems. An innovative market is conforming to the new building target promoting integrated solutions and packages that are now described [46,48].

10.6.1 Overview of Building Envelope, Systems, and Renewables Technologies

The building envelope allows for a considerable reduction in energy needs in a building [49]. There are many innovative materials, components, and elements that give a variety of measures to reduce energy demand in buildings. The envelope includes walls, roofs, basements, and windows surfaces.

Insulation can include the whole building or only a part, such as external walls or roof surface. Measures allow elimination of thermal bridges, air tightness control, and PV façade. A frequent measure within the envelope is the installation of external shading devices on windows. Shadings can be permanent or movable, automatic or manually controlled. The use of these devices to reduce sunlight during summer, and maximize gains during winter is becoming more widespread. Control systems are commonly present in heating, cooling systems and ventilation, and are frequently applied to lighting (e.g. daylight control or occupancy control). Other measures for cooling are the following: ground source heat pumps, free cooling, district cooling, design orientation, selective glazing, and centralized cooling plants with room air-conditioning machines.

Nanotechnology is demonstrating huge potentialities in the synthesis of nanomaterial for the building sector. Insulation materials are able to decrease heat transfer considerably through the envelope. Among these materials, there are fibreglass, polyurethane foam, polystyrene foam, cellulose insulation, and rock wool able to fill or coat walls, roofs, floors, and facades. These materials can allow reaching low U-values (high thermal resistance) of 0.1–0.15 W/m²K. Cool roofs can help minimize solar absorption and maximize thermal emission reducing the incoming heat flow and the energy used for cooling, in addition to reducing heat losses.

Windows give a central contribution to the envelope performance. Double- or triple-glazed windows with low emissivity reduce energy consumption per m² of glazed surface by more than 40%. They are able to reduce thermal transmittance below 2.7 W/m²K, and argon-filled double glazing to 1.1 W/m²K. Argon-filled triple glazing can reach values of 0.7 W/m²K. Films and coatings can be used on existing glazing to limit solar gains.

Building façade technology presents a great opportunity to address energy efficiency. Innovative building façades that integrate different technologies contribute to the overall energy performance of a building.

A minimum exchange of the building volume, expressed as air changes per hour (ACH), is required in a building to maintain indoor air quality. The air-tightness of the envelope can be improved to minimize unwanted air leakage.

In combination with heat recovery ventilation systems, improving air-tightness can provide levels of 0.4–0.6 ACH with an energy efficiency of the installation of over 80%.

Other several techniques can contribute to manage the overall energy performance of a building, integrating materials, and techniques in the envelope. Among these, there are: ventilated facades and windows, solar chimneys, infrared reflective paints, humidity control foils, and solar energy absorbing thermal mass for natural night ventilation. The usefulness of green facades and green walls is also becoming evident in many cities as a tool to mitigate the heat island effect.

The dynamic assessment of the impact of such applications in the overall energy consumption of a building is crucial and demands for the development of specific analysis and simulation techniques.

Efficient and smart building systems also contribute to the energy performance of a building. Different systems can supply the need for heating, cooling, ventilation, and dehumidification [50]. The use of a particular technology usually depends on available energy resources. For example, electricity might be more convenient for heating in one Member State (e.g. France) than in another where more biomass or gas is available (e.g. Finland).

Condensing boilers are an efficient heat-generation system that uses an additional heat exchanger to extract extra heat by condensing water vapor from combustion products. Heat pumps can be used for space heating and hot water, increasing the amount of renewables.

Heat recovery plants can reduce energy consumption of HVAC systems as they use heat exchangers to recover hot or cold air from ventilation exhausts and supply it to the incoming fresh air. Chillers, which produce chilled water rather than cooled air for use in large residential and commercial buildings, can be up to three times more efficient than typical air conditioners [51].

Renewable energy technologies mainly include solar PV, solar thermal, biomass, geothermal, and aerothermal. PV systems are becoming ubiquitous and efficient, stimulating architects to integrate them as a building material. PV can produce electricity to cover a building's energy needs, above all installations and appliances, for direct consumption or for delivery to the grid or local storage for later consumption. Energy storage (i.e., electricity using batteries, compressed air storage, and vehicle-to-grid) represents a cutting-edge technological solution to address the future trend of energy in buildings.

Solar thermal collectors absorb incoming solar radiation and convert it into heat, which is then carried from the circulating fluid either to space heating or hot water equipment or to a thermal energy storage system for later use. Biomass products (e.g. wood logs and pellets) are used in heating systems, and heat pumps (geo- and aerothermal energy) are often used in buildings for ground-coupled and air-to-air heat exchange.

Apart from all the above-mentioned technologies, the energy performance of a building is influenced by several other factors, such as geometry and orientation, as well as building usage (occupancy behavior and auxiliary gains) [52].

Control automation and smart metering devices for interaction with utilities are among the most rapidly developing smart technologies in buildings. These devices allow the control of the energy demand/supply through ICT technologies considerably decreasing energy consumptions. Furthermore, they allow field data to be gathered for use in performance calculations and dynamic simulation modeling. Control systems include daylight responsive control, presence control and motion control.

Requirements related to energy performance, indoor air quality, comfort level, and monitoring are becoming more and more important [53].

Several techniques contribute to manage the overall energy performance of a building, integrating materials and techniques in the envelope. Among these, there are the following techniques: ventilated facades and windows, solar chimneys, infrared reflective paints, humidity control foils, and solar energy absorbing thermal mass for natural night ventilation.

In regards to the connection to the energy infrastructure, many possibilities are available to be connected to utility grids. Among them, district heating is one of the most used systems. Other systems are as follows: electricity grid, and cooling systems, gas pipe network, or biomass and biofuels distribution network. Therefore, buildings have the opportunity to both import and export energy from these grids, thus avoiding on-site electricity storage. While on-grid buildings are connected to one or more energy infrastructures using the grid both as a source and a sink of electricity, off-grid buildings require an electricity storage system in peak load periods or when RES are not available.

Zero-energy districts can effectively overcome physical boundary limitations that are common in the refurbishment of existing buildings at the nearly zero level, such as access to on-site renewable energy generation.

10.7 Conclusion

Improving energy efficiency in buildings represents a great opportunity to decrease energy consumptions and increase renewable production, alleviating energy resources depletion and environmental deterioration. Furthermore, the building sector can contribute to mitigate climate change and at the same time delivering many other benefits. Renovation of European existing buildings would decrease energy needs for gas imports, and related imports costs, and improving energy security. Reducing household energy bills can also alleviate fuel poverty and improve social and territorial cohesion at the European level.

Considering these aspects and the high consumption that characterize the building stock in Europe, buildings are the core of the European 2020 and 2030 strategies of climate and energy targets. Consequently, the attention given to building energy efficiency increased over the last years.

This chapter summarized the main aspects related to the European efforts to improve energy efficiency in buildings. Starting from the overall trends related to energy consumption in residential and non-residential buildings, the European policy framework focused on energy efficiency has been outlined. Data highlights how energy consumptions vary among sectors and energy uses and how, despite the economic crisis, the energy trends started to decrease in European Member States over the last decade. However, despite more stringent building codes, the final energy consumption remains too high especially in non-residential buildings. European Directives and national energy efficiency policies are major driving forces towards energy efficiency in buildings. Further attention to targeted initiatives is still needed in several areas, above all building renovation. Buildings in Europe are more likely to be refurbished than replaced, therefore energy renovation based on energy efficiency in combination with renewables are essential for reaching climate goals in Europe.

As NZEBs are the building target from 2018 onwards, a focus on this building type has been given. The main NZEBs definitions, issues, and status of implementation in European countries have been given considering the cost-optimal approach that is foreseen by the Energy Performance of Building Directive (EPBD). An evaluation is given in relation to the status of NZEBs development and compliance with the EPBD requirements.

The most common barriers and challenges have been described in relation to both retrofit and new buildings. Successful policies and measures are reported identifying best examples in Member States. The achievement of ambitious saving targets is one of the main challenges that Europe will face in near future. The success of energy efficiency policies also depends on the households' involvement as citizens cannot afford the costs of energy renovation. Overcoming these issues is a crucial step for energy renovation and the current policy framework has made relatively little progress towards providing effective solutions to them. Best practices are identified in Member States to highlight good case studies that could be applied to other countries. The analysis highlights that European countries have to further adopt specific actions to exploit the potential energy savings deriving from the building sector. Especially in view of building refurbishment, they should effectively develop detailed strategies both to overcome barriers towards energy efficiency and to guide investment decisions in a forward-looking perspective. Consistent mixtures of policy instruments (policy packages) can provide long-term stability to investors in efficient buildings, including deep and NZEBs renovations. This chapter shows how reliable data to monitor policy impacts are required above all for building stock refurbishment. In some countries with limited solar RES potential (e.g. northern Europe countries), policies that support alternative measures are needed (e.g. biomass). The adoption of roadmaps and indicators would be an additional tool suitable to address specific needs and monitor the implementation.

The chapter also shows how technology plays an important role towards efficiency and how a new market is proposing new solution for a new building concept. It highlights the importance of research projects, innovation, and market uptake in demonstrating NZEBs for a wide implementation into construction practices. Primary energy demand can be reduced through low-energy technologies (e.g. insulation, daylighting, high-efficiency HVAC, natural ventilation, and evaporative cooling) in combination with renewables on-site or off-site. The integration between cost-optimality and high performance technical solutions underpins the deployment of NZEBs. However, reaching buildings requiring nearly zero energy at the lowest cost is not yet reached through Europe at a retrofit level.

It is foreseeable how energy efficiency can further stimulate industrial competitiveness and increase asset values through rental and sales premiums. The economic sector linked to energy efficiency has already begun to involve architects, engineers, and researchers dealing with building physics, economists, environmental analysts, and policy makers. The interdisciplinary nature of energy efficiency in buildings needs further cooperation among all the actors involved in the area. Technological innovation will give a huge input to the creation of new professional figures collaborating and working in the building sector from different perspectives.

References

1. Eurostat. Final energy consumption by sector, 2014, available at: http://epp. eurostat.ec.europa. eu/portal/page/portal/statistics/search/database (Accessed: March 13, 2017).
2. Odyssee-Mure database, available at: http://www.odyssee-mure.eu/ (Accessed: March 15, 2017).
3. Eurostat, European System of Accounts—ESA 1995, *Office for Official Publications of the European Communities*, Luxembourg, Europe, 1996.
4. European Commission. 2014. *Taking Stock of the Europe 2020 Strategy for Smart, Sustainable and Inclusive Growth*, Brussels, Belgium, March 19, 2014.
5. D. D'Agostino, C. Barbara, B. Paolo, Energy consumption and efficiency technology measures in European non-residential buildings, *Energy and Buildings* 153, 2017, 72–86, http://dx.doi.org/10.1016/j.enbuild.2017.07.062.
6. D. D'Agostino, P. Zangheri, L. Castellazzi, Towards nearly zero energy buildings (NZEBs) in Europe: A focus on retrofit in non-residential buildings, *Energies*, 2017, 10, 117, doi:10.3390/en10010117.
7. Odyssee—Mure, Energy Efficiency Trends and Policies in the Household and Tertiary Sectors, An Analysis Based on the ODYSSEE and MURE Databases, 2015, available at: http://www.odyssee-mure.eu/publications/br/energy-efficiency-in-buildings.html (Accessed: March 17, 2017).
8. D. D'Agostino, C. Barbara, B. Paolo, Data on European non-residential buildings, Data in Brief 14, 2017, 759–762, http://dx.doi.org/10.1016/j.enbuild.2017.07.062.
9. Non-Residential Buildings—TABULA Thematic Report N° 3 -Typology Approaches for Non-Residential Buildings in Five European Countries—Existing Information, Concepts and Outlook.
10. EU, Directive 2002/91/EC. European Parliament and of the Council of December 16, 2002 on the Energy Performance of Buildings, *Official Journal of the European Communities*, 2003, 65–71.
11. EU, Directive 2010/31/EU. European Parliament and of the Council of May 19, 2010 on the energy performance of buildings (recast), *Official Journal of the European Union*, 2010, 13–35.
12. EU, 244/2012. Commission Delegated Regulation No244/2012 of January 16, 2012. Supplementing directive 2010/31/EU of the European Parliament and of the Council on the energy performance of buildings by establishing a comparative methodology framework for calculating cost-optimal levels of minimum energy performance requirements for buildings and building elements, *Official Journal of the European Union*.
13. EU, Guidelines accompanying Commission Delegated Regulation (EU) No 244/2012 of January 16, 2012 supplementing Directive 2010/31/EU of the European Parliament and of the Council, *Official Journal of the European Union*.
14. EU, Directive 2012/27/EU. European Parliament and of the Council of October 25, 2012 on energy efficiency, amending Directives 2009/125/EC and 2010/30/EU and repealing Directives 2004/8/EC and 2006/32/EC.

15. EU, Directive 2009/28/EU. European Parliament and of the Council of April 23, 2009 on the promotion of the use of energy from renewable sources and amending and subsequently repealing Directives 2001/77/EC and 2003/30/EC.
16. A. Allouhi, Y. El Fouih, T. Kousksou, A. Jamil, Y. Zeraouli, Y. Mourad. Energy consumption and efficiency in buildings: Current status and future trends, *Journal of Cleaner Production*, 109, 2015, 118–130.
17. EU, Directive 2010/30/EU. Directive of the European Parliament and of the Council of May 19, 2010 on the indication by labelling and standard product information of the consumption of energy and other resources by energy-related products, *Official Journal of the European Union*.
18. European Parliament. 2009, Directive 2009/125/EC. Directive of the European Parliament and of the Council of October 21, 2009 establishing a framework for the setting of eco-design requirements for energy-related products, *Official Journal of the European Union*.
19. Livio Mazzarella Near zero, zero and plus energy buildings: Revised definitions, proceedings of Clima 2016.
20. U.S. Department of Energy. A common definition for zero energy buildings. DOE/EE-1247, September 2015.
21. H. Lund, A. Marszal, P. Heiselbergb. Zero energy buildings and mismatch compensation factors, *Energy and Buildings*, 43 (7), 2011, 1646–1654, doi:10.1016/j.enbuild.2011.03.006.
22. M. Panagiotidou, R. J. Fuller. Progress in ZEBs—A review of definitions, policies and construction activity, *Energy Policy*, 62, 2013, 196–206.
23. J. Laustsen. Energy Efficiency Requirements in Building Codes, *Energy Efficiency Policies for New Buildings*, Organisation for Economic Co-operation and Development/International Energy Agency, Paris, France, 2008.
24. P. Torcellini, S. Pless, M. Deru, D. Crawley. Zero Energy Buildings: A Critical Look at the Definition, National Renewable Energy Laboratory and Department of Energy, Pacific Grove, California, 2006.
25. D. D'Agostino. Assessment of the progress towards the establishment of definitions of Nearly Zero Energy Buildings (nZEBs) in European Member States, *Journal of Building Engineering*, 2015, doi:10.1016/j.jobe.2015.
26. D. D'Agostino et al., Synthesis Report on the National Plans for NZEBs, EUR 27804 EN. doi:10.2790/659611, available at: http://publications.jrc.ec.europa.eu/repository/bitstream/JRC97408/reqno_jrc97408_online%20nzeb%20report(1).pdf (Accessed: March 20, 2017).
27. EC, European Commission, National plans for nearly zero-energy buildings (2013), available: http://ec.europa.eu/energy/efficiency/buildings/implementation_en.htm (Accessed: March 22, 2017).
28. D. D'Agostino, P. Zangheri. Development of the NZEBs concept in Member States, EUR 28252 EN. doi:10.2788/278314.
29. UNI EN 15459. Energy performance of buildings, economic evaluation procedure for energy systems in buildings. 2008.
30. UNI/TS 11300 e Energy performance of buildings. Part 1 (2014): Evaluation of energy need for space heating and cooling. Part 2 (2014): Evaluation of primary energy need and of system efficiencies for space. Part 3 (2010): Evaluation of primary energy and system efficiencies for space cooling. Part (2012): Renewable energy and other generation systems for space heating and domestic hot water production heating, domestic hot water production, ventilation and lighting for non-residential buildings.
31. P.M. Congedo, C. Baglivo, D. D'Agostino, I. Zacà. Cost-optimal design for nearly zero energy office buildings located in warm climates, *Energy*, 91, 2015, 967–982.
32. I. Zacà, D. D'Agostino, P.M. Congedo, C. Baglivo. Assessment of cost-optimality and technical solutions in high performance multi-residential buildings in the Mediterranean area, *Energy Build*, 102, 2015, 250–265.
33. I. Zaca, D. D'Agostino, P. Maria Congedo, C. Baglivo. Data of cost-optimality and technical solutions for high energy performance buildings in warm climate, *Data Brief*. doi:10.1016/j.dib.2015.05.015.

34. C. Baglivo, P.M. Congedo, D. D'Agostino, I. Zacà. Cost-optimal analysis and technical comparison between standard and high efficient mono-residential buildings in a warm climate, *Energy*, 83, 2015, 560–575.
35. P.M. Congedo, C. Baglivo, D. D'Agostino, G. Tornese, I. Zacà. Efficient solutions and cost-optimal analysis for existing school buildings, *Enegies*, 9(10), 2016, 851. doi:10.3390/en9100851.
36. CA EPBD, Concerted action EPBD: Implementing the energy performance of buildings directive (EPBD). Information of the joint initiative of EU Member States and the European Commission, available at: http://www.epbd-ca.eu/themes/cost-optimum (Accessed: March 24, 2017).
37. BPIE (Buildings Performance Institute Europe). Implementing the cost-optimal methodology in EU countries, pp. 1–82. http://bpie.eu/costoptimalmethodology.html (Accessed: March 27, 2017).
38. EU, Commission Recommendation 2016/1318 of July 29, 2016 on Guidelines for the promotion of nearly zero-energy buildings and best practices to ensure that, by 2020, all new buildings are nearly zero-energy buildings.
39. F. Marco, M.G. Almeida, R. Ana, S. M. da Silva. Comparing cost-optimal and net-zero energy targets in building retrofit, *Buildings Research and Information*, 2016, 44, 188–201.
40. Y. Lu, S. Wang, K. Shan. Design optimization and optimal control of grid-connected and standalone nearly/net zero energy buildings, *Applied Energy*, 155, 2015, 463–477.
41. P. M. Congedo, C. Baglivo, I. Zacà, D. D'Agostino. High performance solutions and data for Nzebs offices located in warm climates, data in brief, available online October 13, 2015, doi:10.1016/j.dib.2015.09.041.
42. M. Ferrara, E. Fabrizio, J. Virgone, M. Filippi. A simulation based optimization method for cost-optimal analysis, Energy and Buildings 2014, doi:10.1016/j.enbuild.2014.08.031
43. Buildings Performance Institute Europe (BPIE). Europe's buildings under the microscope. *A Country-by-Country Review of the Energy Performance of Buildings*. BPIE: Brussels, Belgium, 2011.
44. Zebra 2020. Nearly Zero Energy Building Strategy 2020—Strategies for a Nearly Zero-Energy Building Market Transition in the European Union, available at: http://zebra2020.eu/website/wp-content/uploads/2014/08/ZEBRA2020_Strategies-for-nZEB_07_LQ_single-pages-1.pdf (Accessed: March 27, 2017).
45. Taking Stock of the Europe 2020. *Strategy for Smart, Sustainable and Inclusive Growth*. European Commission: Brussels, Belgium, 2014.
46. L. Carmel, K. Anja, S. Kari, W. Annemie. Barriers and challenges in nZEB Projects in Sweden and Norway, *Energy Procedia*, 58, 2014, 199–206.
47. GBPN, Reducing energy demand in existing buildings: Learning from best practice renovation policies, July 2014, available at: http://www.gbpn.org/sites/default/files/08.%20Renovation%20Tool%20Report.pdf (Accessed: March 30, 2017).
48. European Commission, DG Energy: ENER/C3/2012-436—Market study for a voluntary common European Union certification scheme for the energy performance of non-residential buildings.
49. M. Rossia, V. M. Roccoba. External walls design: The role of periodic thermal transmittance and internalareal heat capacity, *Energy and Buildings*, 68, 2014, 732–740.
50. J. Mlakar, J. Štrancar. Overheating in residential passive house: Solution strategies revealed and confirmed through data analysis and simulations, *Energy and Buildings*, 6, 2011, 1443–1451.
51. T. Psomas, P. Heiselberg, K. Duer, E. Bjørn. Overheating risk barriers to energy renovations of single family houses: Multicriteria analysis and assessment, *Energy and Buildings*, 4/1, 2016, 138–148.
52. E. Moretti, E. Bonamente, C. Buratti, F. Cotana. Development of innovative heating and cooling systems using renewable energy sources for non-residential buildings, *Energies*, 6, 2013, 5114–5129.
53. S. Kephalopoulos, O. Geiss, J. Barrero-Moreno, D. D'Agostino, D. Paci. *Promoting Healthy and Energy Efficient Buildings in the European Union, National implementation of related requirements of the Energy Performance Buildings Directive (2010/31/EU)*, EUR 27665 EN, doi:10.2788/396224, LB-NA-27665-EN-N.

11

Advances in Simulation Studies for Developing Energy-Efficient Buildings

Karunesh Kant, Amritanshu Shukla, and Atul Sharma

CONTENTS

11.1   Introduction 210
    11.1.1   Energy Consumption in Buildings 210
    11.1.2   Energy-Efficient Buildings 212
11.2   Energy Balance in Buildings 213
    11.2.1   External Heat Gains 214
    11.2.2   Internal Heat Gains 214
    11.2.3   Transmission Loss 214
    11.2.4   Loss Due to Thermal Bridges 214
    11.2.5   Ground Losses 214
    11.2.6   Filtration Loss, Ventilation Loss 214
11.3   Software for Modeling 215
    11.3.1   Computational Fluid Dynamics 223
        11.3.1.1   COMSOL Multiphysics 224
        11.3.1.2   Fluent 225
    11.3.2   Transient System Simulations 226
    11.3.3   Energy Plus 226
    11.3.4   Open Studio® 227
    11.3.5   Radiance 227
    11.3.6   eQUEST 228
    11.3.7   TRANE TRACE 700 228
    11.3.8   Simergy 229
    11.3.9   FORTRAN 229
    11.3.10   MATLAB 229
        11.3.10.1   Set Point 230
        11.3.10.2   Thermostat 230
        11.3.10.3   Heater 230
        11.3.10.4   Cost Calculator 230
        11.3.10.5   House 232
11.4   Conclusion 232
References 233

11.1 Introduction

The projection of building energy utilization in the world is in the range of 123.1 quad btu [1]. Presently, the energy demand that fulfills the thermal comfort in buildings has increased considerably because of more cooling/heating system set-up in the buildings. The increase in energy consumption, CO_2 emissions, and fuel prices are pushing a new policy to build more sustainable and energy-efficient buildings (EEBs). Thereby, to reduce the energy consumption by private and industrial users is one of the major items that are currently proposed by governments and administrations of various countries. Reducing energy consumption and recycling waste are among the main goals of the majority of countries. Figure 11.1 represents the energy consumption worldwide in different sectors. From the figure, it is obvious that the maximum energy consumption is in the building sector.

11.1.1 Energy Consumption in Buildings

Due to the growth of population, expanding of the economy, and seeking a better quality of life, energy consumption has enlarged, and the growth rates are expected to continue, powering the energy demand further. Augmented energy consumption will lead to additional greenhouse gas (GHG) emissions having serious impacts on the global environment. The advanced urbanization rate with enlarged floor space for both commercial and residential applications has enacted massive pressure on the existing energy sources. Restricted accessibility of energy sources, the existing energy resources, and transient nature of renewable energy sources have enhanced the significance of energy efficiency and conservation in various sectors. With the purpose of the thermovisual relaxation of the residents and according to functionality, lighting systems and, the heating ventilation and air conditions systems, electric motors are the major consumers of energy within buildings. Categorical classification of energy consumption by any end use such as heating, cooling, cooking, and so on, for both residential and commercial buildings (in U.S.) is shown in Figure 11.2 [2].

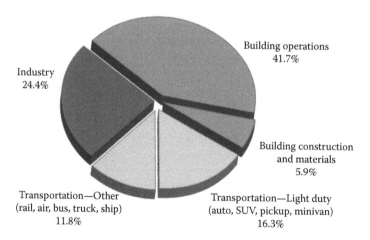

FIGURE 11.1

Energy consumption worldwide in different commercial sectors. (From http://www.architecture2030.org/buildings_problem_why/.)

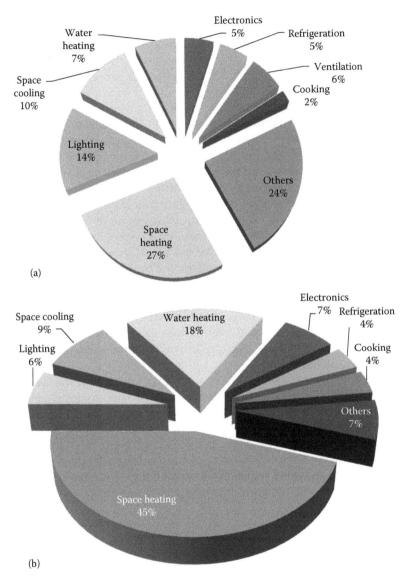

FIGURE 11.2
End-use wise energy consumption in (a) residential and (b) commercial buildings. (From Harish, V.S.K.V., and Kumar, A, *Renew. Sustain. Energ. Rev.* 56, 1272–1292, 2016.)

Energy conservation, which leads to the more proficient use of energy without decreasing comfort levels, does not mean rationing or curtailment or load flaking, however, it is a means of recognizing areas of inefficient use of energy and taking action to decrease energy waste. There are vast opportunities to reduce electricity consumption and increase energy efficiency within buildings. It was estimated that new buildings can decrease energy consumption on an average of 20% and 50% by integrating suitable design interventions in the building envelope, HVAC (20%–60%), lighting (20%–50%), water heating (20%–70%), refrigeration (20%–70%), and electronics and other (e.g., office equipment and intelligent controls, 10%–20%).

11.1.2 Energy-Efficient Buildings

Energy efficiency and carbon management in buildings should be thought of as an ongoing process. The diagram in Figure 11.3 shows how to approach this process in a logical way that empowers you to keep striving for continuous improvement.

Management is the key factor for the energy efficiency in the building. In the building, the technology, as well as management, is important for efficient utilization of energy, without wasting money. The first maxim of energy management is that we can't succeed what we don't measure; consequently, establishing a baseline of energy use should be our top priority as a facilities manager. Get the bills out, check consumption and keep a record of it. If in doubt, check the meters, and reflect how working changes and external elements such as the climate have affected energy consumption.

After the establishment of a baseline, we should then tackle the "no cost" then "low cost" measures suggested in this guide. A no-cost measure is one that is free to the appliance (although it may take an hour or so of your time), and low-cost measures are ones that will pay for themselves within six months. No-cost measures are applicable to both leased and owned buildings as they do not involve changing the building's services.

Once these have been completed, the "medium cost" measures suggested should be considered. These measures should pay for themselves in less than two years. If we are the manager for a leased building, these should be discussed with our landlord using a cost-benefit agreement. Lease renewal negotiations are a perfect time to appeal energy efficiency modifications, as owners may prefer small capital spending to keep a tenant instead of risk an empty building.

It may be that immediate implementation of an action is not appropriate but when any major change is planned for the building portfolio, the decision-making process should be repeated to see if any methods, which were not formerly possible could be implemented during the refurbishment/relocation.

Throughout this process, energy depletion should be controlled and energy savings described to improve the business case for upcoming energy effectiveness plans. The bills should be compared on at least a quarterly basis with data from the previous year to see if unit feeding has decreased. Set up a simple spreadsheet to track monthly consumption and record savings.

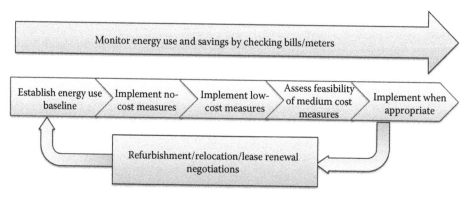

FIGURE 11.3
The energy management decision process.

11.2 Energy Balance in Buildings

Energy balance in the buildings is when the heat enters across a control volume that is equal to the same control volume of heat leaving. The energy enters in the control volume is called energy gain and leaves from control volume are called energy loss in unit time. The energy gain is equal to energy loss. The Figure 11.4 represents the energy balance of heated space.

$$Q_{gains} = Q_{losses}$$

The energy loss and energy gain is described in more details by the following equation:

$$Q_{ext.gains} + Q_{int.gains} = Q_{FL} + Q_{FIL} + Q_{VENT} \pm Q_{ST}$$

where:
  $Q_{ext.gains}$ is the external energy gain
  $Q_{int.gains}$ is the internal energy gain
  Q_{FL} is the fabric loss which includes transmission loss, thermal bridge loss, and ground loss
  Q_{FIL} is the filtration loss, which is related to the air tightness of a building shell
  Q_{VENT} is the energy loss due to mechanical ventilation
  Q_{ST} is the heat which is stored in the building mass

The stored heat is related to the heat capacity and density of the building shell.

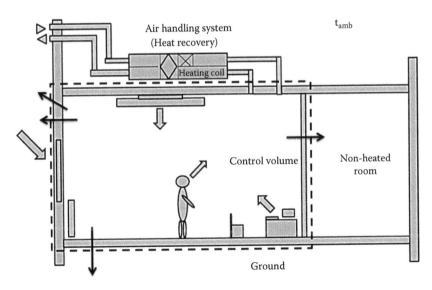

FIGURE 11.4
Energy balance of a heated space.

11.2.1 External Heat Gains

External heat gains, also known as solar gain, refer to the upsurge in temperature in a building space, object, or structure that results from solar radiation. The quantity of solar heat gains upsurges with the intensity of solar radiation, and with the capability of any dominant material to transmit or resist the solar radiation. Entities struck by the sunshine absorb the short-wave radiation from the sunlight and re-radiate the heat at longer infrared wavelengths. The constituents, such as glass and other building materials, lying between the sun and the room entities, which are more transparent to the shorter wavelengths than the longer wavelengths, also result in an upsurge in solar heat gain when the sun is shining.

11.2.2 Internal Heat Gains

Inside the buildings, people, technical equipment, artificial lighting, and even sincere belongings give off heat. High occupancy levels and an extreme quantity of practical devices cause high internal loads, especially in office buildings. The conditions are easily intended by specific office building residence schedules. Domestic buildings are more variable in occupancy and technical equipment. The internal heat gain regimes generally remain lower and therefore much less significant than in office buildings, schools, and so on.

11.2.3 Transmission Loss

Transmission loss of single or multilayer planar building element means conduction and convection. In steady state (constant temperatures of the boundaries), heat flux is constant from and to the boundaries and also constant at each layer.

11.2.4 Loss Due to Thermal Bridges

A thermal bridge is a component, or assembly of components, in a building envelope through which heat is transferred at a substantially higher rate than through the surrounding envelope area; also the temperature is substantially different from the surrounding envelope area.

11.2.5 Ground Losses

Ground loss is also considered as a linear loss. There are two main dependencies: the resistance of the multilayer construction and the relative elevation (which is relative to the ground level around the building).

11.2.6 Filtration Loss, Ventilation Loss

Filtration loss is related to the air tightness of a building shell and ventilation loss is the energy loss due to mechanical ventilation.

11.3 Software for Modeling

The performance of buildings depends on some physical parameter such as solar radiation, solar temperature/outside air temperature, room air temperature, thermophysical properties of construction elements, wind characteristics and precipitation, internal heat gains, sky or cloud conditions ventilation rate, building location (global information), and so on. Some of the parameters are controllable and some are uncontrollable. The list of significant controllable and uncontrollable parameters used in the development of the model is represented in Table 11.1 and software for simulation is represented in Table 11.2. For the modeling of energy efficiency in buildings, various software is available in the market; some of them are described in the following table:

TABLE 11.1

List of the Significant Parameters Used in Development of Model

S. No	Uncontrollable	Application[a]
1	*Solar radiation*: Incident solar radiation is the major thermal load at the building envelope's exterior	R/NR
2	*Sol–air temperature/outside air temperature*: Sol–air temperature is the outdoor air temperature that, in the absence of all radiation changes gives the same rate of heat entry into the surface as would the combination of incident solar radiation, radiant energy exchange with the sky and other outdoor surroundings, and convective heat exchange with outdoor air	NR/R
3	*Thermophysical properties of construction elements*: Thickness	R/NR
4	*Sky or cloud conditions*: Shading, cloudiness of the outdoor weather conditions	NR/R
5	*Building location (Global Information)*: Information about latitude, longitude, time zone, month, day of month, directional orientation of the zone, and zone height (floor to floor)	NR/R
6	*Wind characteristics and precipitation*: Wind speed, wind direction, and terrain roughness	NR
	Controllable	
7	*Room air temperature*: Indoor temperature variations depend on the purpose and occupation of the building	R/NR
8	*Internal heat gains*: Internal heat gains from people, lights, motors, appliances, and equipment can contribute the majority of the cooling load in a modern building Shading,	NR
9	*Ventilation rate*: Flow rate due to intentional introduction of air from the outdoors into a building	NR

[a] R: Residential buildings; NR: Nonresidential (commercial and industrial) building.

TABLE 11.2

List of Building Energy Simulation Programs

S. No.	Building Energy Simulation Software	Applications	Open Source	Simulation Engine	Limitations	References
1	Autodesk green building studio	3D CAD/BIM	Y	DOE-2.2 and Energy Plus	Limited for architectural designing; lead to complex modeling and thus, not suited for control purposes	Autodesk Green Building Studio, (http://www.autodesk.com/Greenbuildingstudio)
2	BSim	Energy, Daylight, thermal and moisture analysis, indoor climate	N	Self	No. No standardized result reports. No possibility batch processing of simulation models at this time—typical simulation time for an average model is though only a few minutes on an up to date computer. No support of geometrical input from CAD tools in IFC file formats —this facility is presently being developed	Danish Building Research Institute. BSim, (http://www.bsim.dk)
3	Building energy analyzer	Heating, On-site power generation, Heat recovery, CHP	N	DOE-2.1 E	General General knowledge of commercial building HVAC systems, and on-site power generation technology is needed to fully utilize Building Energy Analyzer potential	Interagency Software. Building Energy Analyzer, (http://www.interenergysoftware.com)
4	BuildingSim	Thermostat, simulation energy cost	Y	Self	Simulation step size is small, so simulation time takes longer than some competitors	Paragon Robotics. LLC, BuildingSim, (http://paragonrobotics.com/#website/home/free-tools/buildingsim/en-US.html).

(Continued)

TABLE 11.2 (*Continued*)

List of Building Energy Simulation Programs

S. No.	Building Energy Simulation Software	Applications	Open Source	Simulation Engine	Limitations	References
5	COMSOL Multiphysics	Solving 3-D heat PDE	N	Self	For solving coupled systems of ordinary differential equations (ODEs)	COMSOL, Inc. COMSOL Multiphysics, (http://www.comsol.com).
6	DesignBuilder	Building energy simulation, visualization, CO_2 emissions, solar shading, natural ventilation, daylighting, comfort studies, CFD, HVAC simulation, pre-design, early-stage design, building energy code compliance checking, OpenGL EnergyPlus interface, building stock modeling, hourly weather data, heating and cooling equipment sizing	Y	Self	A range of common HVAC systems is available Design-Builder user interface but users requiring a wider range of different HVAC types should consider exporting Energy Plus IDF input files and working in the Energy Plus IDF editor. An advanced HVAC interface to a wide range of Energy Plus HVAC systems will be available early 2007	Design Builder Software Ltd. Design Builder, (http://www.designbuilder.co.uk)
7	Designer's Simulation Toolkit (DEST)	Building simulation, design process, calculation, building thermal properties, natural temperature, graphical interfaces, state space method, maximum load	Y	Self	The speed of exporting reports is a little slow High Expertise required to use it.	Tsinghua University. Designer's Simulation Toolkit (DEST), (http://www.dest.com.cn).
8	DOE-2	Energy performance, design, retrofit Environmental	N	Self	High level of user knowledge	Lawrence Berkeley National Laboratory. DOE-2, (http://simulationresearch.lbl.gov).

(*Continued*)

TABLE 11.2 (*Continued*)

List of Building Energy Simulation Programs

S. No.	Building Energy Simulation Software	Applications	Open Source	Simulation Engine	Limitations	References
9	ECOTECT	Environmental design, environmental analysis, conceptual design, validation; solar control, overshadowing, thermal design and analysis, heating and cooling loads, prevailing winds, natural and artificial lighting, life cycle assessment, life cycle costing, scheduling, geometric and statistical acoustic analysis	N	Energy plus	As the program can perform many different types of analysis, the user needs to be aware of the different modeling and data requirements before diving in and modeling/ importing geometry. For example, for thermal analysis, weather data and modeling geometry in an appropriate manner are important; and appropriate/ comprehensive material data is required for almost all other types of analysis. The ECOTECT Help File attempts to guide/educate users about this and when/how it is important	Centre for Research in the Built Environment. Cardiff University, ECOTECT, (http://www.squ1.com).
10	EnerCAD	Building Energy Efficiency; Early Design Optimization; Architecture Oriented; Life Cycle Analysis	N (Free lower end versions for educational institutions)	Self	Simplified calculation algorithm (monthly heat balance method) Simplified Energy	Univ. of Applied Sciences of Western Switzerland. EnerCAD, (http://www.enercad.ch)
11	Energy expert	Energy tracking, energy alerts, wireless monitoring	N	Self	Energy Expert is used for optimizing the energy performance of existing buildings. It is not used for building design or energy simulation.	Northwrite. Energy Expert, (http://www.energyworksite.com)

(Continued)

TABLE 11.2 (*Continued*)

List of Building Energy Simulation Programs

S. No.	Building Energy Simulation Software	Applications	Open Source	Simulation Engine	Limitations	References
12	Energy plus	Energy simulation, load calculation, building performance, simulation, energy performance, heat balance, mass balance	Y	Self	Text input may make it more difficult to use than graphical interfaces.	U. S. Department of Energy. EnergyPlus, (http://www.energyplus.gov)
13	eQuest	Energy performance, simulation, energy use analysis, conceptual design performance analysis, LEED, Energy and Atmosphere Credit analysis, Title 24, compliance analysis, life cycle costing, DOE 2, PowerDOE, building design wizard, energy efficiency measure wizard	Y	DOE 2.2	Defaults and automated compliance analysis have not yet been extended from California Title 24 to ASHRAE 90.1. It does not yet support SI units (I-P units only). Ground-coupling and infiltration/natural ventilation models are simplified and limited. Daylighting can be applied only to convex spaces (all room surfaces have an unrestricted view of each surface) and cannot be transmitted (borrowed) through interior glazed surfaces. Custom functions in DOE-2.1E (allows users limited customization of source code without having to recompile the code) have not yet been made available in DOE-2.2 or eQUEST.	James J. Hirsch and Associates, eQUEST, (http://www.doe2.com)

(*Continued*)

TABLE 11.2 (*Continued*)

List of Building Energy Simulation Programs

S. No.	Building Energy Simulation Software	Applications	Open Source	Simulation Engine	Limitations	References
14	ESP-r	Energy simulation, environmental performance, commercial buildings, residential buildings, visualization, complex buildings and systems	N	Self	It is a general purpose tool and the extent of the options and level of detail slows the learning process. Specialist features require knowledge of the particular subject. Although robust and used for consulting by some groups, ESP-r still shows its research roots	University of Strathclyde. ESP-r, ⟨http://www.esru.strath.ac.uk/⟩
15	Facility energy decision system (FEDS)	Single buildings, multi-building facilities, central energy plants, thermal loops, energy simulation, retrofit opportunities, life cycle costing, emissions impacts, alternative financing	Y	None	Not a building's design tool	Pacific Northwest National Laboratory. Facility Energy Decision System (FEDS), ⟨http://www.pnl.gov/feds⟩
16	Heat, air, and moisture simulation laboratory (HAMLab)	Heat air and moisture, simulation laboratory, hygrothermal model, PDE model, ODE model, building and systems simulation, MATLAB, Simulink, Comsol, optimization Heat	Y	MATLAB-Simulink, COMSOL	MATLAB, Simulink, and FemLab needed. Some MATLAB expertise needed	Eindhoven University of technology. Heat, Air, and Moisture simulation Laboratory (HAMLab), ⟨http://archbps1.campus.tue.nl/bpswiki/index.php/Hamlab⟩
17	HEAT2	Heat transfer, 2D, dynamic, simulation	No	Delphi	Cartesian coordinates must be used (sloped boundaries are using "steps")	HEAT2, ⟨http://archbps1.campus.tue.nl/bpswiki/index.php/Hamlab⟩

(*Continued*)

TABLE 11.2 (*Continued*)

List of Building Energy Simulation Programs

S. No.	Building Energy Simulation Software	Applications	Open Source	Simulation Engine	Limitations	References
18	ParaSol	Solar protection, solar shading, windows, buildings, solar energy transmittance, solar heat gain coefficient, energy demand, heating, cooling, comfort, daylight Photovoltaic,	Y	None	Only one strategy for controlling sun shades is available	Lund Institute of Technology, ParaSol, (http://www.parasol.se).
19	PVcad	Photovoltaic, facade, yield, electrical; electrical engineers planning grid-connected photovoltaic systems, especially photovoltaic facades; CAD	Y	Self	Only a German-language version available, input of building geometry only via DXF import or with ASCII text containing alist of 3D-coordinates. Weather database not expandable by the user	ISET e V, PVcad, (http://www.iset.uni-kassel.de/pvcad)
20	RIUSKA	Energy calculation, heat loss calculation, system comparison, dimensioning, 3D modeling	N	DOE-2.1E	Only the predefined HVAC systems are available	OlofGranlundOy. RIUSKA, (http://www.granlund.fi)
21	SIMBAD Building and HVAC Toolbox	Transient simulation, control, integrated control, control performance, graphical simulation environment, modular, system analysis, HVAC	N	MATLAB Simulink	The tool does not currently offer a fully multi-zone building	CSTB (French Center for Building Sciences). SIMBAD Building and HVAC Toolbox, (http://ddd.cstb.fr/simbad)

(*Continued*)

TABLE 11.2 (Continued)

List of Building Energy Simulation Programs

S. No.	Building Energy Simulation Software	Applications	Open Source	Simulation Engine	Limitations	References
22	SPARK (Simulation problem analysis and research Kernel)	Object-oriented, research, complex systems, energy performance, short time-step dynamics	N	Self	High level of user expertise in system being modeled required	Lawrence Berkeley National Laboratory. Simulation Problem Analysis and Research Kernel (SPARK), (http://simulationresearch.lbl.gov)
23	TRNSYS	Energy simulation, load calculation, building performance, simulation, research, energy performance, renewable energy, emerging technology	N	Self	No assumptions about the building or system are made (although default information is provided) so the user must have detailed information about the building and system and enter this information into the TRNSYS interface	Thermal Energy System Specialists. TRaNsientSYstem Simulation Program (TRNSYS), (http://sel me.wisc.edu/trnsys)

11.3.1 Computational Fluid Dynamics

Numerical calculation using software is a cost-effective and efficient method to predict thermal behavior and performance of the facade in naturally ventilated buildings among various architectural designs. Simulation approaches for natural ventilation fall into two broad classifications: computational fluid dynamics (CFD) method and building simulation (BS) method. BS can model heat transfer and radiation processes in seconds based on heat balance methods and CFD can predict reliable detailed airflow for outdoor and indoor. CFD simulation offers comprehensive spatial distributions of air velocity, temperature, air pressure, contaminant concentration, and turbulence by numerically solving the governing equations of fluid flows. It is a trustworthy tool for the estimation of thermal environment and contaminant distributions. These results can be used to quantitatively analyze the indoor environment and determine facade system performances.

CFD is the tool by which we can model the heat transfer and fluid flow in the systems. The heat transfer is modeled using diffusion equation while mass and momentum transfer can be modeled using Navier–Stokes Equations [3,4]. CFD requires numerical modeling and analysis algorithm for solving and analyzing the heat transfer and fluid flow problem. The computers are required for the performance of numerical simulations for the interaction of fluid and gas to the solid structure. The flow chart for calculation is shown in Figure 11.5.

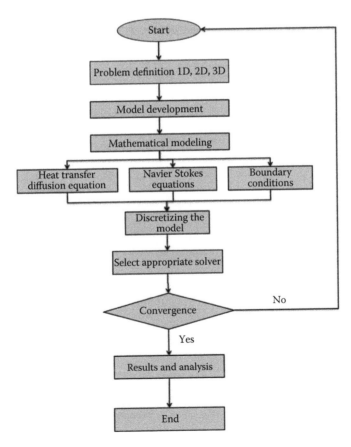

FIGURE 11.5
Flow diagram for CFD analysis.

Various software is applied for the CFD analysis, which is based on finite element analysis. Some of them are as follows:

11.3.1.1 COMSOL Multiphysics

COMSOL Multiphysics collects and solves models using state-of-the-art numerical analysis methods. Numerous methods are used in the add-on modules in this software, including finite element analysis, the finite volume method, the boundary element method, and particle tracing methods, but the emphasis of COMSOL Multiphysics is on the finite element method. Many types of finite elements are available, and fully coupled elements are automatically generated by the software at the time of solving. This patented method of generating finite elements "on-the-fly" is precisely what allows for unlimited Multiphysics combinations. Various studies had been carried out using COMSOL Multiphysics. The window for simulation using COMSOL Multiphysics is shown in Figure 11.6. The models are created using model builder section, where we can define the function, geometry, and so on. Several studies of the building are carried out using COMSOL Multiphysics. Hasse et al. [5] carried out a numerical simulation of phase change materials in honeycomb structure using COMSOL Multiphysics for thermal energy storage in buildings. Jacobs and Schijndel [6] carried out building energy simulation (BES) using BESTEST criteria. In order to predict, enhance and meet a definite set of performance requirements connected to the indoor climate of buildings and the accompanying energy demand, numerical simulation tools are indispensable. In this study, the authors consider two types of numerical simulation tools: Finite Element Method (FEM) and BES. A well-known benchmark case for BES,

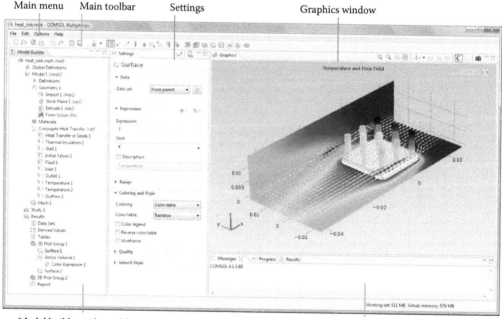

FIGURE 11.6
Heatsink modeling using COMSOL. (From COMSOL Multiphysics User's Guide, *Comsol*, 2015, http://www.comsol.com.)

the so-called BESTEST, was used to verify COMSOL (FEM) 3D simulation results. It was concluded that one of the main benefits of FEM–BES modeling exchange was the likelihood to simulate building energy performances with high spatial resolution.

11.3.1.2 *Fluent*

CFD is the science of envisaging fluid flow, heat and mass transfer, chemical reactions, and related phenomena by solving numerically the set of governing mathematical equations conservation of mass, conservation of mass, momentum, energy, species and effects of body forces. The results of CFD studies are relevant in conceptual studies of new designs, detailed product development, troubleshooting, and redesign. The Fluent is the world leader in the rapidly growing field of design simulation software used to predict fluid flow, heat and mass transfer, chemical reaction, and related phenomena. The Fluent software is used by engineers in corporations globally for the development of the product, design optimization, troubleshooting and retrofitting. With the application of Fluent and CFD, engineers are able to perform faster analysis at a reduced cost with the end result being an improved design, by solving the fundamental heat and mass transfer equations; Fluent offers information on important flow characteristics such as pressure loss, flow distributions, and mixing rates. Fluent modeling overview is represented in Figure 11.7. Liping et al. [3] carried out parametric analysis of facade designs on alignments, window-to-wall ratios and shading expedients were achieved for two usual weeks by coupled simulations between BS ESP-r and CFD (FLUENT). Four ventilation strategies: (1) night-time only ventilation, (2) daytime-only ventilation, (3) full-day ventilation, and (4) no ventilation,

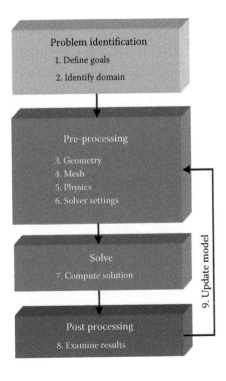

FIGURE 11.7
Fluent modeling overview.

were evaluated for the hot–humid climate. The results indicate that full-day ventilation for indoor thermal comfort is better than the other three ventilation strategies.

Wang et al. [4] discussed coupling strategies between BS and CFD for natural ventilation. The study was performed using two single zone cases to validate coupled simulation with full CFD simulation. The results were indicated that the coupled simulations taking pressure from BS as inlet boundary conditions can provide more accurate results for indoor CFD simulation than taking velocity from BS as boundary conditions.

11.3.2 Transient System Simulations

Transient System Simulations (TRNSYS) is a very flexible graphical user interface (GUI) environment software used for simulating the performance of transient systems. Although the majority of models are focused on evaluating the performance of thermal and electrical energy systems, TRNSYS can similarly be used to model other dynamic systems such as traffic flow or biological processes. TRNSYS is divided into two parts. The first one is an engine also called the kernel which reads and processes the input file, solves the system iteratively, determines convergence, and plots system variables. The kernel also provides utilities that (among other things) determine thermophysical properties, perform linear regressions, invert matrices, and interpolate external data files. The second part of TRNSYS contains a library of components, each of which models the performance of one part of the system. The standard library of TRNSYS includes approximately 150 models ranging from pumps to multi-zone buildings, wind turbines to electrolyzers, weather data processors to economics routines, and basic heating, ventilation, and air-conditioning (HVAC) equipment to cutting-edge emerging technologies. Models are constructed in such a way that users can modify existing components or write their own, extending the capabilities of the environment [8]. Various authors carried out modeling of the building and their components using TRNSYS.

Mei et al. [9] developed a dynamic thermal model of building with integrated ventilated PV façade/solar air collector. In this thermal model, ventilation air is preheated within façade by incident solar radiation. The heating and cooling load with and without facade-ventilated system were calculated and performance was also evaluated with impact on the environment. It was found that the cooling loads are marginally higher with the PV façade for all locations considered, whereas the impact of the facade on the heating load depends critically on location. Bony et al. [10] developed a numerical model for thermal energy storage using phase change materials in buildings. The model had been implemented in an existing TRNSYS type of water tank storage. Liu et al. [11] carried out numerical analysis incorporating the Phase Change Materials (PCM) component into the TRNSYS 16 simulation package [12] that was written in the FORTRAN programming language. The numerical model developed by Liu et al. [13] was written in FORTRAN 6.5 and the FORTRAN codes were linked to use with TRNSYS 16 by filling the codes into a TRNSYS template. Then the investigated phase change thermal storage unit became a TYPE in TRNSYS. Similar to other components (TYPEs) in TRNSYS, it requires a number of constant parameters and time-dependent inputs and produces a number of time-dependent outputs.

11.3.3 Energy Plus

Energy Plus is whole BS program that is applied by engineers, architects, and researchers to calculate energy consumption (for heating, cooling, ventilation, lighting, and plug and process loads) and water use in buildings. Some features of Energy Pulse software are as the following:

- Combined, a concurrent solution of thermal zone conditions and HVAC system response that does not accept that the HVAC organization can change zone loads and can simulate unconditioned and underconditioned spaces.
- Heat balance-based solution of radiant and convective effects that produce surface temperatures thermal comfort and condensation calculations.
- The user-defined time interval for the interface between thermal zones and the environment with automatically varied time steps for interactions between thermal zones and HVAC systems. These allow Energy Plus to model systems with fast dynamics while also trading off simulation speed for precision.
- Combined heat and mass transfer model that accounts for air movement between zones.
- Advanced fenestration models including controllable window blinds, electrochromic glazing, and layer-by-layer heat balances that calculate solar energy absorbed by window panes.
- Illuminance and glare calculations for reporting visual comfort and driving lighting controls.
- Component-based HVAC that supports both standard and novel system configurations.
- A large number of built-in HVAC and lighting control strategies and an extensible runtime scripting system for user-defined control.
- Functional Mockup Interface import and export for co-simulation with other engines.

11.3.4 Open Studio®

Open Studio® is a cross-platform (for Windows, Mac, and Linux) group of software tools to support whole-building energy modeling using Energy Plus and advanced daylight analysis using Radiance. Open Studio is an open source project to facilitate community development, extension, and private sector adoption. Open Studio includes graphical interfaces along with a Software Development Kit (SDK).

11.3.5 Radiance

Radiance is envisioned to aid illumination designers and architects by forecasting the light levels and appearance of a space prior to construction of buildings. The package includes programs for modeling and translating scene geometry, luminaire data, and material properties, all of which are needed as input to the simulation. The lighting simulation itself uses ray tracing techniques to compute radiance values (i.e., the quantity of light passing through a specific point in a specific direction), which are typically arranged to form a photographic quality image. The resulting image may be analyzed, displayed, and manipulated within the package, and converted to other popular image file formats for export to other packages, facilitating the production of hard copy output. Radiance contains many of the features of popular computer graphics rendering programs with the physical accuracy of an advanced lighting simulation. This combination of flexibility and accuracy makes it unique in providing realistic images with predictive power for architects, engineers, and lighting designers. Its flexibility is demonstrated in applications as diverse as forensics (i.e., roadway accident re-enactment by Failure Analysis and Associates, CA) and aerospace (i.e., space station design by NASA, Goddard). Radiance has been compared to other lighting calculations, scale model measurements, and real spaces to validate its capabilities. No other lighting calculation has undergone a more rigorous validation. Many others have attempted to develop a lighting simulation system with similar capabilities, and there have been few notable successes. At least three major European firms (Ove Arup in London, Abacus Simulations in Glasgow, and Siemens Lighting in Germany) have abandoned their own in-house lighting simulations in favor of using Radiance.

11.3.6 eQUEST

eQUEST was originally developed by the US Department of Energy which allows importing building geometry from architectural models or constructing a building envelope within the program. eQUEST can perform simple simulations or very complex models. There are three input wizards in eQUEST that all have differing levels of complexity, or the comprehensive DOE-2 interface can be used. There are several important file types in eQUEST that advanced users can edit these with text editors.

   PD 2—*Building inputs for wizards*: The data in this file is data entered by the user and not the defaults.

   PRD—*Parametric run definitions*: This file contains the options when you run parametric runs, for instance when comparing glass types.

   SIM—*Large output files* (*text*): This can be viewed in eQUEST when we want to view the advanced report.

   INP—Building inputs (Created by wizards).

11.3.7 TRANE TRACE 700

TRACE 700 is a Windows-based software program used to design a virtual building, calculate its air-conditioning loads, and simulate its hourly operation over the course of one year. It can also implement a life cycle cost analysis. TRACE 700 simulates a virtual building, but it does not show a visual image of the building. While this is a commonly requested feature, nearly all surfaces in TRACE are currently entered by two dimensions. A visual image requires additional input and thus more work. TRACE 700 is first a load and then an energy calculation program, while all energy-simulations must first calculate building loads; most energy modeling packages do not precisely determine building design. Therefore, TRANE TRACE 700 can be hand-me-down in both the design phase of a project and the analysis phase. Approximately all kits in TRACE are customizable. Although, the paper will go into more detail later, for now, it will be satisfactory to say that all equipment, materials, and programs are 100% customizable and can be created, edited, and reserved in the TRACE 700 library folder. TRANE TRACE has the capability to model over 33 different airside systems, plus many HVAC plant configurations and control strategies, including thermal storage, cogeneration, fan pressure optimization, and daylighting controls. Customizable libraries and templates simplify data entry and allow greater modeling accuracy.

Documentation of this software comprises comprehensive online help and a printed modeling guide. Skilled HVAC engineers and support experts provide free technical support. Templates provide a fast, easy way to analyze the effects of variations in building loads, air flows, thermostat settings, occupancy, and construction. A widespread library of construction materials, equipment, and weather profiles (nearly 500 locations) enhances the speed and accuracy of your analyses. Choose from seven different American Society of Heating Refrigeration and Air Condition (ASHRAE) cooling and heating methodologies, including the Exact Transfer Function.

11.3.8 Simergy

This software is a GUI software for the simulation using the Energy Plus simulation engine. Both tools were developed by the U.S. Department of Energy. Simergy allows users to import geometry from Building Information Modeling (BIM) software or 2-D CAD, along with the ability to create geometry directly. The conversion of geometry from BIM to Building Energy Models (BEM) has always been a challenge, and Simergy presents an emerging solution for this, as well as allowing added workflows:

Workflow # 1: Industry Foundation Class (IFC) 2.3.2 model view definition for BIM to bring building geometry and supplementary information into Simergy.

Workflow # 2: Build geometry from imported floor plans.

Workflow # 3: Build your own geometry utilizing the Simergy tools to develop models quick and easily from the earliest stages.

Conversion of geometry from BIM to BEM has always been a challenge. Now there are different workflows established on IFC Design concept BIM to bring building geometry and supplementary information into Simergy. In addition, users can build geometry from base floor plans or build their own geometry to develop new designs from the earliest stages.

Whether we choose to construct our HVAC schemes from scratch; use the system templates distributed with Simergy; or use custom system templates, the drag and drop capabilities facilitate your work. The performance characteristics at component level can envisage and integrity of the system design can check at any point by using this software.

11.3.9 FORTRAN

The concurrent partial differential equations describing heat and mass transfer during heating and cooling in the buildings can be modeled and these nonlinear partial differential equations can be solved using the finite difference method (FEM) in FORTRAN. FORTRAN is used for analysis of building energy efficiency that can help to reduce the energy consumption.

11.3.10 MATLAB

The thermal model was developed using MATLAB® Simulink® by Mathworks [14]. This model calculates heating costs for a generic house. When the model is opened, it loads the information about the house from the sldemo_househeat_data.m file. The file does the following:

- Defines the house geometry (size, number of windows)
- Specifies the thermal properties of house materials
- Calculates the thermal resistance of the house
- Provides the heater characteristics (temperature of the hot air, flow-rate)
- Defines the cost of electricity (0.09$/kWhr)
- Specifies the initial room temperature (20°C = 68°F)

Figure 11.8 represents the thermal model of building in Simulink. The thermal model contains the following set of points:

11.3.10.1 Set Point

"Set Point" is a constant block. It specifies the temperature that must be maintained indoors. It is 70°F by default. Temperatures are given in Fahrenheit but then are converted to Celsius to perform the calculations.

11.3.10.2 Thermostat

"Thermostat" is a subsystem that contains a Relay block. The thermostat allows fluctuations of 5°F above or below the desired room temperature. If air temperature drops below 65°F, the thermostat turns on the heater. See the thermostat subsystem in Figure 11.8d.

11.3.10.3 Heater

"Heater" is a subsystem that has a constant air flow rate, "Mdot," which is specified in the sldemo_househeat_data.m file. The thermostat signal turns the heater on or off. When the heater is on, it blows hot air at Temperature T_{Heater} (50°C = 122°F by default) at a constant flow rate of Mdot (1 kg/sec = 3600 kg/hr by default). The heat flow into the room is expressed by the following equation:

$$\frac{dQ}{dt} = (T_{heater} - T_{room}).Mdot.c$$

$\frac{dQ}{dt}$ is the heat flow from the heater into the room

T_{heater} is the temperature of hot air from the heater

T_{room} is the current room air temperature

Mdot is the air mass flow rate through the heater (kg/h)

c is the heat capacity of air at constant temperature

The heater subsystem represents in Figure 11.8b.

11.3.10.4 Cost Calculator

"Cost Calculator" is a Gain block. "Cost Calculator" integrates the heat flow over time and multiplies it by the energy cost. The cost of heating is plotted in the "Plot Results" scope.

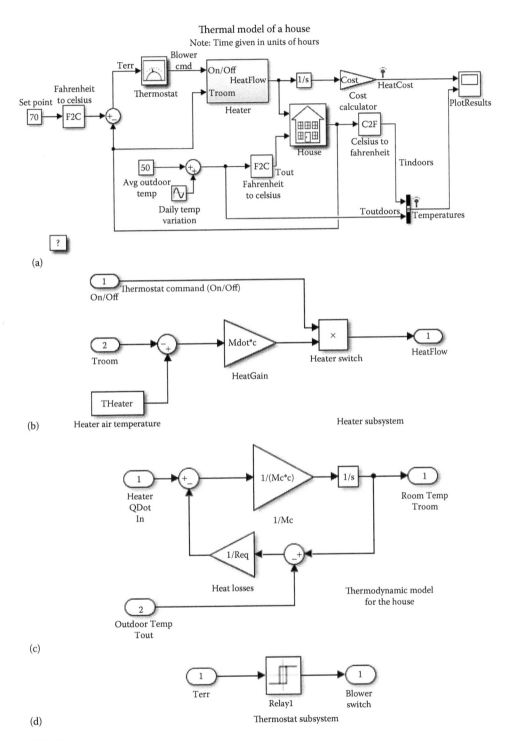

FIGURE 11.8
(a) Thermal model of buildings; (b) the heater subsystem; (c) the house heating model; (d) the "thermostat" subsystem. (From https://in.mathworks.com/help/simulink/examples/thermal-model-of-a-house.html?requested Domain=www.mathworks.com#zmw57dd0e1093.)

11.3.10.5 House

"House" is a subsystem that calculates room temperature variations, which is represented in Figure 11.8c. It takes into consideration the heat flow from the heater and heat losses to the environment. Heat losses and the temperature time derivative are expressed by the following equation:

$$\left(\frac{dQ}{dt}\right)_{losses} = \frac{T_{room} - T_{Out}}{R_{eq}}$$

$$\frac{dT_{room}}{dt} = \frac{1}{M_{air} \cdot c}\left(\frac{dQ_{heater}}{dt} - \frac{dQ_{losses}}{dt}\right)$$

M_{air} is the mass of air inside the house
R_{eq} is the equivalent thermal resistance of the house SPSS

This is an analytical tool for nonlinear regression analysis. SPSS is used for fast and accurate analytical analysis such as coefficient of determination, root mean square error, and analysis of variance, reduced chi-square, and so on. This is a predictive analytics software with advanced techniques in easy-to-use methods:

Statistical analysis and reporting: Address the entire analytical process, planning, data collection, analysis, reporting, and deployment.

Predictive modeling and data mining: Use powerful model-building, evaluation, and automation capabilities.

Decision management and deployment: Analytics with advanced model management and analytic decision management.

Big data analytics: Analyze big data to gain predictive insights.

11.4 Conclusion

Energy modeling in the building using simulation software can enable and improve the design of an EEB. Buildings account for roughly 40% of global energy consumption, and by extension, are major emitters of CO_2 that is the main motive for designing an EEB. In addition, what speaks to most clients, is the motive to reduce a building's operating costs in the face of rising energy prices. Any building design that aims to improve energy performance will benefit from energy and/or airflow simulations.

Along with the development of materials and construction techniques, energy simulation software tools of buildings have also developed over the years. Currently, there are several energy simulation software tools with different levels of complexity and response to different variables. Among the available simulation software tools are the CFD, COMSOL Multiphysics, Fluent, Transient System Simulations (TRNSYS), Energy Plus, Open Studio®, RADIANCE, REQUEST, TRANE TRACE 700, SIMERGY, FORTRAN, and MATLAB.

References

1. EIA. World total primary energy consumption by region. International Energy Outlook 2016, 2016.

2. Harish VSKV, Kumar A. A review on modeling and simulation of building energy systems. *Renewable and Sustainable Energy Reviews* 2016;56:1272–1292.

3. Liping W, Hien WN. The impacts of ventilation strategies and facade on indoor thermal environment for naturally ventilated residential buildings in Singapore. *Building and Environment* 2007;42:4006–4015. doi:10.1016/j.buildenv.2006.06.027.

4. Wang L, Wong NH. Coupled simulations for naturally ventilated rooms between building simulation (BS) and computational fluid dynamics (CFD) for better prediction of indoor thermal environment. *Building and Environment* 2009;44:95–112. doi:10.1016/j.buildenv.2008.01.015.

5. Hasse C, Grenet M, Bontemps AA, Dendievel RR, Sallee H, Sallée H. Realization, test and modelling of honeycomb wallboards containing a phase change material. *Energy and Buildings* 2011;43:232–238. doi:10.1016/j.enbuild.2010.09.017.

6. Jacobs DPM, Van Schijndel AWM. COMSOL Multiphysics for building energy simulation (BES) using BESTEST criteria. *Proceedings of the 2015 COMSOL Conference in Grenoble* 2015;1:1–6.

7. COMSOL Multiphysics User's Guide. *Comsol*, http://www.comsol.com (Accessed: April 1, 2017).; Heat trans.

8. Trnsys. Welcome|TRNSYS: Transient System Simulation Tool 2017, http://www.trnsys.com/ (Accessed: April 1, 2017).

9. Mei L, Infield D, Eicker U, Fux V. Thermal modelling of a building with an integrated ventilated PV façade. *Energy and Buildings* 2003;35:605–617. doi:10.1016/S0378-7788(02)00168-8.

10. Bony J, Citherlet S. Numerical model and experimental validation of heat storage with phase change materials. *Energy and Buildings* 2007;39:1065–1072. doi:10.1016/j.enbuild.2006.10.017.

11. Liu M, Bruno F, Saman W. Thermal performance analysis of a flat slab phase change thermal storage unit with liquid-based heat transfer fluid for cooling applications. *Solar Energy* 2011;85:3017–3027. doi:10.1016/j.solener.2011.08.041.

12. TRNSYS. TRNSYS 16 A transient system Simulation program, 2004, http://www.trnsys.de/download/de/trnsys_kurzinfo_de.pdf (Accessed: April 1, 2017).

13. Liu M, Saman W, Bruno F. Validation of a mathematical model for encapsulated phase change material flat slabs for cooling applications. *Applied Thermal Engineering* 2011;31:2340–2347. doi:10.1016/j.applthermaleng.2011.03.034 (Accessed: April 1, 2017).

14. Thermal model of a house—MATLAB and Simulink example—MathWorks, MathWorks India, https://in.mathworks.com/help/simulink/examples/thermal-Model-of-a-house.html?requestedDomain=www.mathworks.com#zmw57dd0e1093 2017 (Accessed: April 1, 2017).

12

Advances in Energy-Efficient Buildings for New and Old Buildings

A.K. Chaturvedi, Siddartha Jain, Deep Gupta, and Mridula Singh

CONTENTS

12.1 Introduction 236
    12.1.1 What Is Energy Efficiency of a Building? 236
    12.1.2 Energy Consumption in Buildings..236
    12.1.3 Changing Climate—A Challenge 237
12.2 Energy Efficiency in Buildings Methodology..238
    12.2.1 Typical Energy Flow in Buildings 238
    12.2.2 Determining a Building's Energy Performance 239
    12.2.3 Benchmarking Energy Efficiency 240
    12.2.4 Certifying Energy Efficiency 240
    12.2.5 Components of Energy-Efficient Building.......................................241
        12.2.5.1 Automated Controlled Sensors 241
        12.2.5.2 Energy-Efficient Materials..242
        12.2.5.3 Building Orientation..243
12.3 Heating Cooling Solution in Energy-Efficient Building...........................244
    12.3.1 The Heating and Cooling Roadmap Vision 244
    12.3.2 Targets and Assumptions for the Energy-Efficient Buildings.....244
    12.3.3 Technology Development for Energy-Efficient Buildings.............246
    12.3.4 Active Solar Thermal 246
    12.3.5 Combined Heat and Power..247
    12.3.6 Heat Pumps...248
        12.3.6.1 Equipment and Components: Decrease Costs and Increase
                Reliability and Performance through More Efficient
                Components..249
    12.3.7 Thermal Energy Storage..250
12.4 Indian Government Policies and National Missions on Energy...............252
12.5 Case Studies...253
12.6 The Future...254
12.7 Conclusion..255
References...256

12.1 Introduction

12.1.1 What Is Energy Efficiency of a Building?

Energy is needed to meet the demands for cooling, heating, ventilation, and lightning. Normally no attention is paid to the aspect of energy consumed during the construction or to the aspect of energy consumption to run that building during its life span. This approach causes severe depletion of invaluable environmental resources. Thus there is a need to identify areas that need to be addressed to arrive at a solution in which levels of energy and resources consumption can be reduced to meet the occupant's necessity and their thermal comfort without compromising the efficiency and effectiveness of the systems entailed during the construction and during the maintenance phase. Therefore, an idea for energy-efficient design is to be adapted such that the conditions within the building are as close as possible to the comfort zone. To attain the same amount of work for the same duration, the energy efficiency of a building depends on the consumption of a reduced amount of energy during its functioning. Estimation of energy and its consumption has emerged as main concern for the policy-makers/designers. *The ratio of energy input to the calculated or estimated amounts of energy required to cover the various requirements relating to the standardized use of a building serves as the measure of energy efficiency.* As shown by different studies, energy-efficient buildings (EEBs) are capable of saving economic cost as well as greenhouse emissions too. The main sources of energy in today's world are heavily dependent on the availability of oil, coal, and natural gas. These resources not only emit greenhouse gases (GHGs), but they are also nonrenewable; that is, their quantities are limited, and they cannot be replaced as fast as they are consumed. These reasons lend themselves to use of more energy-efficient methods.

Though the words "green" and "energy efficient" are often used interchangeably, there are several differences between them, meaning that a "EEB" is not always "green." A building is green when it helps to reduce the footprint it leaves on the natural environment and on the health of its inhabitants. Green building design includes the use of renewable energy sources such as wind, water, or solar; creating a healthy indoor environment; implementing natural ventilation systems; and using construction materials that minimize the use of volatile organic compounds (VOCs) in the home. On the contrary the use of materials and resources that have low embodied energy, and produce a minimal environmental impact are key elements in an EEB.

12.1.2 Energy Consumption in Buildings

Buildings are currently responsible for more than 40% of global energy and one third of the global GHG emissions. It would also be relevant to note that the emissions from the building sector are on the rise. During the period 1971 and 2004, the carbon dioxide (CO_2) emissions grew at the rate of 2.5% per year for commercial buildings and 1.7% per year for residential buildings (Levine et al. 2007). According to the Intergovernmental Panel on Climate Change's (IPCC's) fourth assessment report, the potential for GHG reduction from buildings is common to both developed and developing countries. This is for the following reasons:

- *Lack of awareness about the low-cost energy efficiency measures (EEM)*: There needs to be a greater awareness among stakeholders about low-cost EEMs, which are proven to be equal to or more effective than the high-cost technologies.

- *Lack of understanding with respect to indicators to measure energy performance in buildings*: Most building occupants have little or no information about the energy-saving potentials of their building. A lack of clear understanding about the indicators to measure energy consumption makes it difficult to gauge savings from energy efficiency improvements. Without a good understanding of the existing type, size, age, construction, and so on, of buildings and what kind of emissions currently are being generated, it is difficult to design and implement suitable reduction policies.

Therefore it can be concluded that with the proven and commercially available technologies, the energy consumption in both new and existing buildings can be cut by an estimated 30%–80% with potential net profit during the building life span (UNEP SBCI 2009).

The energy consumption during the operational phase of a building depends on a wide range of interrelated factors, such as climate and location; level of demand, supply, and source of energy; function and use of building; building design; construction materials; the level of income and behavior of its occupants. The level of GHG emissions from buildings is closely correlated with the level of demand, supply, and source of energy. India had committed to a target of 33%–35% emissions intensity reduction by 2030 ahead of the 2015 United Nations Framework Convention on Climate Change (UNFCCC) climate talks. Clean energy is the centerpiece of India's commitment (UNFCCC 2015). India is committed to strengthening its comprehensive approach based on the National Action Plan on Climate Change (NAPCC) that was released in 2008, through its key missions on energy efficiency and solar energy. Energy efficiency in building sector is a new approach with cleanest and cheapest techniques to meet energy needs. If India could improve energy efficiency in buildings by introducing efficient ways of design, construction technologies and use of correct construction material, then it would be able to save approximately U.S.\$42 billion per year (McKinsey 2009).

12.1.3 Changing Climate—A Challenge

For an EEB professional, there is a fundamental and mandatory part of his training: an understanding of the concept of adaptation with climate change that will play a continuous effort in reducing the GHG emissions from the built up environment while increasing the adaptation strategies (USEPA 2011). Generally climate data is used for informing the different decisions in the design, construction, and operation of the built environment. These decisions include the selection of systems for heating, ventilation, and air-conditioning (HVAC), tree and plant species for landscaping and appropriate building materials.

Many countries are engaging in comprehensive climate adaptation planning. These countries consider various cities of different size and in different geographic locations that are threatened by the natural/manmade hazards like sea level rise in Alaska or an aging infrastructure in some of the larger cities like Boston, Massachusetts, and so on. The following milestones would be a part of the building adaptability with climate change:

1. Thorough study of climate flexibility
2. Adaptation index (related to rise in earth temperature)
3. Develop a climate preparedness plan for most affected cities

Considering various scales for climate adaptation planning at the project level would be an important resource for understanding the local impacts and identifying citywide priorities for increased resiliency. Green building and climate adaptation strategies must be applied to existing buildings as well as new building projects. Climate change will have a significant impact on the effectiveness of regional systems. More resilient regional systems will maintain efficiencies across systems and limit conflict.

Climate changes have a great impact on energy systems; it will affect both energy supply (generation, transmission, and distribution) and demand. The increased variability with global warming, rise in temperature, and climate change could result in more efficient utilization of energy resources (Amato et al. 2005). Climate projections indicate warming trends during both winter and summer seasons. As a result, effects on energy demand will include small reductions in space heating but substantial increases in space cooling (USCCSP 2008).

12.2 Energy Efficiency in Buildings Methodology

12.2.1 Typical Energy Flow in Buildings

The energy flow of a building refers to the process of the energy transfer between a building and its surroundings. The energy flow is estimated for two types of buildings: conditioned building, in which flow is estimated for the heating and cooling load; and for a nonconditioned building; energy flow is calculated for the temperature variation inside the building over a specified time. These quantifications enable us to determine the effectiveness of the design of a building and help in evolving improved designs for EEBs with comfortable indoor conditions.

It may be noted that the heat flows by conduction through various building elements such as walls, roof, ceiling, floor, and so on. Heat transfer also takes place from different surfaces by convection and radiation. Besides, solar radiation is transmitted through the transparent windows and is absorbed by the internal surfaces of the building (Figure 12.1). There may be evaporation of water resulting in a cooling effect. Heat is also added to the space due to the presence of human occupants and the use of lights and equipment. Due to metabolic activities, the body continuously produces heat, part of which is used as work, while the rest is dissipated into the environment for maintaining body temperature. The body exchanges heat with its surroundings by convection, radiation, evaporation, and conduction. Movement of air affects the rate of perspiration, which also affects body comfort. Therefore it can be summarized that the thermal performance of a building depends on a large number of factors. Some of these are the following:

1. Design variables (geometrical dimensions of building elements such as walls, roof and windows, orientation, shading devices, etc.)
2. Material properties (density, specific heat, thermal conductivity, transmissivity, etc.)
3. Weather data (solar radiation, ambient temperature, wind speed, humidity, etc.)
4. A building's usage data (internal gains due to occupants, lighting and equipment, air exchanges, etc.)

The influence of these factors on the performance of a building can be studied using appropriate analytical tools. Several techniques are available for estimating the performance of buildings. They can be classified under Steady-State methods, Dynamic methods, and Correlation methods.

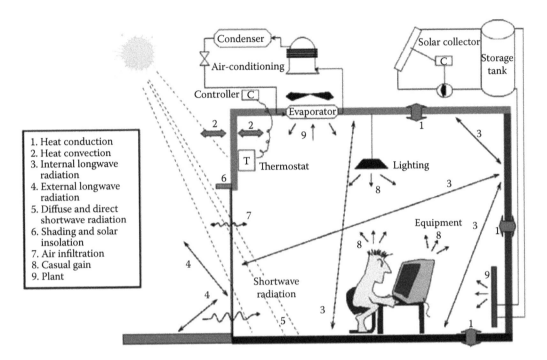

1. Heat conduction
2. Heat convection
3. Internal longwave radiation
4. External longwave radiation
5. Diffuse and direct shortwave radiation
6. Shading and solar insolation
7. Air infiltration
8. Casual gain
9. Plant

FIGURE 12.1
Energy flow in buildings.

12.2.2 Determining a Building's Energy Performance

The major concern is how to measure energy efficiency? A wide range of building calculation and modeling tools exist, some with multiple levels of application and others designed for particular areas of analysis. Advancements in science, technology, and industry (as well as growing awareness of the need to respect and preserve environmental resources) have contributed to the development of tools that calculate, model, and simulate various aspects of building performance.

The Department of Energy (DoE) of the United States of America offers a directory on their website highlighting a continuously expanding list of building performance software. Currently 332 tools on the site are available; ranging from those that work at the whole building level to those that analyze specific materials, components, equipment, and systems. Programs vary in their levels of accuracy, required effort, and cost.

When looking at energy efficiency at the whole building level, one is essentially interested in thermal performance. In considering efficiency, designers and engineers increasingly make use of energy programs that model and predict annual building energy consumption in terms of energy units (e.g., BTUs), financial cost, or environmental impact.

Several simulation tools are available to estimate energy efficiency of a building:

1. Therm Version 1.0 (Thermal Evaluation Tool for Buildings) which evaluates the thermal performance of a passive or partly air-conditioned building, by calculating the hourly floating inside air temperature.
2. TADSIM (Tools for Architectural Design and Simulation) has been developed as a computer interface between building design and simulation software. The basic

philosophy is to ensure that architects can make use of simulation tools quickly and efficiently.

3. *TRNSYS*: Is a dynamic simulation tool for estimating the performance of any solar thermal system, that is, it can estimate the performance of a building, a solar photovoltaic system, and solar domestic hot water system. It is one of the most widely used commercially available tools for building simulation.

4. *DOE-2.1E*: Predicts the hourly energy use and energy cost of a building.

5. *Energy plus*: Is a modular, structured software tool based on the most popular features and capabilities of BLAST and DOE-2.1E. It is primarily a simulation engine; input and output are simple text files.

6. *EQUEST*: Is an easy-to-use building energy use analysis tool that provides professional-level results with an affordable level of effort. This is accomplished by combining a building creation wizard, an EEM wizard, and a graphical results display module, with an enhanced DOE-2.2-derived building energy use simulation program

12.2.3 Benchmarking Energy Efficiency

Benchmarking energy efficiency is an important tool to promote the efficient use of energy in commercial buildings. Benchmarking is the practice of comparing the measured performance of a device, process, facility, or organization to itself, its peers, or established norms, with the goal of informing and motivating performance improvement. When applied to building energy use, benchmarking serves as a mechanism to measure energy performance of a single building over time, relative to other similar buildings, or to modeled simulations of a reference building built to a specific standard (such as an energy code). Benchmarking models are mostly constructed in a simple benchmark table (percentile table) of energy use, which is normalized with floor area and temperature.

Benchmarking is useful for state and local government property owners and facility operators, managers, and designers. It facilitates energy accounting, comparing a facility's energy use to similar facilities to assess opportunities for improvement, and quantifying/verifying energy savings. A planned approach to benchmarking helps create a more practicable and utilizable benchmarking system. The plan should agree on the purpose for the benchmarking program and the intended audience for the program results. Some of the steps which provide a framework for designing a benchmarking plan are the following: establishing the goal for benchmarking; securing buy-in from leadership; building a benchmarking team; identifying output metrics; identifying data inputs; selecting a benchmarking tool; determining the collection method; considering a data verification process; evaluating analysis techniques; communicating the plan and finally planning for a change.

12.2.4 Certifying Energy Efficiency

The Energy Efficiency Certification Scheme (EECS) gives assurance that the best energy efficiency outcome for your building has been planned. It certifies professionals that have the skills and experience to lead and manage all types and scale of building energy upgrades, up to and including an Integrated Building Energy Retrofit (IBER) and to work effectively with their clients. A Building Energy Efficiency Certificate (BEEC) puts out the energy efficiency rating of a building or an area of a building, which consists of two parts:

Part 1 consists of a National Built Environment Rating System (NBERS) for offices rating for the building —it provides for offices rating information on the building's energy efficiency. *Part 2* consists of a CBD Tenancy Lighting Assessment (TLA) for the area of the building that is being sold, leased or subleased. The TLA is an assessment of tenancy lighting that measures the power density of the installed general lighting system. Australia has a very successful program on the subject in the form of National Australian Built Environment Rating System (NABERS), which is worth emulating.

12.2.5 Components of Energy-Efficient Building

12.2.5.1 Automated Controlled Sensors

One of the techniques, to achieve the efficiency of the building is to develop a system that controls a process by highly automated means using electronic devices and reducing human intervention to a minimum. Automation has been achieved by various means including mechanical, hydraulic, pneumatic, electrical, electronic devices and computers, usually in combination. Building intelligence starts with monitoring and controlling information services known as Smart Building Automation System (SBAS). Smart building automation project is an integrated building solution system that facilitates lighting control, heating, air-conditioning (HVAC), and access control to share information and strategies with an eye to reduce energy consumption, improve energy efficiency management, provide value-added functionality, and make the building easier to operate.

The objective of an integrated design process is to achieve a cost-effective, energy-efficient design for a building. An integrated system can provide a level of occupant control.

Intelligent building, with the use of automated control system such as SBAS, enables both building owners and occupants to enjoy the benefits of financial gain and enhance comfortable accommodation/management quality. Building automation and control systems rely on many sensors and actuators placed at different locations throughout a building. Reducing the power consumption of a modern building requires continuous monitoring of various environmental parameters inside and outside the building.

SBAS includes a compilation of sensors that determine the situations or position of parameters that need to be controlled, such as lighting, temperature, relative humidity, and pressure. The SBAS enables intelligent features like air flow control. This regulates the flow of air as per occupancy of the room. For example, the flow of air increases when the thermostat senses the increase in temperature due to increased higher occupancy within the room. The sensor is connected to a duct which will be opened and closed on the basis of output of the thermostat. This may also regulate the fan speed with respect to increase and decrease in temperature. In a similar fashion, the lighting can also be controlled using occupancy and motion through specially installed detection sensors. These automated controlled systems have a large impact on energy saving and cost reduction of running expenditure. Also, these systems are easy to monitor and easy to maintain when installed. However, installation cost of such systems may be initially more, but in view of the reported lower failure rates and with the improvement in the availability of a responsive service network, quick and effective service can be provided to occupants and users.

Electronic equipment embedded with modern technology such as Internet of things (IoT) is a need for the EEBs (Figure 12.2). Use of IoT, into the inbuilt embedded systems or equipment can be controlled, which leads to greater efficiency. Two different approaches using automated sensors and smart materials can be utilized. The former is totally technology driven while the latter is controlling the physical parameters of the sources of energy associated with building.

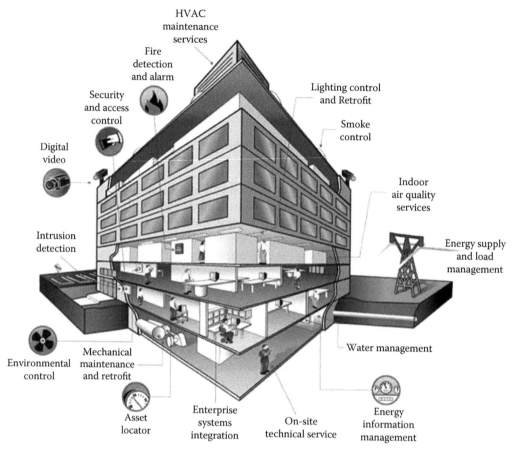

FIGURE 12.2
Segments of smart buildings driven by automated sensors.

12.2.5.2 Energy-Efficient Materials

The construction field has the largest potential for delivering long-term, significant, and cost-effective GHG emissions (Omer 2003). An energy-efficient home is now no longer built by merely using bricks and mortar; they are constructed from a variety of different materials. Every homeowner should take advantage of the new eco-friendly technological advances in home building, because they're affordable and more efficient. Below are some of the energy-efficient materials:

1. *Vacuum insulation panels (VIP)*: The Vacuum insulation panels comprise a textured silver rectangle that encloses a core panel in an airtight envelope. Due to this framework there will be minimum heat loss.

2. *Structural insulated panels (SIP)*: These panels are manufactured from an insulated layer of foam which is sandwiched between plywood or cement panels. It is fire resistant and suitable for floors, basements, foundations, as well as load bearing walls.

3. *Cool roof*: The ancient technology of improving heat dissipation and lowering temperatures in summers by increasing the reflectiveness of the roof surface is an effective method of energy optimization.

4. *Insulating concrete forms*: This technique uses poured concrete between two insulating layers and left in place. It can be used for free-standing walls and building blocks.

5. *Plant-based polyurethane foam*: The natural form of insulating material obtained from plant-based polyurethane such as bamboo, hemp, and kelp, offers high resistance to moisture and heat and protects against mold and pests.

6. *Recycled steel*: Recycling being one the most cost effective methods in any energy design and management process can be used for the construction industry too. If recycled steel is used, it will take just 6 scrap cars to serve the same purpose. Steel is a very durable material and particularly useful in areas where there are earthquakes and high winds.

7. *Reflective indoor and outdoor coatings*: By reflecting light better than normal paints, these coatings maximize the feeling of space and illumination. This allows reduction in the amount of energy used for artificial lighting and/or increases the perceived illumination by natural light.

8. *Phase change materials (PCM)*: The phase change materials are those which are used such that it enables walls and ceilings to absorb and store excessive heat during the day in order to dissipate that excessive heat during the night when air temperatures have gone down or vice versa. PCM basically increases the thermal inertia of the wall and ceilings, making them behave like the old-fashioned thick stone walls found in buildings of hundreds of years ago.

12.2.5.3 Building Orientation

Building orientation refers to the way a building is situated on a site and the positioning of windows, rooflines, and other features. A building oriented for solar design takes advantage of passive and active solar strategies. Passive solar strategies use energy from the sun to heat and illuminate buildings. Building orientation and building materials also facilitate temperature moderation and natural day lighting. Active solar systems use solar collectors and additional electricity to power pumps or fans to distribute the sun's energy. Heat is absorbed and transferred to another location for immediate heating or for storage for use later. Passive solar heating makes use of the building components to collect, store, distribute, and control solar heat gains to reduce the demand for fossil fuel powered space heating. Passive solar heating strategies also provide opportunities for day lighting and views to the outdoors through well-positioned windows. The goal of passive design is to maximize solar gain while minimizing conductance. Passive cooling removes or rejects heat from the building, keeping temperatures cool. Avoiding any mechanical operations to moderate temperature achieves energy and cost savings by alleviating the cooling load demanded. Shading devices can also reduce unwanted solar gains by blocking the sun during the summer months, while natural ventilation, which relies on natural airflow and breezes, can reduce the need for mechanical cooling when the building is occupied (see Glare and Heat Gain Reduction strategy).

12.2.5.3.1 How to Optimize Building Orientation

It is best to incorporate passive solar systems into a building during the initial design. Passive solar systems utilize basic concepts incorporated into the architectural design of the building. They usually consist of the following: rectangular floor plans elongated on an east-west axis; glazed south-facing wall; thermal storage medium exposed to the solar

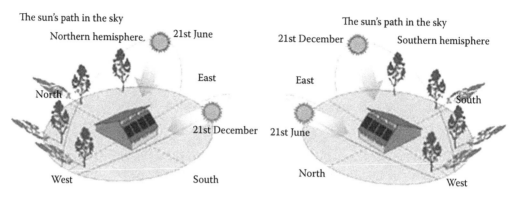

FIGURE 12.3
Optimization of orientation of building for different geographical areas and seasons.

radiation; light shelves/overhangs or other shading devices which sufficiently shade the south-facing elevation from the summer sun; south elevation overhangs that are horizontal while east and west elevations usually require both horizontal and vertical overhangs; windows on the east and west walls, and preferably none on the north walls (Figure 12.3). A correctly orientated building can save a lot of money which will not be required any longer while doing heating and cooling costs expenditure—in effect the building itself maintains a comfortable environment for the occupant with little additional costs. This is especially relevant with rising fuel bills and the increasing costs of electricity. By simply building this way, a house can reduce its heating and cooling costs by almost 85%.

12.3 Heating Cooling Solution in Energy-Efficient Building

12.3.1 The Heating and Cooling Roadmap Vision

For an energy-conscious regulatory authority in the building sector, it is highly important that a vision for future is stated. Such an approach will help the authority to work out a framework of policies, which lend themselves to right sizing the optimal utilization of resources to achieve conservation of energy without resulting into reduction of comfort levels and without causing excessive increase in the cost of construction and subsequently maintenance of the structures. One suggested vision could be as follows:

To accelerate the widespread adoption of energy-efficient heating and cooling technologies worldwide between now and 2050 in order to achieve significant reductions in energy demand, CO_2 and other pollutant emissions, and energy bills with a view to shift the buildings sector to a more sustainable future.

12.3.2 Targets and Assumptions for the Energy-Efficient Buildings

Energy efficiency options are available in the buildings sector that can reduce energy consumption and CO_2 emissions from heating and cooling equipment, lighting, and appliances rapidly and at low cost. But achieving deep cuts in energy consumption and CO_2 emissions in the building sector will be much more expensive and faces significant barriers. It will require an integrated approach, with much more ambitious policies on building shells than are currently foreseen, particularly in the existing stock of buildings in OECD countries, and on de-carbonizing the energy sources used. The most cost-effective approach to the transition to a sustainable buildings sector will involve three parallel efforts:

- The rapid deployment of existing technologies that are energy-efficient (including designing and building better building shells to minimize overall energy demand) to low-cost applications and the use of low/zero carbon technologies. R&D into new technologies will need to be increased and existing technologies optimized for new applications in the building sector.
- The deployment of existing technologies into applications with higher abatement costs, along with efforts to adapt the existing building stock in OECD countries, and the deployment of emerging technologies at modest scale.
- Maximizing the deployment of energy-efficient technologies, substantially renovating 60% of the OECD building stock by 2050 and ensuring the widespread deployment of new technologies, particularly those that decarbonize energy use—electric, hydrogen, and solar in the buildings sector (in the BLUE Map scenario).

While an essential first step, energy efficiency alone will not be sufficient to meet ambitious climate change goals which also require a significant shift in fuel use to low-carbon energy sources. The consumption of electricity, district heat, heat from building-scale Combined Heat and Power (CHP), and solar is higher in 2050 than in 2005 in the BLUE Map scenario. Solar grows the most, accounting for 11% of total energy consumption in the building sector, as its widespread deployment for water heating (30%–60% of useful demand today depending on the region) and, to a lesser extent, space heating (10%–35% of useful demand today depending on the region) helps to improve the efficiency of energy use in the building sector and to reduce CO_2 emissions.

The increased deployment of heat pumps for space and water heating as well as the deployment of more efficient heat pumps for cooling account for 63% of the heating and cooling technology savings. Solar thermal systems for space and water heating account for about 29% of the savings. CHP plays a small but important role in reducing CO_2 emissions and account for 8% of the savings and also assists in the balancing of the renewables-dominated electricity system in the BLUE Map scenario by adding increased electricity generation flexibility.

These CO_2 emissions reductions stem from a dramatic transformation in the markets for these technologies which will take them from small-scale deployment (with the exception of heat pumps for air-conditioning) to large-scale, mass-market technologies that are the incumbent technologies for heating and cooling from 2030 onward. Developing scenarios for the future is an inherently uncertain exercise. To explore the sensitivity of the results to different input assumptions, several variants of the BLUE Map scenario have been analyzed. They are the following:

- *BLUE heat pumps*: This scenario looks at ultra-high efficiency heat pump air conditioners (COP of 9) for cooling and humidity control, and faster cost reductions for space and water heating applications. Heat pumps in this scenario save 2 Gt CO_2 in 2050.
- *BLUE solar thermal*: This scenario explores the situation where low-cost compact thermal storage is available by 2020 and system costs come down more rapidly in the short term. Solar thermal in this scenario saves 1.2 Gt CO_2 in 2050.

- *BLUE buildings CHP*: This scenario explores the impact of more rapid declines in the cost assumptions for fuel-cell CHP units using hydrogen and their potential contribution to a higher penetration of distributed generation. CHP in this scenario saves 0.5 Gt CO_2 in 2050. The main distinction between these scenarios is that in each case a specific technology is assumed to achieve significant cost reductions earlier than in the BLUE Map scenario. This technology, therefore, gains a higher share of installations than competing options. In the BLUE Solar Thermal and BLUE Heat Pumps scenarios, each of these technologies becomes the dominant technology in 2050 for space heating and hot water. In the BLUE Buildings CHP scenario, the share of useful energy for space and water heating provided by small-scale CHP in the buildings sector doubles.

12.3.3 Technology Development for Energy-Efficient Buildings

Although many of the energy-efficient and low/zero carbon technologies for heating and cooling are commercially available in many applications, a significant number of improvements can be expected with increased R&D efforts, particularly in terms of cost reductions and in optimizing systems for a wider range of applications. These improvements are needed to achieve the energy savings and CO_2 emissions reductions envisioned in this roadmap in a timely manner.

12.3.4 Active Solar Thermal

Mature solar thermal technologies are commercially available, but further development is needed to provide new products and applications, reduce the cost of systems, and increase market deployment. Depending on location, new buildings constructed to low-energy or passive house standards could derive all of their space and water heating needs from solar thermal by 2030 at reasonable cost. Solar thermal renovations resulting in a solar coverage of well over 50% should become a cost-effective refurbishment option for single- and multifamily houses and smaller-scale commercial buildings (Figure 12.4).

These goals are ambitious but realistic if the right mix of R&D, industry development, and consistent market deployment programs are applied. These goals can be reached with development of the following technologies:

- *Integration of solar collectors in building components*: Building envelopes need to become solar collectors themselves, so both the performance of collectors and their direct integration into buildings needs to be improved. This should lead to the development of multifunctional building components which act as elements of the building envelope and as solar collectors.

- *Alternative materials*: The development of new components for use in collectors—such as polymers or plastics, the coating of absorbers (optimized to resist stagnation temperatures) and new materials to tackle deterioration resulting from UV exposure—could help to reduce the cost and improve the economics of solar thermal systems.

- *Solar cooling systems*: Solar thermal systems will require more compact thermally-driven cooling cycles (sorption chillers and desiccant systems) with higher coefficients of performance, operating at lower temperatures. This will require R&D into new sorption materials, new coatings of sorption materials on heat exchange

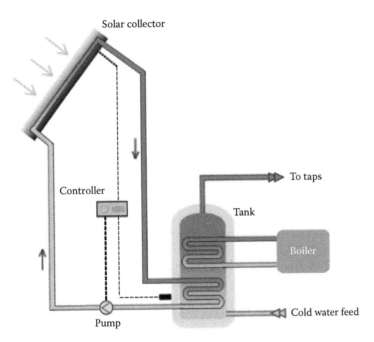

FIGURE 12.4
Active solar thermal system.

surfaces, new heat and mass transfer concepts, and the design of new thermodynamic cycles. This will need to be complemented by design guidelines and tools specifically developed for solar cooling systems and applications.

- Low-cost Compact Thermal Energy Storage will be critical to AST, meeting a larger proportion of space and water heating and cooling requirements.
- Intelligent control systems that communicate with building energy management systems will increase the useful solar energy available. These centralized and integrated control systems need to be able to benchmark and self-diagnose problems, while facilitating the integration of complementary systems (e.g., hybrid solar thermal/heat pump systems) and communicating upstream to utilities.
- Improving the automation of manufacturing will help to reduce initial system costs and expand the economic application to a wider range of customers, particularly for retrofitting existing buildings.

12.3.5 Combined Heat and Power

CHP includes technologies such as fuel cells, micro turbines, and Sterling engines that have yet to be widely deployed in buildings and have, in some cases, significant opportunity for reducing costs and improving performance (Barr et al. 2005).

The key challenges are optimizing components and lowering costs through more R&D but also through large-scale, high-volume production. In addition, further R&D into flexibility of operation and variable heat/ electricity balance would improve their economics.

Similarly, R&D and demonstration will be required on micro-CHP integration into smart grids and real time data exchange with the network. For engines, the current upsurge in work on emissions reduction must continue and overcome several significant engineering and chemical engineering challenges.

The following areas need to be addressed by increased R&D efforts:

- Reciprocating engines are a mature technology, but incremental improvements in efficiency, performance, and costs should be possible. The US DoE's Advanced Reciprocating Engine Systems program (ARES) aims to deploy an advanced natural gas-fired reciprocating engine with higher electrical efficiency, reduced emissions and 10% lower delivered energy costs. Manufacturers of liquid fuel-fired reciprocating engines are incorporating design modifications and new component technologies to improve performance and reduce emissions.

- Micro turbines and gas turbines: Technology development for micro turbines is focused on improving efficiency (through higher temperatures and pressures, new materials such as ceramics and thermal barrier coatings), advanced blade design, recuperators (to boost electrical efficiency at expense of overall efficiency), and ultra-low emissions (through lean-pre-mix dry low-emission combustors and reduced-cost SCR systems) BSRIA Ltd. (2009). Gas turbines are more mature, but there could be modest declines in capital and maintenance costs, while recuperated dry low-emission combustors could meet very low emissions standards without the need for exhaust gas clean-up.

- Sterling engines are at the market introduction stage and R&D to reduce their costs and improve their electrical efficiency is required. This can be achieved by increasing the working hot-end temperature by using high-temperature materials in the hot-end components; these exist today, but their costs need to come down. To help identify the best applications for Sterling engines, more demonstration programs are required. The development of a wider range of systems will also help expand the range of applications in which Sterling engines can compete.

- Fuel cells: The R&D priorities are to reduce costs and improve durability and operational lifetimes. Better fuel–cell system design, new high-temperature materials, and an improved understanding of component degradation and failure could considerably enhance the durability of fuel cells. Fuel cells and their balance of plant will need to have an operating life of 40,000–80,000 hours to be competitive in buildings; current designs are expected to meet the lower end of this range, but further progress is needed. Improving polymer electrolyte membrane (PEM) fuel cells (PEMFC) tolerance to impurities is a priority, while the development of a wide range of commercial solid oxide fuel cell (SOFC) and molten carbonate fuel cell (MCFC) designs is required. Lower cost catalysts, membranes, bipolar plates, and gas diffusion layers all need to be developed further. Balance-of-plant system costs can be lowered by reducing the costs of power conditioning systems (inverters) and the fuel pre-treatment system. Another important goal is to increase net system efficiencies by reducing parasitic loads.

12.3.6 Heat Pumps

R&D priorities for heat pumps are to continue improving the components and systems of existing technologies, and design systems that maximize COPs across a wide range of applications, climates, and operator behavior to widen their potential market. This will require improving the design and sizing of systems, their integration with the building design, and in their operation and control (Figure 12.5). The development of hybrid systems

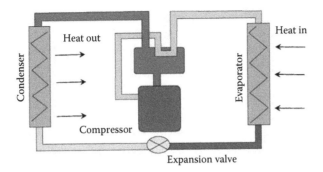

FIGURE 12.5
Basic heat pump configuration.

(e.g., heat pump/solar thermal systems) offers the potential for very high year-round COPs. R&D also needs to focus on developing packaged integrated heat pump systems capable of providing cooling and space and water heating simultaneously for small-scale applications. These goals will also require extensive demonstration programs to refine designs and optimize systems for different applications and customers.

Improved performance is important, but efficiency will increase more slowly now that highly efficient systems are available. Just as important is the technology effort to reduce costs of systems, so that they are competitive in a wider range of applications. Research is needed on the following technical areas:

12.3.6.1 Equipment and Components: Decrease Costs and Increase Reliability and Performance through More Efficient Components

The key component areas are

- Heat exchangers
- Compressors
- Expansion devices/valves
- Fans, circulators and drives
- Heat pump cycles
- Variable speed compressors
- Defrosting strategies
- Advanced system design (including for colder climates)
- Smart controls
- *Systems/Applications*: Optimize component integration and improve heat pump design and installations for specific applications to achieve higher seasonal efficiency in wider capacity ranges. Improve optimization with ventilation systems in larger applications.
- *Control and operation*: Develop intelligent control strategies to adapt operation to variable loads and optimize annual performance. Develop automatic fault detection and diagnostic tools. Improve communication with building energy management systems and upstream to smart energy grids.

- *Integrated and hybrid systems*: Develop integrated heat pump systems that combine multiple functions (e.g., space-conditioning and water heating) and hybrid heat pump systems that are paired with other energy technologies (e.g., storage, solar thermal, and other energy sources) in order to achieve very high levels of performance.

- *Improved design, installation, and maintenance methodologies*: Develop and promulgate information defining and quantifying benefits for good design, installation, and maintenance of systems in order to realize the full efficiency potential of the heat pumps. In parallel, improvements in building design and operation that reduce the temperature lift performed by the heat pump will increase the average operating efficiency (the seasonal or annual performance factor). A combined heating cooling solution for an EEB is shown in Figure 12.6.

12.3.7 Thermal Energy Storage

R&D for thermal energy storage should focus on reducing costs and improving the ability to shift energy demand —for electricity, gas, and so on, —over hours, days, weeks, or seasons and facilitating the greater use of renewable energy. Both central and decentralized energy storage systems are likely to play a role. Ongoing R&D into the underlying science of thermal energy storage and the integration and optimization of storage with heating and cooling technologies still needs to be perfected. This optimization relates both to the storage itself (size, materials, etc.) as well as the operation and control of the overall system, including storage and interaction with the building occupancy profile.

A high number of charging and discharging cycles is critical for most TES applications, so the stability of materials in the systems is very important—not only the storage medium itself but also materials used in systems components such as containers, heat exchangers and pipes. Once thermal energy storage technologies have reached the level for prototype or demonstration, further improvements will be necessary to bring them to market. Better materials are the most promising way to achieve this, but cost barriers may prevent otherwise effective solutions from being implemented.

More R&D into the real boundary conditions of material properties in TES systems is also necessary, given that it is essential to have very high confidence in the stability of materials and the number of charge/discharge cycles that systems can achieve. Operating conditions and hence performance are sometimes quite different from the assumptions made during early development.

Worldwide R&D activities on novel materials for PCMs and thermochemical approaches are insufficiently linked at the moment, and this needs to change. Many projects are focused on material problems related to one specific application and potentially miss wider opportunities for material applications in storage.

Over the last few years, the emphasis of co-operative R&D efforts has shifted toward storage technologies that improve the manageability of energy systems or facilitate the integration of renewable energy sources.

Meeting the strategic goals for the BLUE Map scenario will require research focused on key areas of technical advancement:

- Phase-change materials and other material advancements
- Stability of materials and system components over lifetime charge/discharge cycles

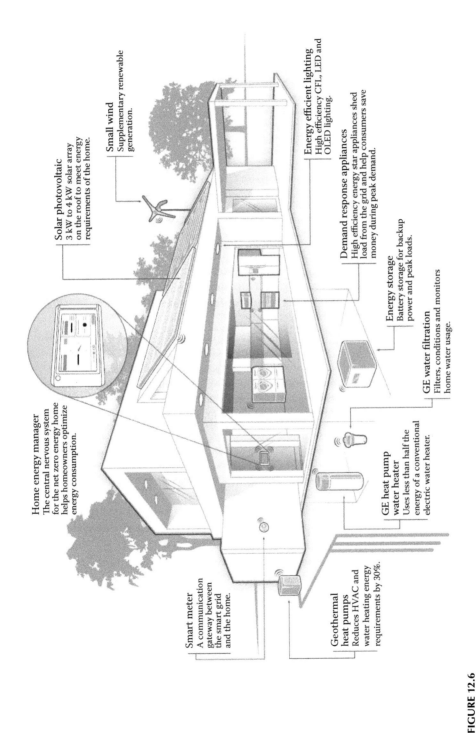

FIGURE 12.6
Combined heating cooling solution for energy efficiency.

- Analysis of system-specific storage parameters for different applications
- Optimized control and operation
- High-temperature energy stores

The key performance expectation for the household sector is that low-cost compact thermal energy storage will become available for small-scale applications in heating and cooling systems by 2020–2025. This will allow initial deployment between 2020 and 2025 and large-scale deployment from 2030. The most promising areas of R&D are in PCMs and thermochemical stores, with hybrid systems (combining PCM and sensible heat systems) likely to allow early deployment of systems at a reasonable cost.

12.4 Indian Government Policies and National Missions on Energy

The government of India realized the potential of energy efficiency in building and made it an integral part of the energy conservation (EC) act, 2001. The EC resulted in formation of the Bureau of Energy Efficiency (BEE). The broad objectives of BEE are to exert leadership and provide policy recommendation and direction to national energy conservation and efficiency efforts and programs (1). This also includes establishing systems and procedures to measure, monitor, and verify energy efficiency results in individual sectors as well as at a macrolevel.

The National Mission for Enhanced Energy Efficiency (NMEEE) is one of the eight missions under the NAPCC. NMEEE aims to strengthen the market for energy efficiency by creating conducive regulatory and policy regime and has envisaged fostering innovative and sustainable business models to the energy efficiency sector. Major considerations for EEBs for better optimization of energy need to be considered. These are enumerated by categories under major headings and then subheadings of each are discussed latter.

1. *Efficient use of energy*: This depends on climate responsiveness of buildings; good urban planning and architectural design; good housekeeping and design practices, passive design and natural ventilation; Use landscape as a means of thermal control; energy efficiency lighting; energy efficiency air conditioning; energy efficiency household and office appliances; heat pumps and energy recovery equipment; combined cooling systems; Fuel cells development.

2. *Utilize renewable energy*: Photovoltaics; wind energy; small hydros; waste-to-energy; landfill gas; biomass energy; biofuels.

3. *Reduce transport energy*: Reduce the need to travel; reduce the level of car reliance; promote walking and cycling; use efficient public mass transit; alternative sources of energy and fuels.

4. *Increase awareness*: Promote awareness and education; encourage good practices; environmentally sound technologies; overcome institutional and economic barriers; stimulate energy efficiency and renewable energy markets.

5. *Perform achieve and trade scheme (PAT)*: A regulatory instrument to reduce specific energy consumption in energy intensive industries, with an associated

market-based mechanism to enhance the cost effectiveness through certification of excess energy saving which can be traded.

6. *Market transformation for energy efficiency (MTEE)*: For accelerating the shift to energy-efficient appliances in designated sectors through innovative measures to make the products more affordable.

7. *Energy efficiency financing platform (EEFP)*: For creation of mechanisms that would help finance demand side management programs in all sectors by capturing future energy savings.

8. *Framework for energy-efficient economic development (FEEED)*: For development of fiscal instruments to promote energy efficiency.

12.5 Case Studies

Energy efficiency in building is one of the major growing technologies, and it has been already implemented on several buildings of India and abroad. Some of the case studies are listed here. The initial criteria used to select high-performance buildings is that it is necessary that the buildings were either near net zero energy new construction or deep energy retrofits for existing buildings. Resilient design, replication potential, community benefits, and market transformation were considered important secondary criteria.

- *ECBC complaint building*: "L" shape design with main entrance toward north; Longer axes along NE and NW directions; by "L" shape configuration, the width of the floor Plate is reduced for the same amount of floor plate area thereby allowing natural light to penetrate deep into the interior spaces. It ensures that part of the façade is always shaded

- *CSR House, NSW (AUSTRALIA)*: CSR is Australia's leading building products supplier, driving innovation in the building and construction industry through continued investment in research and development to further improve building systems and solutions. Thermal Comfort Star Ratings heating and cooling energy makes up around 40% of the overall energy use in a home. Adequate levels of insulation in walls, ceilings and floors; draft proofing the building envelope; effective ventilation to control internal temperatures and air quality makes it the biggest home energy requirement by far.

- *181 Fitzroy Street, St Kilda*: The building includes five stories and a four-level underground car park. The building and services are in good working condition with glazing to the north and south only. On the basis of the assessment and subsequent works, the building had a 3.5 star NABERS rating.

- *Anthony Ziem, Team Leader Facilities Maintenance, City of Kingston*: Upgrades to reduce energy use by 25%. The upgrades implemented include the following: installation of high-efficiency chillers; fitting of variable speed drives on all pumps and fans; installation of new lifts; CO_2 sensors; new BMS controls installed and commissioned; BMS controls five year preventative maintenance; solar heating; new standalone air-conditioning units in one room; building tuning including;

changes to temperature settings; replacement of sensors in Variable Air Volume (VAV) boxes; installation of carbon monoxide sensors in the car park so ventilation runs only when it is needed; envelopment of a plan for upgrading the building façade with window shadowing; installation of LED lighting with an organic response lighting control system.

- Infosys, Pocharam, Hyderabad: Efficient air-conditioning consists of highly efficient day lighting systems.
- *Infinity Benchmark, Kolkata*: The building is furnished with CO_2 monitor sensors; rainwater harvesting; wastewater recycling system; and humidification controls. The exterior of the building is made of brick wall block while the roof comprises of deck thick polyurethane foam for better insulation
- *San Francisco public utilities commission (SFPUC) building*: The building uses the IBMS to integrate data from various building systems. The IBMS monitors and manages the data with analysis and control, and regarding energy management, it monitors and manages these various systems: elevators, lighting, HVAC, power monitoring, solar energy collector metering, wind energy power generator metering, interior and exterior shade control
- Olympia Tech Park, Chennai: It reacts to changing temperatures and automatically and adjusts to the most efficient energy-saving method available to heat or cool a home. It is a hybrid-HVAC system.

There are many other examples of EEB around the world. All that is required is to create a general awareness among people, so that these methods can be efficiently implemented in each of the residential and commercial buildings.

12.6 The Future

The future of energy efficiency heavily depends on the description and the current approaches to encourage energy efficiency in building codes for new buildings. Different building codes for energy efficiency exist for different countries, which depends on regions, its climatic conditions, and type of building (residential, commercial, or complicated high-rise buildings). Integrating different regions and their standards is good practice to successfully address the energy efficiency statement. Energy efficiency improvement can be followed through initiatives such as efficiency labeling or certification, very best practice buildings with extremely low- or no-energy consumption and other policies to raise buildings' energy efficiency beyond minimum requirements.

When referring to the energy-saving potentials for buildings, several studies around the world use the analysis of recent IEA publications, including the World Energy Outlook 2006 (WEO) and Energy Technology Perspective (ETP). The major goal is to recommend policies which could be used to realize large and feasible energy-saving potentials in new buildings, and the use of building codes by renovation or refurbishment.

Improvement of buildings' efficiency at planning stage is relatively simple while improvements after their initial construction are much more difficult, that is, decisions made during a building's project phase will determine consumption over much, if not all, of a building's lifetime. Some measures to improve efficiency are possible only during construction or by major refurbishment, likely to happen only after several decades. To realize the large potential for energy conservation in new and existing buildings, governments must surmount the barriers to energy efficiency in the building sector. Policies and measures to improve buildings' efficiency include the following: All governments, states, or regions should set, enforce, and regularly update requirements for energy efficiency in new buildings. These requirements can appear independently or within building codes. Requirements for efficiency should be based on least costs over 30 years. Best practice and demonstration buildings such as Passive Houses and Zero Energy Buildings should be encouraged and supported to help these buildings penetrate the market. Financial restraints for new buildings preventing energy efficiency should be removed. Energy efficiency for buildings should be made visible in the market place to give building owners a real choice. This could be by certification, labeling, or other declaration of energy consumption. Governments should set up a package of initiative to address the barriers for energy efficiency in both new and existing buildings including the mentioned recommendations above. Most importantly, special outreach activities should be taken for the fast-developing countries such as China and India where most of the world's new buildings are constructed.

12.7 Conclusion

Due to increasing population and energy demand, the energy consumption has been increased significantly. Technological improvements in building design and appliances offer new opportunities for energy savings. Low-energy design of urban environment and buildings in densely populated areas requires consideration of a wide range of factors, including urban setting, transport planning, energy system design, and architectural and engineering details.

Different studies have shown that EEBs offer opportunities to save money while reducing GHG emissions too. Main sources of energy in today's world are heavily depending upon the availability of oil, coal, and natural gas. However, these resources not only emit GHGs, but they are also nonrenewable, therefore, quantities are limited, and they cannot be replaced as fast as they are consumed. These reasons require use of more energy-efficient methods.

Significant barrier in energy saving and optimization is the lack of information on energy consumption trends. An efficient policy for energy effectiveness in buildings should outline the need and benefits of energy efficiency in buildings and estimate potential savings both in terms of energy use and reduction of capacity. There are still large gaps in performance of commercial equipment and their theoretical limits. An appropriate institutional framework is necessary to set achievable targets and timelines, its outline, and monitoring.

These energy-efficient methods need to be monitored and verified periodically, and it should be ensured that building systems are functioning desirably. More alternative methods need to be formulated for better energy efficiency management. Installing high-performance mechanical systems and appliances and applying life cycle assessment to the trade-offs between capital and operating costs manages the energy consumption in the building.

References

Amato, A.D., Ruth, M., Kirshen, P., and Horwitz, J. (2005). Regional energy demand responses to climate change: Methodology and application to the commonwealth of Massachusetts. *Climatic Change* 71:175–201.

Barr, S., Gilg, A., and Ford, N. (2005). The household energy gap: Examining the divide between habitual- and purchase-related conservation behaviors. *Energy Policy* 33:1425–1444.

BSRIA Ltd. (2009). Press release on the BSRIA. http://www.bsria.co.uk/news/global-airconditioning-sales-reach-us70-billion-in-2008.

Bureau of Energy Efficiency. (2007). www.beeindia.gov.in.

Contribution of Working Group II to the Fourth Assessment Report of the IPCC. M. Parry, O. Canziani, J. Palutikof, P. van der Linden and C. Hanson (Eds.), Cambridge, UK: Cambridge University Press.

Crawley, D.B., Pedersen, C.O., Lawrie, L.K., and Winkelmann, F.C. (2000). EnergyPlus: Energy simulation program. *ASHRAE Journal* 42(4):49.

Darling, D. (2005). The Encyclopedia of alternative energy and sustainable living. Retrieved October 28, 2005.

Discovery Insights, LLC. (2006). Commercial and industrial CHP technology cost and performance data analysis for EIA's NEMS, Discovery insights, Bethesda, MD.

DOE/EERE (United States Department of Energy, Office of Energy Efficiency and Renewable Energy). (2003). Buildings energy data book, DOE/EERE, Washington, DC.

DOE/EERE. (2009). Buildings energy data book, DOE/ EERE, Washington, DC.

ECES (Energy Conservation through Energy Storage) Implementing Agreement, submission to the IEA by Andreas Hauer and Astrid Wille.

Ehlers, G.A., Howerton, R.D., and Speegle, G.E. (1996). Engery management and building automation system. U.S. Patent No. 5,572,438. November 5, 1996.

ESTTP (European Solar Thermal Technology Platform). (2007). Solar heating and cooling for a sustainable energy future in Europe, ESTTP, Brussels.

EU (European Union). (2009). Directive 2009/28/ EC of the European Parliament and of the Council of April 23, 2009 on the Promotion of the Use of Energy from Renewable Sources and Amending and Subsequently Repealing Directives 2001/77/EC and 2003/30/EC. *Official Journal of the European Union* L140, June 5, 2009.

http://www.eesolutionsinc.net

http://www.regelgroup.com

IEA (International Energy Agency). (2007). Mind the gap, OECD/IEA, Paris.

IEA Heat Pump Programme. (2010). Submission to the IEA from Monica Axell.

IEA Solar Heating and Cooling Programme. (2010). Submission to the IEA from Esther Rojas.

IEA. (2008). Combined heat and power-evaluating the benefits of greater global investment, OECD/ IEA, Paris.

IEA. (2010). Money matters: Mitigating risk to spark private investments in energy efficiency, OECD/ IEA, Paris.

IEA. (2010a). Energy technology perspectives, OECD/IEA, Paris.

IPCC. (2007). Climate change 2007: Impacts, adaptation and vulnerability contribution of working group II to the fourth assessment report of the intergovernmental panel on climate change. M.L. Parry, O.F. Canziani, J.P. Palutikof, P.J. van der Linden and C.E. Hanson. Cambridge, UK: Cambridge University Press.

Japan Gas Association. (2010). Submission to the IEA from Motomi Miyashita.

Laustsen, J. (2008). Energy efficiency requirements in building codes, energy efficiency policies for new buildings. *International Energy Agency (IEA)* 477–488.

Levine, M., Urge-Vorsatz, D., Blok, K., Geng, L., Harvey, D., Land, S., Levermore, G. et al. (2007). Residential and commercial buildings, Climate change 2007: Mitigation, contribution of working group III to the fourth assessment report of the intergovernmental panel on climate change. B. Metz, O.R. Davidson, P.R. Bosch, R. Dave, and L.A. Meyer (Eds.). Cambridge, UK: Cambridge University Press.

Majumdar, M. (Ed.). (2001). *Energy-efficient Buildings in India*. New Delhi: TERI Press.

Marcogaz. (2009). Submission to the IEA via Motomi Miyashita (Japan Gas Association).

Mckensy & Company Inc. (2009). Environmental and energy sustainability: An approach for India (2009). http://www.mckinsey.com/

Mowris, R. (2004). Evaluation measurement and verification report for the time of-Sale Home Inspection Program #180-02, Robert Mowris & Associates, Olympic Valley, CA.

Murtishaw, S. and Sathaye, J. (2006). Quantifying the effect of the principal-agent Problem on US residential energy use, Lawrence Berkeley National Laboratory, San Francisco, CA.

Navigant Consulting Inc. (2007). EIA–Technology forecast updates–Residential and commercial building technologies, Navigant Consulting, Washington DC.

NEDO website. (2009). http://www.nedo.go.jp/.

Omer, A.M. (2010). Low energy building materials: An overview, environment 2010: Situation and perspectives for the European Union May 6–10, 2003. Porto, Portugal.

Prindle, B. (2007). Quantifying the effects of market failures in the end-use of energy, American Council for an Energy-Efficient Economy (ACEEE), ACEEE Report Number E071, Washington, DC.

Rognon, F. (2008). The role of the state and the economy in the promotion of heat pumps on the market. *Paper Presented at the 9th IEA Heat Pump Conference 2008*, Zurich, Switzerland.

Rosenfeld, A.H. and Hafemeister, D.W. (1988). Energy-efficient buildings. *Scientific American* 258(4):78–87.

Roth, K., Zogg, R., and Brodrick, J. (2006). Cool thermal energy storage. *ASHRAE Journal* 48:94–96.

U.S. EPA. (2011). EPA Green Building. http://www.epa.gov/greenbuilding/.

UNEP SBCI. (2009). Buildings and climate change, summary for decision-makers. United Nations Environment Program.

UNFCCC. (2015). India's intended nationally determined contribution. www4.unfccc.int/submissions/INDC/Published Documents/India/1/INDIA INDC TO UNFCCC.pdf.

USCCSP. (2008). Impacts of climate change and variability on transportation systems and infrastructure: Gulf coast study, Phase I A report by the U.S. Climate Change Science Program and the Subcommittee on Global Change Research. M.J. Savonis, V.R. Burkett and J.R. Potter (Eds.), Systematics. Washington, DC: Department of Transportation.

VHK. (2007). Eco-design of hot water heaters: Task 4 Report, VHK, Delft, the Netherlands.

Weiss, W. et al. (2010). Solar heat worldwide: Market and contribution to the energy supply 2008. IEA Solar Heating and Cooling Programme.

Zhang, Y., Zhou, G., Lin, K., Zhang, Q., and Di, H. (2007). Application of latent heat thermal energy storage in buildings: State-of-the-art and outlook. *Building and Environment* 42(6):2197–2209.

13

Role and the Impact of Policy on Growth of Green Buildings in India

Manish Vaid and Sanjay Kumar Kar

CONTENTS

13.1 Introduction 259
13.2 Green Buildings Can Support Sustainable Development Goals 260
13.3 The Case of Green Buildings in India 261
13.4 What Is Green Building? 263
13.5 Urban Missions 264
    13.5.1 Atal Mission for Rejuvenation and Urban Transformation (AMRUT) 265
    13.5.2 Smart Cities Mission 265
    13.5.3 Pradhan Mantri Awas Yojana—Urban 266
13.6 Green Building Growth Prospects in India 267
13.7 Policy Impact of Green Buildings 267
13.8 Conclusion 269
References 269
Endnote 271

13.1 Introduction

India, being a dynamic and developing economy, and, having a diverse society, is driven by robust urbanization. While every sixth person in India is urbanized, 11–12 million people are added to its annual urban population growth. Because of this urbanization trend, India is projected to become world's fourth-largest economy by 2022 and third-largest by 2030, overtaking [1] Germany ($4.38 trillion) and Japan ($6.37 trillion), respectively.

According to NITI Aayog [2] (NITI), Indian economy will expand annually at an average of 8% to ₹469 lakh crores, or $7.25 trillion by 2030, which would be a threefold expansion from ₹137 lakh crores, or $2.1 trillion in 2015–2016. These economic prospects, according to former NITI Aayog vice chairman Arvinda Panagariya [2], "Make India's future bright because of a very large GDP base and a projection of average 8% growth over the next 15 years."

However, rising population in India, which is expected to overtake that of China in 2022 [3], has already put tremendous pressure on current and prospective resources in urban centers, wherein availability of energy, water, land, and other basic amenities (such as housing and air quality) have been severely hampered. These resources, if continued to be mismanaged, can aggravate the existing resource crisis, creating chaos among urban dwellers.

With an annual growth rate of around 6%, the decadal population growth rate during the period 2001–2011 is recorded at 17.64%, during which India's urban centers [4] grew from 5161 classified towns and 384 urban agglomerations in 2001 to 7935 classified towns and 475 urban agglomerations in 2011.

In 2011, India also entered a different demographic trajectory [5] with the net increment to urban population exceeding the net increment to rural population, signifying increased migration to urban centers.

Having a population spread of 377.16 million in 7933 cities and towns, as recorded by Census 2011 [6], most of its projected population increase, as per United Nations (2014) [7], would take place in urban areas between 2015 and 2030; this will add another 164 million people to its urban base.

Urbanization, which contributes to around 60% of its GDP, needs to be managed well. To accommodate the vast population base and rejuvenate the urbanization process, the government has launched several programs including, Atal Mission for Rejuvenation and Urban Transformation (AMRUT), Smart Cities Mission (SCM), and Pradhan Mantri Awas Yojana (PMAY). This is mapped under Sustainable Development Goals (SDG)-11 on "Making cities and human settlements inclusive, safe, resilient and sustainable [8]." This is set to transform India's building (comprising of both residential and services sectors)[1] and infrastructure sector, being critical to Indian economy.

With a strong push through reform measures, India is expected to be the third largest construction market globally by 2030 [9]. The construction sector is expected to contribute 15% of India's GDP. With this in mind, there is an urgent need to push for sustainable growth through green building under these missions. Effective implementation of these missions would drive investments in the real estate sector and create millions of direct and indirect jobs.

These missions aim to ensure access to adequate, safe and affordable housing, basic services, upgrade slums, improve air quality and waste management. They also provide financial and technical assistance in building sustainable and resilient buildings and utilizing local materials.

Interestingly, incorporating the concept of green buildings into these missions would facilitate in achieving other SDGs, which can go beyond its existing mapping of SDG-11, as has been perceived by the World Green Building Council (WGBC) [10].

The WGBC sees green buildings beyond an inanimate structure that goes beyond saving energy, water, and curbing carbon emissions, to include more education, creation of jobs, strengthening communities, improving health and well-being. This approach can steadfast and direct green building growth in a more inclusive manner to cover a larger section of the society, offering wider and unwavering benefits of adopting green buildings.

13.2 Green Buildings Can Support Sustainable Development Goals [10]

WGBC has suggested a meaningful link between green building and the following nine out of 17 SDGs:

1. *Goal # 3 (Good Health and Well-Being)*: Green building can ensure healthy life while promoting well-being for all and at all ages by offering some of its features like improved lighting, better air quality and greenery.

2. *Goal # 7 (Affordable and Clean Energy)*: Green building can ensure access to affordable, reliable, sustainable, and modern energy for all by saving energy through energy-efficient means and renewable energy, such as, solar energy.

3. *Goal # 8 (Decent Work and Economic Growth)*: Contrary to conventional building, green building can further promote inclusive and sustainable economic growth, employment, and decent work for all, benefitting from the life cycle of the green building itself.

4. *Goal # 9 (Industry Innovation and Infrastructure)*: Green building can also help developing resilient infrastructure, promote sustainable industrialization, and foster innovation, particularly in an era of changing global climate, considering the net zero emissions future.

5. *Goal # 11 (Sustainable Cities and Communities)*: Green buildings can make cities inclusive, safe, resilient and sustainable, compared to conventional buildings, which consume significant amount of energy and water besides emissions and cannot sustain livelihood in the future.

6. *Goal # 12 (Sustainable Consumption and Production)*: Green buildings also ensure sustainable consumption and production patterns by reducing and recycling of waste under "circular economy" principles that extract maximum value, unlike a traditional linear economy, that follows "make, use, and dispose" principles.

7. *Goal # 13 (Climate Action)*: Green buildings can go a long way in combating climate change and its adverse impacts. The building sector is the largest contributor to global greenhouse gas (GHG) emissions, emitting 30% of GHG.

8. *Goal # 15 (Life on Land)*: Green buildings can manage forests, combat desertification, halt and reverse land degradation, halt biodiversity loss, given the material that make up a green building is used in a more sustainable way. In this regard, green building certification tools become a useful component which can recognize the need to reduce water use, the value of biodiversity, and the importance of ensuring it is protected.

9. *Goal # 17 (Partnerships for the Goals)*: Green building has a major role to play in revitalizing the global partnership for sustainable development. In 2015, a significant milestone was achieved when WGBC, United Nations Environment Programme (UNEP), the French government and several other organizations came together to host the first ever "Buildings Day" as part of the official COP21 agenda and to launch the Global Alliance for Building and Construction [11].

Furthermore, green buildings can also help in achieving SDG #6 on "Clean Water and Sanitation" to ensure availability and sustainable management of water and sanitation for all, making it an essential element of green building composition.

Thus, India can push its urban missions through broader framework of SDGs to promote the green building concept, thereby supporting housing sector, which has a significant economic and social impact on development.

13.3 The Case of Green Buildings in India

According to UNEP, the building sector is the largest contributor to GHG emissions, emitting 30% of GHG emissions, while using 40% of the global energy. In India, building accounts for 40% of the total energy consumption, of which residential real estate accounts for over 60% of it. This has resulted in environmental degradation, impacting natural resources (including air, soil, and water) and increase in toxic waste due to rapid urbanization. According to the Intergovernmental Panel on Climate Change (IPCC), the building sector has the highest potential to reduce energy use (and resultant carbon dioxide emissions) at the lowest cost.

With the current stock [12], green buildings contribute only at 5% of the total buildings in India, which suggests the massive market potential of green buildings. However, continuation of inefficient use of energy, water, and embedded materials would result in approximately 10 times increase in GHG emissions [13] in the construction sector by 2050.

According to India Energy Outlook 2015 [14], three-quarters of the projected increase in energy demand in residential buildings comes from urban areas. With most of the 2040 building stock yet to be constructed, India has an opportunity to expand and tighten efficiency standards on building sector to ensure future demand for energy services is met without putting undue strain on energy supply. With an additional 315 million people expected to live in Indian cities by 2040, India's urbanization would be the key driver of energy consumption.

According to the U.S. Green Building Council's (USGBC) Dodge Data & Analytics World Green Building Trends 2016, supported well by environmental regulations and demand for healthier neighbourhoods, by 2018, the green building industry in India will grow by 20% [15]. Currently, India has about 2.68 billion sq. ft. of registered green building space [16] across 3000 projects (second largest in the world), of which 600 are certified and fully functional.

India ranks third [17] among the top ten counties for Leadership in Energy and Environmental Design (LEED) rating system devised by the USGBC to evaluate the environmental performance of buildings through sustainable design.

LEED [18] is a certification program which works for all buildings, covering all phases of development that are based on points system across several areas of sustainability issues. Based on the number of points earned, the project attains one of four LEED rating levels, namely, Certified, Silver, Gold, and Platinum [18]. In 2016, nearly 650 projects in India earned LEED certification.

According to The Energy and Resources Institute (TERI), [19] if all buildings in urban areas adopt green building concepts, India could save more than 8400 MW of power, which is enough to light half of Delhi or 5.5 lakh homes a year.

In a presentation [20] made by a group of secretaries to the Prime Minister on energy efficiency earlier in 2017, buildings are acknowledged to be the fastest growing energy consuming sector. It was therefore proposed to target for 30% penetration for energy-efficient buildings (EEBs); this can help India save approximately 2500 MW of power.

Paris Climate Agreement has prompted several governments to place greater emphasis on low carbon development of their cities. India too has pledged to cut its carbon emissions per unit of gross domestic product by 33%–35% by 2030 from 2005 level [21]. To this end, it has formulated several mitigation strategies under different sectors in its Nationally Determined Contribution submitted for Conference of Parties—21 (COP) of United National Framework Convention on Climate Change (UNFCCCC) to develop climate resilient urban centers.

In addition to curb energy intensity of the Indian economy, the National Mission for Enhanced Energy Efficiency (NMEEE) has been introduced to replace all incandescent lamps with Light-Emitting Diode (LED) bulbs in the next few years leading to energy savings of up to 100 billion kilowatt hours (kWh).

Further, to implement energy efficiency, the Bureau of Energy Efficiency (BEE) has introduced the Energy Efficiency Building Code (ECBC) [22] to improve the energy performance of buildings and curb emissions. This code sets a minimum energy efficiency performance standard for buildings and has already been adopted and notified by eight states/union territory (UT) such as Rajasthan, Odisha, Uttrakhand, Punjab, Karnataka, Andhra Pradesh, and UT of Puducherry, making more than 300 new buildings ECBC compliant. Design Guidelines for Energy Efficient Multi-story Residential buildings have also been launched.

To stimulate the large-scale replication of EEBs, India has also developed its own building energy rating system, namely, Green Rating for Integrated Habitat Assessment (GRIHA) based on 34 criteria like site planning, conservation, and efficient utilization of resources, and so on.

The National Urban Housing and Habitat Policy (NUHHP, 2007), which was formulated with the goal of "Affordable Housing for All" with an emphasis on the vulnerable sections of the society, including urban poor, sought to earmark land for economically weaker and low-income households in the new housing projects. The provisions of building codes and by-laws have also been modified along with the provisions for green buildings, natural disaster resilience, and inclusive design for the elderly and the physically challenged. Therefore, the NUHHP [23] not only considers affordable housing for weaker sections of the society but also lays emphasis on the use of modern technology in the housing sector for enhancing energy and cost efficiency, productivity, and quality through the concept of "green" and "intelligent" buildings. Policy interventions such as the three urban missions, namely, AMRUT, SCM, and PMAY are in line with NUHHP, wherein the role of green buildings has to play a major role. These interventions are formulated to provide cleaner, healthy, and sustainable living to all the urban dwellers, which would further shape future growth of India's green buildings. The following section elaborates the urban missions undertaken by the government in detail.

13.4 What Is Green Building?

According to the WGBC [24], a "green" building is a building that, in its design, construction, or operation, reduces or eliminates negative impacts and can create positive impacts on our climate and natural environment. Green buildings preserve precious natural resources and improve our quality of life. The following features can make a building "green." These include:

- Efficient use of energy, water, and other resources
- Use of renewable energy, such as solar energy
- Pollution and waste reduction measures and the enabling of re-use and recycling
- Good indoor environmental air quality
- Use of materials that are non-toxic, ethical, and sustainable
- Consideration of the environment in design, construction, and operation
- Consideration of the quality of life of occupants in design, construction, and operation
- A design that enables adaptation to a changing environment

The features mentioned above can make any building a green building, whether it's a home, an office, a school, a hospital, a community center, or any other type of structure.

However, green buildings in different countries and regions could vary based on distinctive climatic conditions, unique cultures and traditions, diverse building types and ages, or wide-ranging environmental, economic, and social priorities.

According to UN-Habitat [25], green building can include energy efficiency along with features such as healthy materials, water efficiency, sustainable waste management, resource efficiency, and land use. Often materials and methods used to incorporate green aspects have synergistic properties providing energy efficiency and other benefits such as resource efficiency. For instance, a low-flow shower can reduce both water consumption and energy to heat that water.

Further, there are vernacular approaches [26] to green building design that advance the sustainable aspect in housing going beyond environmental sustainability to include social, cultural, and economic sustainability. Vernacular schools of building design are also deeply embedded in the traditional wisdom that enhances the cultural settings of India's built environment. Under this design [26], each project addresses an integrated approach to design with a special emphasis on climatology; solar passive architecture; bio-climatic design and low- energy architecture to achieve appropriate human comfort, low-energy, low-cost community development; use of recycled municipal/domestic waste as building material; and a financial model that may be implemented for successful promotion of sustainable building design principles.

Green building is synonymous with "high-performance buildings," "green construction," "sustainable design and construction" as well as other terms that refer to a comprehensive approach to design and construction [27]. Green building maximizes the owner's return on investment by sustained savings of energy (40%–50%), water savings (20%–30%), and a smaller payback period on initial investment [27]. Thus, green building design and construction is an opportunity to use resources more efficiently, while creating healthier and more efficient homes.

The building sector in any developing country is a big consumer of resources such as energy, water, and materials and a generator of pollutants during the construction, operation, and demolition phase [28]. Consumption of these resources by building increases with the level of economic development, population growth, urbanization patterns/level, shift from biomass to commercially available energy carriers, especially electrification, income (a strong determinant of the set of services and end-uses), level of development, cultural features, level of technological development, individual behaviors, availability, and financial aspects of technologies/materials [28]. As a result, there is a severe pressure on its resources due to rising population in its urban space. Therefore, there is a stringent need to shift toward green building concept from the existing commercial building idea of energy and resource guzzlers. To meet this objective, the government of India has undertaken a holistic path by launching several programs such as, AMRUT, SCM, and PMAY, that have the potential to address the core challenges of building sectors and provide better sustainable environment to urban settlements and its dwellers.

13.5 Urban Missions

On June 25, 2015, Prime Minister Narendra Modi launched three mega urban schemes, namely, "AMRUT," "SCM" and "PMAY" for urban transformation and to enable better living while eventually driving the economic growth.

These schemes, focusing on transforming urban centers, are directed to address the rising urbanization challenges in India with emphasis on the need to improve the basic services such as water supply, sewerage, and urban transport to further improve the quality of life based on the spirit of cooperative and competitive federalism. These Missions are also in line with the 74th Constitutional Amendment of 1992, which was a landmark legislation in decentralizing governance in its true spirit.

All the three missions, inter-alia aims to improve the quality of life for all especially the poor and disadvantaged in respective cities/towns as identified in each of these missions, thereby improving their housing and working environment.

The role of green buildings would play a significant role, both directly and indirectly to this housing and working environment by offering sustainable environment to all urban dwellers.

Given India's deteriorating ecological resources in the urban centers, it is important to move toward the green architecture model by promoting increased numbers of green buildings which can integrate development with sustainability. Therefore, green growth in building sector aims to minimize environmental consequences due to construction activities by focusing on energy conservation, energy efficiency, integration of renewables, less consumption of water and sustainable waste management.

Further, to provide sustainability, such a growth also aims at building resilience to climate change impacts and extreme events without compromising the comfort of inhabitants.

Before delving upon the possible links with urban missions and green buildings, the following paragraphs describes the three urban missions.

13.5.1 Atal Mission for Rejuvenation and Urban Transformation (AMRUT)

Through AMRUT [29], the government would ensure access to tap with assured supply of water and a sewerage connection; increase the amenity value of cities by developing greenery and well maintained open spaces, including, parks and switching to public transport and promote walking and cycling to curb pollution. As a national priority, this mission aims to provide water supply, sewerage, urban transport and other basic services to households and build amenities in cities to improve the quality of life for all, including, the poor and disadvantaged, as its main area of focus [30].

The mission would include [31] all cities and towns with a population of over one lakh with notified municipalities, including Cantonment Boards. Out of 4040 estimated urban zones in India, 500 cities are to be taken up under its ambit. In this regard, the total estimated outlay [32] is ₹50000 crores for the period from FY2015-16 to FY2019–2020. The mission may continue thereafter in the light of evaluation done by the MoUD.

13.5.2 Smart Cities Mission [33]

The objective of the SCM is to promote sustainable and inclusive cities that provide core infrastructure while offering a decent quality of life to its citizens with a clean and sustainable environment though "Smart" solutions. This initiative is meant to create a replicable model catalyst to the creation of similar Smart Cities with core infrastructure elements to include, adequate water supply; assured electricity supply; sanitation and solid waste management and good governance along with citizen participation.

This mission plans to cover 100 cities and with a duration of five years from FY 2015–2016 to FY 2016–2017 which could be extended thereafter in the light of an evaluation to be done by the MoUD and incorporating the learnings from the mission. The total number of 100 smart cities have been distributed among the States and Union Territories (UT) on the basis of an equitable criteria [34], giving equal weightage (50:50) to urban population of the State/UT and the number of statutory towns in the State/UT. The implementation of SCM at the city level will be done by a Special Purpose Vehicle (SPV), having core elements namely, smart energy and environment, smart governance, smart citizens, smart buildings and homes, smart infrastructure and smart mobility.

The strategic aspect [35] of this mission would be undertaken under two main approaches: area-based development including retrofitting (city improvement), redevelopment (city renewal) and greenfield development (city extension); and pan-city development, which applies Smart Solutions. Retrofitting will introduce planning to make the existing area more efficient and liveable, covering more than 500 acres. Under redevelopment a replacement of the existing built-up environment and co-creation of a new layout with enhanced infrastructure using mixed land use and increased density would be undertaken covering more than 50 acres. Under a greenfield development with Smart Solutions a previously vacant area covering more than 250 acres will be introduced.

The paradigm shift in the approach to urban development of this mission is the introduction of the nationwide "Smart City Challenge" approach to make cities competitive with the total planned investment of around ₹1 trillion ($15 billion), including around ₹500 billion ($7.5 billion) to be allocated by the Central Government.

13.5.3 Pradhan Mantri Awas Yojana—Urban

The mission of PMAY or Housing for All aims to provide housing to all citizens by the year 2022 by providing central assistance to implementing agencies through the States and UTs.

This Mission [36] is being implemented as a Centrally Sponsored Scheme except for the component of credit linked subsidy that will be implemented as a Central Sector Scheme. The mission effective from June 17, 2015 will cover all 4041 statutory towns as per Census 2011 with focus on development of 500 Class-I cities in the following three phases, covering 100 cities in Phase I (April 2015–March2017); 200 cities in Phase II (April 2017–March 2019) and remaining cities in Phase III (April 2019–March 2022) selected from States/UTs. However, ministry will have flexibility regarding inclusion of additional cities in earlier phases in case there is a resource backed demand from States/UTs.

Inter-alia, the Technology Sub-mission [36] would also be set up to facilitate adoption of modern, innovative and green technologies and building material for faster and quality construction of houses. The mission will also facilitate preparation and adoption of layout designs and building plans suitable for various geo-climatic zones, besides assisting States/Cities in deploying disaster resistant and environment friendly technologies.

The implementation methodology [37] of this mission would be based on four pillars, namely, in-situ slum redevelopment, affordable housing in partnership, subsidy for beneficiary led individual house construction or enhancement and affordable housing credit through credit linked subsidy. Thus, government can push to promote the development of green buildings through all the three missions.

However, one of the biggest barriers to green building growth in India is the notion that green buildings are expensive and not financial viable. Therefore, there is a need to unfold the complex relationship [38] between sustainability and low-income housing in cities. Accordingly, low-income urban housing cannot be made sustainable unless the themes of ecology and energy are taken into consideration. Sustainability of a house can be improved by efficient and equitable provision of sanitation, safe water, and collection of solid waste, in addition to the introduction of technology in pro-poor settlements. For instance, depending upon contextual conditions, environmentally friendly, yet durable and affordable construction materials may be produced locally, based on relatively simple technologies. Bamboo, timber, adobe bricks, compressed earth blocks and interlocking stabilized soil blocks are some of such materials that may be used in self-managed housing and in low-cost housing schemes [39].

Most of the urban housing schemes, which are largely Center or State-regulated, are implemented by private construction companies as a result, the objective of stimulating macro-economic growth [39] overrides the goal of providing housing to urban poor. This could defeat the holistic vision of several such housing schemes, wherein the marginalized and poor are left out of the needed benefits. Even the middle and upper middle class fell short of reaping the benefits of those housing schemes which are planned away far from their working place, making it difficult and expensive for prospective residents to get and do their jobs, shop, and pay social visits to their near and dear ones. Consequently, those houses remain out of reach to these prospective residents, resulting in large vacancies of those houses. The case in point is the faulty housing policies of China, which has resulted in creation of many ghost towns.

In addition, housing under urban schemes should meet the foremost objectives of reduction of greenhouse gas emissions, both during the construction of house as well as throughout its life cycle.

13.6 Green Building Growth Prospects in India

As per the SmartMarket Report [40], green building construction on a global scale is doubling every three years. India, an emerging country with robust urbanization taking place, is very much a part of this trend. This report, based on the opinions of 1026 survey respondents from 69 countries and 13 country-specific profiles, including India, noted that in 2015, green building accounts for 37% of the total building project activity and is projected to increase to 57% among all the countries included in the survey. This suggests a strong market for green building products and services. 52% of Indian respondents considered environmental regulation as the single most significant factor triggering India's green building activity for future, while 28% of respondents considered a healthier neighborhood as another factor which can boost green building growth in India. However, lack of public awareness and concerns about corruption are the biggest challenges preventing growth in green buildings in India.

13.7 Policy Impact of Green Buildings

Government of India has several policy initiatives even before the launch of the above three urban missions to mainstream green building practices and the 12th Five-Year Plan had faster adaptation of Green Building Codes as one of the twelve focus areas. Energy Conservation Building Codes (ECBC) was launched in May 2007, specifying energy performance requirements of commercial buildings in India, having a connected electrical load of 100 kW or more. ECBC has been developed by the BEE under the provision of the Energy Conservation Act of 2001.

The scheme of Energy-Efficient Solar/Green Buildings was also introduced for implementation during 2013–2014 and rest of the 12th FYP period with a budgetary allocation of ₹10 crore. The main objective of the scheme is to promote the widespread construction of energy-efficient solar/green buildings in the country through a combination of financial and promotional incentives and other support measures, so as to save a substantial amount of electricity and other fossil fuels apart from having peak load shavings in cities and towns.

The NUHHP of 2007 endeavors to devise sustainable strategies for development of cities and towns in an environmentally friendly manner with concern about dealing with solid waste disposal and drainage. While advocating for accelerating the pace of development of housing and related infrastructure, this policy emphasized the need of using technology for modernizing the housing sector for enhancing energy, cost efficiency, productivity and quality. The concept of "green" and "intelligent" buildings was also planned to put in place on the ground.

The National Action Plan on Climate Change introduced in 2008 also highlights energy efficiency in the building sector, the National Mission on Sustainable Habitat aims to make the built environment more sustainable through improvements in energy efficiency, and the government will promote the ECBC as an integral component of urban planning.

India has also developed its own building-energy-rating system named GRIHA [41] which is based on 34 criteria such as site planning, conservation, and efficient utilization of resources and Building Operation and Maintenance, and Innovation points. All buildings more than 2500 sq m, (except for industrial complexes) that are in the design stage, are eligible for certification under GRIHA. Buildings include offices, retail spaces, institutional buildings, hotels, hospital buildings, healthcare facilities, residences, and multi-family high-rise buildings.

Further, there were several institutional initiatives such as the Indian Green Building Council (IGBC), which is a part of the Confederation of Indian Industry (CII)-Godrej GBC to promote Green Building Concept in India. This council serves as a single point solution provider and key engine to facilitate all Green Building activities in India. IGBC has brought in green building rating systems which help to achieve building energy savings of 30%–50% with the investment generally paid back in over just two-three years. As a part of indigenization of the LEED rating system, IGBC has been working on LEED India for the past decade with the launch of LEED India for New Construction in January 2007.

The impacts of policy intervention in green building growth have been widespread. For instance ICICI Bank Data Centre [42] in Hyderabad became India's first platinum rated project to be rated under IGBC Green Data Centre Rating System. This system would provide both tangible benefits by 20%–25% reduction in Power Usage Effectiveness and 25%–30% reduction in water consumption. Intangible benefits include enhanced air quality, excellent daylighting, health, and well-being of the staff operating such facility.

The impact of policy interventions is also being felt on airports [43] as green airports are gradually becoming a necessity all around the country due to a growing carbon emission footprint resulting from booming traffic. India is already among the top five fastest growing countries for plane passengers. In the fiscal 2015–2016, Indian airports had a total of 223.6 million passengers (almost 55 million international passengers and 169 million domestic ones). Cochin International Airport is one of such green airports with 46,000 solar panels are installed. This panel, which produces between 50,000–60,000 power units per day, saves 300,000 tonnes of carbon emissions, making the airport "absolutely power neutral." In October 2016, PM Modi inaugurated India's second green airport in Vadodara's Harni Airport. This airport has rainwater harvesting systems and energy-saving cooling mechanisms and was designed and built by the Airports Authority of India.

13.8 Conclusion

To conclude, it can be said that while initiatives on shaping India's green building growth started over a decade ago with the launch of the ECBC by the BEE in 2007, it has not picked up to the desired level primarily because of lack of capacity building exercises on the part of the government and institutions responsible for taking up these initiatives. Consequently, a notion that green buildings are expensive and financial unviable developed. According to Syed Beary, who chairs the Bengaluru chapter of the IGBC urged to break this myth stating, that "green building may cost 3%–5% more than the conventional building while being constructed, but the extra cost is more than offset in the first 3 to 5 years itself as operation cost are considerably less." We support the argument and we are sure that green building is the future of India's construction space.

Green buildings should not be seen as a prerogative only of the rich if one has to transform urban settlements and its standard of living of the people living in these urban spaces.

The idea should be to make urban spaces habitable and inclusive to make them more sustainable, and developers along with the government and institutions should come up with attractive financing solutions for green buildings to provide greater space for their growth.

The green building can also provide to facilitate SDGs such as affordable, clean, and sustainable energy (SDG#7); providing clean water and sanitation to ensure availability and sustainable management of water and sanitation for all (SDG#6); inclusive and sustainable economic growth (SDG#8); fostering industry innovation and infrastructure (SDG#9); sustainable cities and communities (SDG#11); sustainable consumption and production (SDG#12); combating climate change and its impact (SDG#13); sustainably manage forests, combat desertification, halt and reverse land degradation, halt biodiversity loss by introducing green building sustainability tools (SDG#15) and revitalize the global partnership for sustainable development (SDG#17).

Therefore, meeting of these SDGs with the help of green building could be possible through urban missions, which could further spur growth of green building in India.

References

1. Ray, R. K. India's economy to become third largest, surpass Japan, Germany by 2030, *Hindustan Times*, April 28, 2017.
2. ET Bureau. Economy to grow over 3-fold to $7.25 trillion by 2030: Arvind Panagariya, *The Economic Times*, April 24, 2017.
3. Population forecasts—The world's biggest country, *The Economists*, August 13, 2015.
4. Abbu, N. et al. Volume 1, *Urban Green Growth Strategies for Indian Cities*. New Delhi, India: ICLEI—Local Governments for Sustainability, 2015, p. 10.
5. Ministry of housing and urban poverty alleviation, government of India. India Habitat III, National Report 2016, New Delhi, India: MoHUPA, p. 31.
6. Ministry of housing and urban poverty alleviation, government of India. India Habitat III, National Report 2016, New Delhi, India: MoHUPA, p. 30.
7. Ministry of housing and urban poverty alleviation, government of India. India Habitat III, National Report 2016, New Delhi, India: MoHUPA, p. 21.
8. National Institution for Transforming India (NITI Ayog). Sustainable Development Goals (SDGs), Targets, CSS, Interventions, Nodal and other Ministries (As on June 8, 2016). Available at http://niti.gov.in/writereaddata/files/SDGsV20-Mapping080616-DG_0.pdf.

9. Pandey, N. India to be 3rd largest construction market globally by 2030: Report, *The Hindu Business Line* (As on August 19, 2016). Available at http://m.thehindubusinessline.com/news/real-estate/india-to-be-3rd-largest-construction-market-globally-by-2030-report/article9008113.ece.

10. World Green Building Council. Green building: Improving the lives of billions by helping to achieve the UN Sustainable Development Goals (As on March 23, 2017). Available at http://www.worldgbc.org/news-media/green-building-improving-lives-billions-helping-achieve-un-sustainable-development-goals.

11. Available at http://globalabc.org/.

12. PRNewswire. India Green Building Market Opportunity Outlook 2020, PRNewswire, (As on December 7, 2016). Available at http://www.prnewswire.com/news-releases/india-green-building-market-opportunity-outlook-2020-300374805.html.

13. TERI. Green financing is not the only tool to unlock the potential of the green real estate sector: GRIHA, TERI, (As on March 3, 20177). Available at http://www.teriin.org/files/PressRelease-GRIHASummit2017_Day2.pdf.

14. International Energy Agency. India Energy Outlook—World Energy Outlook Special Report. Available at https://www.iea.org/publications/freepublications/publication/IndiaEnergyOutlook_WEO2015.pdf.

15. Express New Service. Green buildings to grow by 20% in India by 2018: Report, The Indian Express, (As on February 20, 2016). Available at http://indianexpress.com/article/india/india-news-india/green-buildings-to-grow-by-20-in-india-by-2018-report/.

16. UNFCCC. India's Intended Nationally Determined Contribution: Working Towards Climate Justice. Available at http://www4.unfccc.int/submissions/INDC/Published%20Documents/India/1/INDIA%20INDC%20TO%20UNFCCC.pdf.

17. Sawit, Amanda. Infographic: Top 10 Countries for LEED in 2016, USBC.ORG, (As on December 14, 2016). Available at http://in.usgbc.org/articles/infographic-top-10-countries-leed-2016.

18. About LEED. Available at http://in.usgbc.org/leed.

19. Bershilia, K. Innovation in design: A study of green buildings, *International Journal of Scientific and Research Publications*, 4(5), (As on May 2014). Available at http://www.ijsrp.org/research-paper-0514/ijsrp-p29128.pdf.

20. Available at http://www.teriin.org/files/PressRelease-GRIHASummit2017_Day2.pdf.

21. India's Intended Nationally Determined Contribution.

22. Bureau of Energy Efficiency. Available at https://beeindia.gov.in/content/ecbc.

23. National Urban Housing and Habitat Policy, 2007. Available at http://mhupa.gov.in/writereaddata/NUHHP_2007.pdf.

24. World Green Building Council. Available at http://www.worldgbc.org/what-green-building.

25. UN-HABITAT and UNEP. Green Building Interventions for Social Housing, [UN-Habitat, 2015]. Available at https://unhabitat.org/books/green-building-interventions-for-social-housing/.

26. UNEP. The "State of Play" of Sustainable Buildings in India, UNEP, 2010. Available at http://staging.unep.org/sbci/pdfs/State_of_play_India.pdf.

27. Tathagat, D. and Ramesh D. D. Role of green buildings in sustainable construction—Need, challenges and scope in the Indian scenario, *IOSR Journal of Mechanical and Civil Engineering*, 12(2):4.

28. TERI. *Green Growth and Buildings Sector in India*. New Delhi, India: The Energy and Resources Institute, 2015, p. 5.

29. AMRUT—The Mission. Available at http://amrut.gov.in/writereaddata/The%20Mission.pdf.

30. AMRUT—Thrust Area. Available at http://amrut.gov.in/writereaddata/Thrust%20Areas.pdf.

31. AMRUT—Coverage. Available at http://amrut.gov.in/writereaddata/Coverage.pdf.

32. AMRUT—Fund allocation. Available at http://amrut.gov.in/writereaddata/Fund%20Allocation.pdf.

33. SMART CITIES. What is Smart City. Available at http://smartcities.gov.in/upload/upload-files/files/What%20is%20Smart%20City.pdf.

34. How many smart cities in Each State/UT? Available at http://smartcities.gov.in/upload/uploadfiles/files/No_%20of%20Smart%20Cities%20in%20each%20State.pdf.
35. Smart Cities Mission—Strategy. Available at http://smartcities.gov.in/content/innerpage/strategy.php.
36. PIB. Housing for All by 2022" Mission—National Mission for Urban Housing, (As on June 17, 2015). Available at http://pib.nic.in/newsite/PrintRelease.aspx?relid=122576.
37. MoHUA. Pradhan Mantri Awas Yojana. Available at http://pmaymis.gov.in/.
38. Smets, Peer and Paul van Lindert. Sustainable housing and the urban poor, *International Journal of Urban Sustainable Development*, 8, 2016. Available at http://dx.doi.org/10.1080/19463138.2016.1168825.
39. Bredenoord, J. Sustainable building materials for low-cost housing and the challenges facing their technological developments: Examples and lessons regarding bamboo, earth-block technologies, building blocks of recycled materials, and improved concrete panels, *Journal of Architectural Engineering Technology*, 6(1), 2017. Available at https://www.omicsonline.org/open-access/sustainable-building-materials-for-lowcost-housing-and-the-challengesfacing-their-technological-developments-examples-and-lessonsr-2168-9717-1000187.pdf.
40. Dodge Data & Analytics. World Green Building Trends 2016: Developing Markets Accelerate Global Green Growth—SmartMarket Report. Available at http://fidic.org/sites/default/files/World%20Green%20Building%20Trends%202016%20SmartMarket%20Report%20FINAL.pdf.
41. GRIHA—GRIHA Rating. Available at http://www.grihaindia.org/index.php?option=com_content&view=article&id=87.
42. ICICI Data Centre gets India's first CII-IGBC Platinum rating, Telangana Today, (As on April 19, 2017). Available at https://telanganatoday.com/icici-data-centre-india-platinum-rating.
43. Dembinski, Stanislas. Green Airports Taking Off, Media India Group, January-March 2017. Available at http://mediaindia.eu/feature/green-airports-taking-off/.

Endnote

1. The services sector includes, among others, public buildings, offices, shops, hotels, and restaurants.

14

Energy-Efficient Buildings: Technology to Policy and Awareness

Saurabh Mishra

CONTENTS

14.1 Introduction 273
14.2 Policy and Regulations...275
14.3 Awareness-Need for Community and Technology Partnership 276
14.4 Conclusion..279
Bibliography 280

14.1 Introduction

Infrastructure development is important to root the growth and development of any nation. Developing countries like India throughout the world are trying in earnest to develop and strengthen their infrastructure. Equally important is to realize the need for empowering the thrust in this area coupled with Energy-Efficient Building (EEB) concept. Buildings form a very important component of any nation's plan to guarantee a better quality of life to its citizens. It is not only a material parameter of growth, development, and economic well-being, but it also manifests a sense of emotional and social security and satisfaction among the stake holders. The cost that this infrastructural development bears is phenomenal with economic perspective being more discussed and deliberated than the energy perspective. Inferring upon this further it is important to ponder that both pre- and post-building energy consumption efficiency need to be looked into.

According to an estimate, the building construction's energy consumption around the world is far more than any other individual sector, harnessing nearly 42% of the total available resource (EEB Africa report). As more rural area gets transformed to urban consolidation, the electricity consumption registers a steep spike in the consumption graph. This is despite the fact that most of the infrastructurally poor countries do lack in power generation too. There is a yawning gap in the demand and supply of power. This raises the concern that development is being done at a questionable cost. The buildings become liability creation rather than asset formulations.

Urbanization pressure is mounting and the core and fundamental challenge is to focus on constructing buildings that are energy efficient. This is as India is shifting the crux from being rural oriented to being more urban centric. The heart of the matter becomes even more serious when exodus of rural population to urban areas is considered in context with fewer urban resources being burdened with heavy populace pressure and high energy demands. Between 2001 and 2011 census, India's urbanization went up from 27.8% to 31.1%, which doesn't seem too impressive in terms of growth challenges, but the fact lies in hazardous human capital accumulation or what many prefer to call "Hidden Urbanization" (Kumar, 2016). According to the latest data between 2015 and 2031, the pace of urbanization, as per compounded annual growth rate (CAGR) of 2.1%, is estimated to be almost double the growth rate of China (The Hindu). Appending to claims, Arvind Panagariya, the chairman of NITI Aayog (2016), had commented that urbanization in India is slated to increase to over 60% in the next 30 years, assuming a 7%–9% rate of economic growth. The challenge of disorganized urbanization haunts the entire growth pattern. The "Smart Cities Mission" program by Government of India envisions to deliver from a variety of bottlenecks: energy efficiency, waste management, environmental sustenance, and so on.

Urban populations are predicted to grow all over the globe. In China and India, the urban dwelling is predicted to swell by over one billion by 2050 (Figure 14.1). India is predicted to overtake Japan in energy consumption by 2030 (WBCSD, 2009) (Figure 14.2). Hence, with building energy projections anticipating volumes in energy consumption in India, EEB form a mandatory supplement for containing and rejuvenation of the unorganized urbanization in the country. The energy resource crunch confronting the supply sector can still be assuaged by designing and developing future buildings on the concepts of energy efficiency and sustainability (The Energy and Resources Institute (TERI).

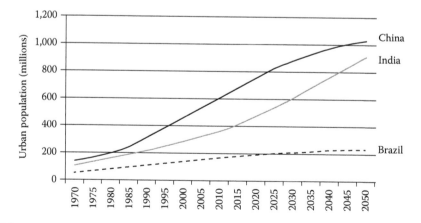

FIGURE 14.1
The rising urban population in developing countries (China, India, and Brazil). (From WBCSD, Energy efficiency in buildings: Transforming the market. Report online accessed on May 4, 2017: Available at http://wbcsdservers. org/wbcsdpublications/cd_files/datas/business-solutions/eeb/pdf/EEB-TransformingTheMarket.pdf, 2009.)

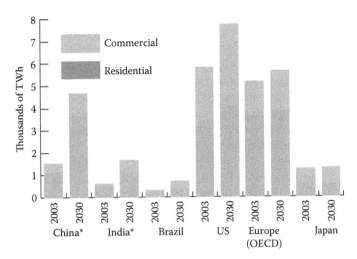

FIGURE 14.2
Building energy projection by regions in 2003 and 2030. (From WBCSD, Energy efficiency in buildings: Transforming the market. Report online accessed on May 4, 2017. Available at http://wbcsdservers.org/wbcsdpublications/cd_files/datas/business-solutions/eeb/pdf/EEB-TransformingTheMarket.pdf, 2009.)

14.2 Policy and Regulations

In India, the Bureau of Energy Efficiency (BEE) had taken the initiative back in 2004 by launching the Energy Conservation Building Code (ECBC). This was done to give the energy poverty struck nation structures that demand less energy and also help to save energy. ECBC would ensure "cross-check for building designs and specifications" (Times of India, 2014). Apart from BEE, other agencies have also been working toward making Indian building energy efficient; most of them partaking in a role through Green Building initiative viz. National Building Code (NBC), Indian Green Building Council (IGBC) and ECBC. Leadership in Energy and Environmental Design (LEED) formulated by the United States Green Building Council is an internationally accepted certification system for Green Buildings. LEED has specially designed certification as per need of Indian scenario. Green Rating Integrated Habitat Assessment (GRIHA), a TERI initiative, is the national rating system developed for India. Being an issue belonging to various states, many have opted to formulate their own policies along the guidelines as prescribed by the center. As per ECBC, "While the Central Government has powers under the EC Act 2001, the state governments have the flexibility to modify the code to suit local or regional needs and notify them. Presently, the code is in voluntary phase of implementation" (BEE). Much of the impetus is lingered and left dangling in wake of clear cut mandatory guidelines. A comprehensive energy efficiency policy in India has not yet seen the light of the day; about 22 states are at various stages of mandating ECBC although states like-Rajasthan, Odisha, Uttarakhand, and Andhra Pradesh have already notified the code.

"Energy Efficiency Improvements in Commercial Buildings" is a collaborative project of UNDP and Ministry of Power of th Government of India. The project aims at strengthening the capacities of State Designated Agencies (SDAs) by helping them with demonstration of energy-efficient commercial buildings with reference to various climatic zones of India. This would aid in compliance and expansion of the ECBC across the Indian states.

Primary energy demand in India has shown immense growth from about 450 million tons of oil equivalent (Mtoe) in 2000 to about 770 Mtoe in 2012. Furthermore, it is estimated to increase from about 1250 to 1500 (estimated in the Integrated Energy Policy Report) Mtoe in 2030 (Ministry of Power, Govt. of India). The prime factors contributing to the mammoth growth in energy demand are building constructions and post construction consumptions. This growth features the current low levels of energy supply in India. In India, "the scope for energy efficiency improvements in existing buildings is immense. Energy Audit Studies have revealed a savings potential to the extent of 40% in end use such as lighting, cooling, ventilation, refrigeration, and so on," (BEE).

As intuited by the Ministry of Power, no country in the world has been able to achieve a Human Development Index of 0.9 or more without an annual energy supply of at least 4 Mtoe per capita. For developing countries like India, energy production is as big a challenge as efficient utilization of present energy availability. This would help in two ways: first, by using efficient technologies, the current energy demands would be more amicably met, and second, the future energy demand pressure would also be depreciated.

UN Resident Coordinator and UNDP Resident Representative, Yuri Afanasiev of India opined that, "Given that India is urbanizing faster than most countries in the world, the need for both new residential and commercial buildings across the country, especially in cities, is escalating. India is expected to increase the building stock by nearly three times the existing building stock by 2030. Hence, there is no better time than the present for India to integrate energy efficiency and conservation into its efforts towards sustainable development." The economist intelligence unit of the "Global Buildings Performance Network" has ascertained that by 2050, India will see an unprecedented escalation of floor area of around 400%. Much of his would be contributed from the residential buildings sector, approximately 75%, and hence the Indian policy makers should start to focus on residential as compared to large scale commercial buildings.

14.3 Awareness-Need for Community and Technology Partnership

The EEB (2008) project commissioned research indicates "awareness" as an important component of fostering EEBs throughout the world (Figure 14.3). The figure shows a three-stage analysis (awareness, consideration, involvement) wherein it is clearly evident that though a high percentage of professionals in the building industry are aware about the concept of EEB, this number witnesses a drop when the consideration of the plan is upfront, finally stooping to a staggering low percentage actually involved in the act of realizing EEB. For India, comparing the present figure is a dismal show with 64% professionals being aware of the EEB and only 13% of those consider it, and the final conversion rate of those actually involved is a mere 05%. Falling only ahead of Japan where the actual involvement rate is high (5%–5%). The problem of conversion rate is prime and even bitter for developed countries like France and Spain.

The problems of increasing awareness and conversion rate must be dealt with differently. Although most of the solutions are hedged at targeting the two through superimposing each other, this leads to lower rates of involvement, even where there are higher rates of awareness as is the case with France, Spain, and other countries (Germany being the exception). Awareness could constitute simple advertising to begin with, but beguiling the

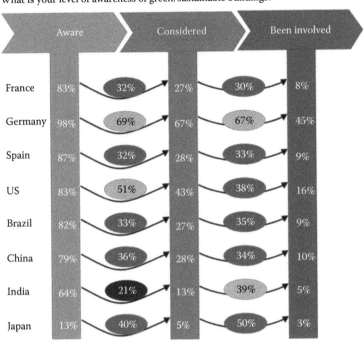

"What is your level of awareness of green/sustainable buildings?"

FIGURE 14.3
Depicting "Awareness" and involvement of building professionals. Note: Figures between the blue columns represent the proportion of the previous column number; for example, in France 32% of those who were aware had considered, and 30% of those had been involved.

same to involvement and actual action certainly requires better information. It calls for a strong community and technology partnership.

Policy awareness is an important component of object-oriented proliferation and energy sustainability. Awareness is generally dealt with dissemination of relevant data to the stakeholders at large. Most of the efforts are targeted to reach Meta audiences. Audiences are heterogeneous in nature and hence require different approaches and options. Similar kind of data in regular and repetitive pushing makes the entire process of awareness look mundane. Too much data pounding has also been known to be linked with development of apathy in the target audiences. This does not just end there, but it breeds in the coming generations. This then makes the intended message/communication being treated as a fallout effort on governmental or non-governmental front—a simple waste of time and money. There is also a significant suo moto of annoyance that percolates within the stakeholders; this leads them to assume there is nothing new here.

Learning needs empathy, motive, and effect to be imbibed by audiences before they embark on the journey to go from being perspective shareholders to operational stakeholders, to becoming real contributors. Improving energy efficiency is a process of change as it requires changing attitudes to energy use (Srinivas et.al. 2015). For special programs like EEBs, there is also a certain level of empathy that needs to be inculcated within the target audiences. This needs to be done from various perspectives as the Meta audience is approach sensitive. The chapter proposes a lotus model of policy awareness in this regard.

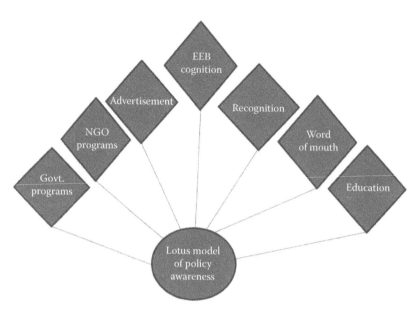

FIGURE 14.4
Depicts lotus model of policy awareness.

The Lotus Model (Figure 14.4) suggests favorable conditioning to arouse contribution from the various stakeholders. The model proposes to handle audiences with choice approaches. The model is based on the belief that a sense of social achievement is the prime and conditioned desire of every human being. The model is divided into three subsections:
   On the Right side (Individual Level) of the model are components:

   a. *Education*: Any policy and awareness related to it appends an important component of understanding and pursuance for the same. Education is known to implant the confluence of both. Awareness and Education are to be treated as synonyms with little knowing that they are supplementary activities rather than substitutes. In India, there are several myths associated with EEB itself; most confusing it with green building and low carbon emissions etc.
      Educating individuals: Primary, secondary, senior secondary, and graduate levels would ensure the encapsulation of the basic need for EEB and help in norm culturing. It would also help to sensitize the young minds and future bearers about the necessity of the same.

   b. *One to one publicity (word of mouth)*: The government should aim at imparting knowledge and training people. Helping them gain hands-on experience on the latest guidelines, technologies, and governmental schemes with regards to EEB. The Indian civic governmental system may identify Block Development Officers (BDOs), Zila Panchayat members, and other officials at Nigam levels for this purpose. These people should then be awarded with the task to approach the society by disseminating and encouraging in person. Person-to-person mapping proves to be pretty impressive in dealing with the nuances of individuals and it also helps to set up general beliefs and mounting community pressure and social approvals.

c. *Recognition, awards, and accolades*: Help to set a prudent and recurring ripple of sense of positive achievement within the members of the civic societies. It mobilizes people to act in a definitive way as proposed by the government.

This helps to reach the helm of the model from right hand side by way of EEB Cognition.

On left hand side (Meta) of the model are components:

a. *Government programs/policy*: Governmental role in incubating the impetus is for EEB program. Government has already launched several programs that are supplemental to EEB. Documentaries prove a versatile source of imparting and adding to the knowledge bank of the general masses, as has been proven in various governmental campaigns. Grants in the form of loans and advances and easy EMI solutions for the public can be another way which can help to induce real action on the ground. Public meeting and messages are also known to impact people and persuade them by building mass acceptability. Organizing competitions and awareness week at various levels could also help to mobilize opinion and action.

b. *Role of nongovernmental organizations (NGOs)*: Nongovernmental Organizations have been known to work in tandem with governmental policies as this has strengthened various campaigns. NGOs often have deeper penetration levels in places where governmental programs have not been taken seriously. Impeccable and designed nature of outreach viz. nukkad nataks, local dialectal slogans, themed rhymes have been NGOs favored approaches. These have helped to cater community awareness, understanding, and support.

c. *Advertisement/slogan/punchlines*: The post popular mode of approaching the stakeholders is via advertisement. Although it is an important mode of making the EEB effect happen, it should still be practiced with much caution. Too much dependence on this singular mode can withhold the desired impact. Hatched in proper connotation with the earlier two components, it can work wonders. Hence, it cannot be excluded as a component of the model.

These three components of Meta reach to the helm of the model from right hand to EEB Cognition. The proposed model simplistically tends to the problems of only Meta approach. It also helps to recognize and attend to the importance of individual sensitization through norm generation and empathy embankment.

14.4 Conclusion

In an interview with a national daily, the Director of the Sustainable Habitat Division of TERI, Ms. M. Majumdar, suggested that Buildings constitute a lot to our daily lives. These buildings utilize a large amount of energy which comes from resources that produce significant amount of carbon footprint; therefore, there is an urgent need to maximize the performance of these buildings. (The Hindu). A report by The Economist Intelligence Unit (2013), suggests that the major hurdle in enhancing building efficiency is awareness of energy-efficient technologies and their benefits and awareness programs for various

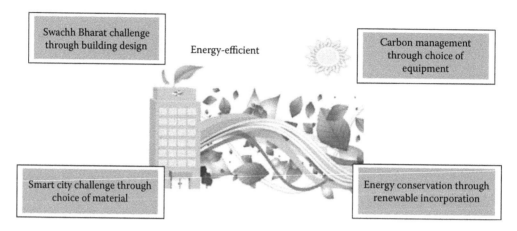

FIGURE 14.5
Energy-efficient buildings as a role manager for various state challenges.

stakeholders should be encouraged to make people more comfortable and familiar with the latest technology trends.

Building Energy Efficiency (BEE) implementation is not just a unilateral program with singular impact and results, it is a multilateral program that would help to mitigate through various state challenges setting in the domino effect. India, as a developing nation, is faced with a variety of developmental and environmental challenges. Government of India has framed programs which are aimed to resolute transformation, such as Swachh Bharat, Smart Cities Mission, Pradhan mantri Awas Yojna, Sulabh Awas Yojna, State Energy Conservation Fund (SECF) Scheme, National Mission for Enhanced Energy Efficiency (NMEEE), Super Efficient Equipment Program (SEEP), Energy Efficiency Financing Platform (EEFP), and so on. The BEE program excogitates an automated response for such state challenges and aids the various governmental programs as shown in Figure 14.5. It also aims to address environmental concerns by reducing carbon emissions, which is in line with the Green Building Program inducing waste management, environmental preservation, water management, rain water harvesting, and so on.

Government's ability to monitor and enforce energy efficiency implementation has observed to be generally weak. Policy formulations should be should be stepped up and engrossed by developing the statewide code in India, as was planned to be implemented mandatorily by the end of 12th five-year plan (2012–2017). Stakeholders are the main builders and should be approached with an aim to sensitize and empathize; approached with messages of building energy efficiency both at Meta and Individual levels.

Bibliography

Abdul, N. S. (2015). Market assessment of energy-efficient building materials. In *Proceedings of International Conference on Energy Efficiency in Buildings*, Passau, Germany, December 17–18.

Achieving scale in energy-efficient buildings in India A view from the construction and real estate sectors. (2013). A report from The Economist Intelligence Unit. Available at http://www.gbpn.org/sites/default/files/06.EIU_INDIA_Casestudy.pdf

Alam, M. Sathaye, J., and Barnes, D. (1998). Urban household energy use in India. Efficiency and policy implications. *EnergyPolicy*, 26 (downloaded on 13.05.2013)

Al-Shemmeri, T. (2011). *Energy Audits—A Workbook for Energy Management in Buildings*, Wiley-Blackwell Hoboken United States.

Altomonte, S., Reimer, H., Rutherford, P., and R Wilson (2013). Towards education for sustainability in university curricula and in the practice of design. *Proceedings of the PLEA*. Munich, Germany.

Andrzij, Z. (2013). *Energy Systems for Complex Buildings*, Springer-Verlag London.

Annual Progress Report. (2015). UNDPGEF- BEE Project Energy Efficiency Improvements in Commercial Buildings (July 1, 2014 to June 30, 2015).

A ProDoc [Project Document]. (2012). Under GEF 4 cycle for implementation by United Nations Development Programme and Bureau of Energy Efficiency. *Energy Efficiency Improvements in Commercial Buildings*. Online accessed on May 12, 2017: Available at http://www.undp.org/content/dam/india/docs/energy_efficiency_improvements_in_commercial_buildings_project_document.pdf

A Report on Energy Efficiency and Energy Mix in the Indian Energy System. (2030). Using IndiaEnergy Security, Scenarios, 2047, NITI Aayog, 2015.

Bentler, P. M. (1990). Fit indexes, Lagrange multipliers, constraint changes and incomplete data in structural models. *Multivariate Behavioral Research*, 25(2): 163–172.

Beteille, A. (2001). The Indian Middle Classes, Times of India 5 February (downloaded on June 6, 2013). CREED (n.d.). Available at http://www.hs-owl.de/creed/.

Buildings Energy Data Book. (2015). U.S. Department of Energy's Office of Energy Efficiency and Renewable Energy. Online accessed on May 4, 2017: Available at https://www.nrel.gov/docs/fy17osti/66591.pdf

Building Energy Efficiency Opportunities, Retroficiency, Boston, MA.

BSEEP (2009), UNDP-GEF, Project Document of Buildings Sector Energy Efficiency Project (BSEEP). Online accessed on May 4, 2017: Available at http://www.my.undp.org/content/malaysia/en/home/operations/projects/environment_and_energy/building-sector-energy-efficiency-project.html

Bureau of Energy Efficiency (BEE). Available at https://www.beeindia.gov.in/content/buildings.

EEICB. (2011). UNDP-GEF Project Document of Energy Efficiency Improvements in Commercial Buildings. Online accessed on May 12, 2017: Available at http://www.indiaenvironmentportal.org.in/files/energy%20efficiency%20in%20commercial%20buildings.pdf

Energy Efficient Buildings: Europe, Navigant Research, London.

Energy Efficiency in Buildings Business Realities and Opportunities. (2008).WBCSD. Report online accessed on May 4, 2017. Avaiable at http://wbcsdservers.org/wbcsdpublications/cd_files/data/ busines-solutions/eeb/pdf/EEB-Facts&Trends-FullReport.pdf

Energy Performance of Buildings Directive, The Directorate-General for Energy, European Commission, Brussels, Belgium.

Fangzhu, Z. and Philip C. (2010). *Green Buildings and Energy*. Cardiff University, Cardiff, Wales.

ICEEB. (2015). Implementing energy efficiency in buildings. A compendium of experiences from across the world, December 17 – 18, 2015, p.7. UNDP. New Delhi, India.

IEA. Modernizing Building Energy Codes to Secure our Global Energy Future (2013), Policy Pathway, IEA-UNDP. (2013)a. Available at https://www.iea.org/publications/freepublications/publication /Policy PathwaysModernisingBuildingEnergyCodes.pdf

IEA. Technology Roadmap: Energy Efficient Building Envelopes. (2013)b. Available at https://www.iea.org/publications/freepublications/publication/TechnologyRoadmapEnergyEfficientBuildingEnvelopes.pdf.

Investing in energy efficiency in Europe's buildings—A View from the Construction and Real Estate Sectors (2013). The Economist Intelligence Unit Limited, Global Buildings Performance Network (GBPN).

Jouhara, H. et. al. (2013). *Building Services Design for Energy Efficient Buildings*, Milton Park, UK and New York, Routledge.

Krishna, K. (2012). Building Construction—Design Aspects of Leakage and Seepage Free Buildings, McGraw Hill, New Delhi, India.

Kubba, S. (2012). *Handbook of Green Building Design and Construction*, Butterworth-Heinemann. Woburn, United States.

Kumar, S., Kapoor, R., Rawal, R. et.al. (2010). *Developing and Energy Conservation Building Code Implementation Strategy in India*, Summer Study, American Council for an Energy Efficient Economy, Washington, DC.

Leiserowitz, A. and Thaker, J. (2012). Climate change in the Indian mind. Yale Project on Climate Change Communication (downloaded on May 09, 2013).

Mawdsley, E. (2004). India's middle classes and the environment. *Development and Change* 35(1): 79–103. (downloaded on May 09, 2013).

Mercom Capital Group (2011). India Renewable Energy Awareness Survey. Executive Summary 2011 (downloaded on 09.05.2013).

Meyer, C.H. and Birdall, N. (2011). New Estimates of India's Middle Class. Technical Note. Centre of Global Development. Avaiable at http://www.cgdev.org/doc/2013_MiddleClassIndia_MHA (2011). Census of India 2011. Ministry of Home Affairs. Govt. of India. Available at http://censusindia. gov.in (downloaded on 22.05.2013).

Mili Majumdar. (2001). *Energy-Efficient Buildings in India*, The Energy and Resources Institute (TERI), New Delhi, India.

Ministry of Power. Avaiable at http://www.powermin.nic.in/en/content/energy-efficiency Global Buildings Performance Network. Achieving scale in energy-efficient buildings in India: A view from the construction and real estate sectors. Available at http://www.gbpn.org/reports/achieving-scale-energy-efficient-buildings-india-view-construction-and-real-estate-sectors.

Reddy, B. S. (2004). Economic and Social Dimensions of Household Energy Use: A case study of India. In Ortega, E. and S. Ulgiati (Eds.). *Proceedings of IV Biennial International Workshop. Advances in Energy Studies*. Campinas: pp. 469–477. Available at http://www.unicamp.br/fea/ortega/energy /Reddy.pdf (downloaded on 02.10.2013).

Shabnam, B. (2015). Benchmarking of commercial buildings in India. In *Proceedings of International Conference on Energy Efficiency in Buildings*, December 17–18. New Delhi.

Shukla, R. (2005). India Science Report. Science Education. Human Resources and Public Attitude towards Science and Technology. NCAER. New Delhi, India (downloaded on June 26, 2013).

Shukla.Y., Rawal, R., and Shnapp, S. (2015). *Residential Buildings in India: Energy Use Projections and Savings Potentials, Summer Study, European Council for an Energy Efficient Economy*, Hyères, France.

Srinivas, S. N., Butchaiah, G., Sanjay, S., and Vittalkumar, D. (2015). *Towards "Building" Energy Efficiency International Conference on Energy Efficiency in Buildings (ICEEB 2015)* December 17–18. United Nations Development Programme, New Delhi, India.

Steg, L. and Vlek, C. H. (2009). Encouraging pro-environmental behavior: An integrative review and research agenda. *Journal of Environmental Psychology* 29: 309–317.

Stern P. C., Dietz, T. H., Abel, T., Guagnano, G. A., and Kalof, L. (1999). A value-belief-norm theory of support for social movements: The case of environmentalism. *Human Ecology Review*, 6 (2): 81–97. TechnicalNote_CGDNote.pdf. (downloaded on May 21, 2013).

The Energy Conservation Building Code (ECBC). (2007). *Ministry of Power*, Government of India. New Delhi.

TOI (2016). India's urbanization likely to be 60% in 3 decades: Panagariya. Available at https://timesofindia.indiatimes.com/business/india-business/Indias-urbanization-likely-to-be-60-in-3-decades-Panagariya/articleshow/52008570.cms%20Apr%2027

UNEP SBCI|TERI, Background paper on sustainable buildings and construction for India: Policies, practices and performance, *Influence of Indian Buildings on Climate Change*. Online accessed on May 4, 2017: Available at http://www.teriin.org/eventdocs/files/sus_bldg_paper_1342567768.pdf

Vlek, C. H. and Steg, L. (2007). Human behavior an environmental sustainability. Problems. driving forces, and research topics. *Journal of Social Issues* 63(1): 1–19.

Waters, J.R. (2013). *Energy Conservation in Buildings: A Guide to Part L of the Building Regulations*, Blackwell. Oxford, United Kingdom.

WBCSD. (2009). Energy efficiency in buildings: Transforming the market. Report online accessed on May 4, 2017: Available at http://wbcsdservers.org/wbcsdpublications/cd_files/datas/business-solutions/eeb/pdf/EEB-TransformingTheMarket.pdf

World Green Building Council. (2009). Perspectives on green building, *Renewable Energy Focus* Nov/Dec. 2009.

Index

Note: Page numbers followed by f and t refer to figures and tables respectively.

181 Fitzroy Street, St Kilda. *See* Case studies

A

AAC blocks, 22, 75–76
Absorption of solar energy, 74
Active solar systems, 243
Active solar thermal, 246–247, 247f
Active thermal storage systems, 135
Aerogel, 76
Aerogel blanket needle glass fibers
　chemical characteristics, 155–156, 156f
　hydric properties, 156–159, 157f–158f
　texture, 153–155, 154f–155f
　thermal properties, 159–160, 160f
Aerogel blankets
　elaboration of, 153f
　method, 152–153
　needle glass fibers, 153–160, 154f–160f
　specific heat capacity, 160–163,
　　161f–162f, 162t
Aerothermal energy, 203
Air changes per hour (ACH), 203
Air-source heat pump (ASHP), 172–173
Albedo, 66
Aluminum coating, 66
Aluminum powder, 75
American Society for Testing and Materials
　(ASTM) standards, 42
Amorphous silicon solar cell, 115, 116f
AMRUT, 265
Anthony Ziem, 253–254
Anthropogenic greenhouse gas (GHG)
　emissions, 6f
Architectural integration PVs, 121
Arup's Sustainable Project Appraisal Routine
　(SPeAR), 14, 17f
*ASHRAE Handbook of HVAC Systems and
　Equipment*, 42, 174
Atal Mission for Rejuvenation and Urban
　Transformation (AMRUT), 265
Attenuated total reflectance (ATR) mode, 155
Autoclave aerated concrete (AAC) blocks, 22,
　75–76
Autodesk green building studio, 216t
Automated controlled sensors, 241, 242f

B

Balance-of-plant system costs, 248
Benchmarking energy efficiency, 240
BESTEST, 225
Biomass, 203–204
BIPVs. *See* Building integrated photovoltaics
　(BIPVs)
Black carbon emissions, 6
Block Development Officers (BDOs), 278
BLUE Map scenario, 245–246, 250, 252
BREEAM (United Kingdom), 11, 60
Brunauer-Emmett-Teller method (BET), 155
Brundtland Commission, 60
BSim, 216t
Building Energy Efficiency Certificate (BEEC),
　240–241
Building-energy-rating system, 268
Building energy simulation (BES), 216t–222t, 224
Building envelope glazing materials and
　technologies
　electrochromic devices, 64–65
　micro-blinds, 65
　polymer dispersed liquid crystal devices, 65
　suspended particle devices, 64
　switchable glazing, 64
　vacuum glazing, 64
Building envelope roofs
　domed roofs, 66
　evaporative cooling roofs, 69
　green roofs, 67
　photovoltaic roof envelopes, 68
　reflective roofs, 66–67
　thermal insulation, 68–69
　ventilated roofs, 66
Building envelopes
　building simulation software/programs,
　　69–70
　defined, 60
　elements, 73
　frames, 65
　glazing materials and technologies, 64–65
　maintenance, 70
　roofs, 66–69
　systems, and renewables technologies,
　　202–204

Building envelopes (*Continued*)
  thermal mass, 69
  thermal resistance (R-value), 61
  walls, 61–62
  windows and doors (fenestrations), 63–64
Building envelope walls
  latent heat storage, 62
  lightweight concrete, 62
  passive solar walls, 61–62
  ventilated or double skin walls, 62
Building Information Modeling (BIM)
    software, 229
Building integrated photovoltaics (BIPVs),
    68, 119
  building attached photovoltaic
    products, 120
  codes and standards, 126
  facade integration of photovoltaic, 124–125
  foil products, 119
  installation, 120–121
  market limitations, 126
  module products, 119
  performance, 125–126
  price, 125
  roof integration of photovoltaic, 121–124
  for smart cities, 127
  solar cell glazing products, 120
  tile products, 119
Building management, energy efficiency
    in, 212
Building Performance Simulation (BPS), 35
Building Research Establishment's
    Environmental Assessment Method
    (BREEAM), 11, 60
Buildings. *See also* Energy-efficient
    buildings (EEBs)
  automation and control systems, 241, 242f
  cooling load, 24, 174
  determining energy performance of,
    239–240
  development concepts, sustainability, 7f
  energy analyzer, 216t
  energy codes implementation, 10f
  energy efficiency policies for, 12f
  façade technology, 203
  integration PVs, 121
  orientation, 243–244, 244f
  thermal inertia, 21
  typical energy flow in, 238, 239f
Building sector, growth of, 3–4
BuildingSim, 216t
Bureau of Energy Efficiency (BEE), 113,
    262–263, 275

C

Cadmium telluride (CdTe) solar cell, 116, 116f
Campus district cooling system, Cairns Campus,
    James Cook University, 146–147
Carbon dioxide (CO_2), 90
  emissions, 4–6, 26
Carbon monoxide, 80
CASBEE (Japan), 11
Cellular polystyrene, 69
Case studies, 253
  181 Fitzroy Street, St Kilda, 253
  Anthony Ziem, 253–254
  CSR House, NSW (AUSTRALIA), 253
  ECBC complaint building, 253
  Infinity Benchmark, Kolkata, 254
  Infosys, Pocharam, Hyderabad, 254
  Olympia Tech Park, Chennai, 254
  San Francisco public utilities commission
    (SFPUC) building, 254
Cement stabilized rammed earth (CSRE), 98
CFD. *See* Computational fluid dynamics (CFD)
Chemical vapor deposition (CVD), 63
Chilled-ceiling systems, 176t
China
  commercial energy in, 5
  construction market in, 4
CHP. *See* Combined heat and power (CHP)
Climate change, challenge of, 237–238
Climatic conditions, 63
Coal, consumption, 2–3
Coefficient of performance (COP), 171–172
Combined heat and power (CHP), 245, 247–248
Comfort-to-comfort systems, 174
Commercial energy consumption, 4–5
Commercial sector delivered energy
    consumption, 5t
Compliance cycle, 13f
Composite solar walls, 61
Compounded annual growth rate (CAGR), 274
Compound parabolic concentrator (CPC), 79
Computational fluid dynamics (CFD), 223–226
  COMSOL Multiphysics, 224–225, 224f
  flow diagram for, 223f
  Fluent, 225–226, 225f
  Navier–Stokes equations, 223
  simulation approaches, 223
COMSOL Multiphysics, 217t, 224–225, 224f
Concrete-structured buildings, 97
Concrete walls, 62
Concrete with microencapsulated PCM, 143f
Condensed refrigerant, 42
Condensing boilers, efficient buildings, 203

Confederation of Indian Industry (CII), 112, 268

Conference of Parties—21 (COP), 262

Control automation, 204

Cooling with PCM integrated into ceiling, 143f–144f

Cool roofs, 242. *See also* Solar reflective roofs

COP, 171–172

Correlation methods, buildings performance, 238

Cost Calculator gain block, 230–231

Cost effective energy-efficiency concepts, 18–21

    glazed components, shading of, 21–22

    green roof concept, 22–25

    opaque components, insulation of, 22

    ventilation, improved, 21

Cost-optimality, 191–193, 192f, 193t

CSR House, NSW (AUSTRALIA). *See* Case studies

D

Daylighting, 205

Decentralized phase change material cooling system, 145

Density diffusivity, 83

Desiccant cooling systems, 176t

DesignBuilder, 217t

Designer's Simulation Toolkit (DEST), 217t

DGNB (Germany), 11

Direct evaporative cooling (DEC) systems, 170, 175t

DOE-2, 217t

DOE-2.1E, 240

Domed roofs, 66

Domestic electricity consumption, 2f

Domestic hot water, 138, 138f

Double- or triple-glazed windows, efficient buildings, 203

Double skin façades (DSF), 75

    thermal performance, 75

Double skin roofs, 66

Drought, 2

Dry green roof, 79

Duct-free split heating and air-conditioning system, 168

Dynamic energy efficiency policies, 11

Dynamic methods, buildings performance, 238

Dynamic thermal simulation, 34

E

Earth to air heat exchangers (EATHE), 83

ECBC. *See* Energy Conservation Building Codes (ECBC)

ECBC complaint building. *See* Case studies

ECOTECT, 218t

Ejector cooling systems, 176t

Electrochromic devices, 64–65

Embodied energy, 84–85

    construction materials and buildings, 92–98

    defined, 92–93

    intensities for common building materials, 94t, 95f

    significance, 91–92

EnerCAD, 218t

Energy balance in buildings

    energy gain, 213

    energy loss, 213

    external heat gain, 214

    filtration loss, 214

    ground losses, 214

    heated space, 213f

    internal heat gains, 214

    loss due to thermal bridges, 214

    transmission loss, 214

Energy conservation, 134

Energy Conservation Building Codes (ECBC), 8–9, 18, 267, 275

Energy consumption, building, 2–3

    changing climate, 237–238

    commercial, 4–5

    current levels of, 236–237

    in different commercial sectors, 210, 210f

    end-use wise energy consumption and, 211f

    energy conservation and, 211

    energy demand and, 132–134

    energy efficiency in building sector and, 237

    by energy-efficient buildings, 4–6, 212, 212f

    by energy source, 3f

    GHG emissions and, 237

    global, 2–3, 110f, 133f, 210f

    Intergovernmental Panel on Climate Change's (IPCC's) fourth assessment report on, 236–237

    National Action Plan on Climate Change (NAPCC) and, 237

    in residential *vs.* commercial buildings, 4, 211, 211f

    trends in, 2f

Energy consumption in Europe

    European climate and energy package, 181

    greenhouse gases (GHGs) emissions, 181

    non-residential sector, 186–187, 187f

    overall trends in buildings, 183–184, 183f–184f

    primary energy consumption, 182f

    residential sector, 185–186, 185f–186f

    savings at European level, 182f

Energy demand and consumption, 132–134
Energy efficiency, 193–194
    in buildings methodology, 204, 238–244,
        239f, 242f, 244f
    challenge of climate change and, 237–238
    current state of, 237–238
    defined, 236
    economic crisis and, 194
    in efficient buildings, barriers and
        challenges for, 196–198, 196t–198t
    energy conservation and, 134
    EU Projects for Building Stock
        Renovation, 195t
    implementation, 194–196
    member states, policies, measures, and best
        practices, 198–202, 199t–201t
    NZEBs target, 194
    regulations, 85
    renovation levels, 194
Energy Efficiency Building Code (ECBC),
        262–263
Energy Efficiency Certification Scheme (EECS),
        240–241
Energy Efficiency Financing Platform (EEFP),
        253, 280
Energy-efficient buildings (EEBs), 1–3, 6–7,
        60–61, 85, 89, 128, 210, 212, 253–254
    automated controlled sensors, 241, 242f
    awareness and need for community and
        technology partnership, 276–279
    benchmarking, 240
    building envelope, systems, and renewable
        technologies in, 202–204
    building orientation, 243–244, 244f
    building sector and, 3–4
    building sustainability and, 6–8
    cost effective energy-efficiency
        concepts, 21–25
    definition, 236
    energy conservation building codes
        and, 8–18
    energy consumption and, 4–6
    Energy Efficiency Certification Scheme
        (EECS), 240–241
    energy-efficient materials, 242–243
    energy flow in buildings and, 238, 239f
    energy management decision process
        and, 212f
    GHG emissions in, 4–6
    green building rating tools, 8–18
    heating cooling solution in, 244–252
    heating, ventilation, and air-conditioning
        (HVAC) systems in, 177, 177t–178t

    in high performing building categories,
        188–189
    NZEBs, 189–190, 191t
    performance of, 239–240
    policy and regulations, 275–276
    as role manager, 280f
    simulations of energy consumption by,
        212, 212f
    society, benefits and impact on, 25–26
Energy-efficient heating systems, 35
Energy-efficient materials, 242–243
Energy-Efficient Solar/Green Buildings, 267–268
Energy expert, 218t
Energy flow
    in buildings, 238, 239f
    typical, 238, 239f
Energy gain, 213
Energy loss, 213
Energy management decision process, 212f
Energy Performance Index (EPI), 113
Energy Performance of Building Directive
        (EPBD), 205
EnergyPlus software, 35, 219t, 226–227, 240
Energy resources, 132
Energy savings (ES), 18
    in BAU, 19
    night ventilation rate (ACH_N), 21, 21f
    potential of green roof, 24t
Energy-saving strategies for HVAC systems
    building behavior, effect of, 174, 175t–176t
    evaporative-cooled air-conditioning system,
        171–172
    evaporative cooling systems, 170, 172f
    ground-coupled HVAC technology, 172–173
    heat recovery systems, 174
    thermal energy storage (TES) systems,
        173–174, 173f
Energy storage, 37, 204
Energy sustainability, 60–61
Energy Technology Perspective (ETP), 254
Environmental certification, 89
Environmental degradation, minimization, 7
Environmental impact, 14
eQUEST software, 219t, 228, 240
ESP-r, 220t
European policies for energy efficiency, 187–188
Evaporative cooling, 205
    air conditioning systems utilizing,
        171–172, 175t
    roofs with, 69
    systems for, 170, 172f
Evaporative roofs, 77
Evapotranspiration, 79

Expanded polystyrene (EPS), 148
Extensive green roofs, 67
External heat gain, 214

F

Facade integration of phototvoltaic, 124–125
Facility energy decision system (FEDS), 220t
Faucets and showerheads, 7
Fenestrations, 63–64
Filtration loss, 214
Finite Element Method (FEM), 224–225
Floating pavilion, Rotterdam, The
    Netherlands, 147–148
Fluent software, 225–226, 225f
Fluid inventories, 14
Fluorinated tin oxide (SnO_2:F), 63
Fly ash, 75
Foil products, 119
FORTRAN software, 226, 229
Frames, 65
Framework for energy-efficient economic
    development (FEEED), 253
Fuel cells, 248
Fumed silica (SiO_x), 80

G

Gallium arsenide solar cells, 114, 115f
Geothermal energy, 203
GHGs. *See* Greenhouse gases (GHG) emissions
Glazing
    materials and technologies for, 64–65
    passive and low energy buildings, 76–77
    shading of components in, 21–22
    solar heat gain coefficient (SHGC), 76
Glazing materials and technologies, 64
    electrochromic devices, 64–65
    micro-blinds, 65
    polymer dispersed liquid crystal devices, 65
    suspended particle devices, 64
    switchable glazing, 64
    vacuum glazing, 64
Global Construction 2030, 3
Global energy potential, 133f
Global warming, 1
Graphical user interface (GUI), 226
Green buildings, 6–7, 111
    concepts and rating systems, 8–18
    construction, status, 90–91, 91f
    definition, 263–264
    drivers, conceptual framework for, 15t
    growth prospects in India, 267

in India, 261–263
    LEED India approach, 112
    policy impact of, 267–268
    reducing negative impacts, benchmark, 13f
    sustainable development goals, 260–261
    urban missions, 264–267
Green buildings rating system in India, 111
    Bureau of Energy Efficiency (BEE), 113
    Green Rating for Integrated Habitat
        Assessment (GRIHA), 11, 26, 92f, 112
    Indian Green Building Council, 112–113
Green building technology, 145
Green construction, 264
Green Data Centre Rating System, 268
Greenhouse gases (GHG) emissions, 1, 4–6, 6f,
    33–34, 74, 89, 132
    current state of, 236–237
    definition of energy efficiency and, 236
    energy consumption and additional,
        210–211, 210f–211f
Green Mark (Singapore), 11
Green Rating for Integrated Habitat Assessment
    (GRIHA), 11, 26, 92f, 112, 263, 275
Green roofs, 67
    concept, 22–25
    energy saving potential of, 24t
Green Star (Australia), 11
Ground cooling system, 83
Ground-coupled HVAC technology,
    172–173, 176t
Ground granulated blast slag (GGBS), 98
Ground loss of heat, 214
Ground source heat pumps (GSHP), 34,
    172–173
Ground source water to air heat pump of ZØE
    lab, 43f
Gypsum, 75
    board with phase change materials, 141
    plaster wallboards, 82

H

HEAT2, 220t
Heat, Air, and Moisture Simulation Laboratory
    (HAMLab), 220t
Heat and energy balance, 213–214, 213f
Heated space, 213f
Heater subsystem, 230
Heat gains
    external, 214
    internal, 214
Heating and air-conditioning split system,
    167–168

Heating and cooling solutions in EEBs
  active solar thermal, 246–247, 247f
  combined heat and power (CHP), 247–248
  heat pumps, 248–250, 251f
  roadmap vision, 244
  targets and assumptions for, 244–246
  technology development for, 246
  thermal energy storage, 250–252, 251f
Heating, ventilation, and air-conditioning
      (HVAC) systems, 60, 73, 81, 99, 166,
      167f, 205
  components, 169–170
  diagram for ZØE, 43f
  duct-free split heating and air-conditioning
      system, 168
  energy consumptions in, 169
  in energy-efficient buildings, 177, 177t–178t
  heating and air-conditioning split system,
      167–168
  hybrid heat split system, 168
  packaged heating and air-conditioning
      system, 168–169
Heating, ventilation, and air-conditioning
      (HVAC) systems, energy-saving
      strategies for
  building behavior, effect of, 174, 175t–176t
  evaporative-cooled air-conditioning system,
      171–172
  evaporative cooling systems, 170, 172f
  ground-coupled HVAC technology,
      172–173
  heat recovery systems, 174
  thermal energy storage (TES) systems,
      173–174, 173f
Heat pumps, 35, 248–250, 251f
Heat recovery systems, 174, 176t
Heat release rate (HRR), 80
Heat transfer, 61, 76
  fluid, 139
Heat waves, 2
High density polyethylene pipes (HOPE), 42
High-embodied energy, 101
High-performance retrofits, 18–20, 20f
High performing buildings, 264
  categories of, 188–189
High-performing constructed systems, 16f
HOBO sensors, 49–50
HOMER, 34
HOMESKIN European project, 152–153
Hominal thermal features, 49
House subsystem, 232
Humidity, aerogel blanket, 156–157

Hurricanes, 2
HVAC systems. *See* Heating, ventilation, and
      air-conditioning (HVAC) systems
Hybrid heat split system, 168
Hydro-NM-Oxide, 81

I

Ignition time, 80
Illumination, 60
Indian Government policies, 252–253
Indian Green Building Council (IGBC),
      112–113, 268
Indirect evaporative cooling (IEC)
      systems, 175t
Infinity Benchmark, Kolkata. *See* Case studies
Infosys, Pocharam, Hyderabad. *See* Case
      studies
INP, building inputs, 228
Insulated Trombe walls, 61
Insulating concrete forms, 243
Insulation, 202–203, 205
  materials, 7
  opaque components, 22
Integrated Building Energy Retrofit (IBER),
      240–241
Integrated ceiling and wall, 143
Integrated Energy Policy Report, 276
Intensive green roofs, 67
Inter-alia Technology Sub-mission, 266
Intergovernmental Panel on Climate Change
      (IPCC), 1, 236, 262
Internal heat gains, 214
International Building Codes, 126
Internet of things (IoT), 241

K

KNAUF PCM smartboard, 142f

L

Latent heat storage system (LHTES), 62,
      135–136, 136t, 138
Leadership in Energy and Environmental
      Design (LEED)™ system, 11, 60, 111,
      262, 268
  building in India, 113t
LEED-NC 2009, 11, 26
LENSES (Living Environments in Natural,
      Social and Economic Systems), 14, 16f

Life cycle energy (LCE), 91
    of buildings, 98–103
Life cycle energy analysis (LCEA), 98, 99f
Light-emitting diode (LED) bulbs, 262
Lighting arrangements, 73
Lightweight cement-based materials, 75
Lightweight concrete (LWC), 62, 75
Liquid pressure amplification (LPA)
        systems, 175t
Lithographic processes, 65
Loss due to thermal bridges, 214
Lotus Model, 278, 278f
Low-cost compact thermal energy
        storage, 247
Low-e glass windows, 35, 39t

M

Magnetron sputtering, 63
Market transformation for energy efficiency
        (MTEE), 253
Masonry-based walls, 61
Material utilization, 14
MATLAB software, 229–232, 231f
    Cost Calculator, 230–231
    Heater, 230
    House, 232
    Set Point, 230
    thermal model of buildings, 230, 231f
    Thermostat, 230
Mega urban schemes, 264–265
Metal-based walls, 61
Metallic reflective coatings, 66
Micro-blinds, 65
Microchannel systems, 14
Micro-generation government incentive
        scheme, 34
Micro/nanoencapsulated plasters and concrete,
        141–142
Micro turbines and gas turbines, 248
Modeling software, 216t–222t
    computational fluid dynamics, 223–226,
        223f–225f
    Energy Plus, 226–227
    eQUEST, 228
    FORTRAN, 229
    MATLAB, 229–232, 231f
    Open Studio, 227
    Radiance, 227–228
    significant parameters used in development
        of model, 215, 215t
    Simergy, 229

TRANE TRACE 700, 228–229
    Transient System Simulations (TRNSYS), 226
Moisture condensation, 61
Molten carbonate fuel cell (MCFC), 248
Monolithic aerogel, 76
MTEE, 253
Multilayer roofs, 66

N

Nanocrystal-impregnated window panes, 65
Nanotechnology, 203
Nansulate® coating, 81
National Action Plan on Climate Change
        (NAPCC), 237, 268
National Australian Built Environment Rating
        System (NABERS), 241
National Building Code (NBC), 275
National Built Environment Rating System
        (NBERS), 241
National Mission for Enhanced Energy
        Efficiency (NMEEE), 252–253, 262, 280
National Urban Housing and Habitat Policy,
        The (NUHHP), 263
Natural ventilation, 41
Navier–Stokes Equations, 223
Near-infrared (NIR) transmission, 63
Nearly net-zero energy homes (NNZEHs), 35
Nearly zero energy buildings (NZEBs), 189–190,
        191t, 205–206
Needle glass fibers aerogel blankets
    chemical characteristics, 155–156, 156f
    contact angle, 157, 158f
    hydric properties, 156–159, 157f–158f
    microstructure, 153–154, 154f–155f
    specific surface, porosity, and density, 155
    texture of, 153–154, 154f–155f
    thermal conductivity, 159–160, 159t, 160f
    water uptake, 156–157, 157f
    water vapor transmission, 158–159, 158f
Net-zero energy buildings (NZEB), 7, 33–34, 101.
        See also Zero-energy buildings (ZEB)
Net-zero energy homes (NZEHs), 35
Night cooling, 21
Night ventilation, 84
Night ventilation rate (ACH_N), 21, 21f
Nippon Sheet Glass (NSG), 77
NMEEE. *See* National Mission for Enhanced
        Energy Efficiency (NMEEE)
Nongovernmental organizations, 279
Non-insulated buildings, 79
Non-toxic interior paints, 7

O

Off-the-grid applications, 44
Olympia Tech Park, Chennai. *See* Case studies
On-site renewable energy generation, 204
Opaque components, insulation of, 22
Open Studio® software, 227
Organic solar cells, 116–117, 117f
Organization for Economic Co-operation and
        Development (OECD), 5, 110
Orientation, windows, 63

P

Packaged heating and air-conditioning system,
        168–169
ParaSol, 221t
Paris Climate Agreement, 262
Paris Climate Conference (COP21), 3
Passive and low energy buildings, 73–74
    embodied energy, 84–85
    glazing, 76–77
    ground cooling, 83
    life cycle energy demand, 85f
    night ventilation, 84
    roof, 77–79
    thermal insulation, 80–82
    walls, 74–76
Passive houses, 255
Passive solar heating, 243
Passive solar walls, 61–62, 74
Passive thermal storage systems, 135
PCM. *See* Phase change materials (PCM)
PD 2, building inputs for wizards, 228
Peak heat release rate (PHRR), 80
Perform achieve and trade scheme (PAT),
        252–253
Perkins+Will framework, 14, 17f, 18
Person-to-person mapping, 278
Phase change materials (PCM), 81, 226, 243
    integrated floor cooling, 144–145, 144f
    microencapsulated, 82
    TES, 138
    Trombe walls, 62
Photovoltaic (PV) technology, 114
    amorphous silicon solar cells using, 115
    architectural integration with, 121–124
    building envelopes using, 61
    building integration with, 121, 124–125
    cadmium telluride solar cells using, 116
    cells efficiencies in, 118
    gallium arsenide solar cells using, 114
    organic solar cells using, 116–117

roof envelopes, 68
roofs using, 77, 122f–124f
roof tiles, 123f
single-crystal silicon solar cells using, 114
solar cells efficiency in, 117–118
systems using, 25, 34
thin-film solar cells using, 114–115
Photovoltaic Trombe wall (PV–TW), 62
Physiological sensors, 49f
Plant-based polyurethane foam, 243
Plaster with microencapsulated PCM, 142f
PMAY. *See* Pradhan Mantri Awas Yojana
        (PMAY)
Policy awareness, 277–278, 278f
Policy interventions, 263
Pollution emissions, 33
Polyisobutylene (PIB), 39
Polymer dispersed liquid crystal
        devices, 65
Polymer electrolyte membrane (PEM) fuel
        cells, 248
Polystyrene, 22
Portland cement, 75
Pradhan Mantri Awas Yojana (PMAY), 260,
        266–267, 280
PRD, parametric run definitions, 228
Process-based approach, 93
Process-to-comfort systems, 174
Process-to-process systems, 174
Propylene glycol mixture (PGM), 39
PV/T/D (Photovoltaic/Temperature/
        Daylighting) system, 79

Q

Quick lime, 75

R

Radiance software, 227–228
Radiant floor heating system, 140f
Radiation control strategies, 21
Rainwater filtration system, 45
Reciprocating engines, 248
Recycled steel, 243
Reflected light color roof, 40
Reflective electrochromic materials, 65
Reflective indoor and outdoor coatings, 243
Reflective roofs, 66–67
Reflective-transmittive barrier, 68
Reflectivity, 66
REGEN, 14, 16f
Renewable electricity, 35

Renewable energy, 7, 101, 110
Replacement factor, 95, 96t
Residential dwellings, 3–4
Retrofit, 37
    high-performance, 18–20
RIUSKA, 221t
Roof technologies
    domed roofs, 66
    evaporative cooling roofs, 69
    green roofs, 67
    insulation, 77
    integration of photovoltaic, 121–124
    passive and low energy building, 77–79
    photovoltaic roof envelopes, 68
    reflective roofs, 66–67
    thermal insulation, 68–69
    top garden, 78f
    ventilated roofs, 66

S

San Francisco public utilities commission
        (SFPUC) building. *See* Case studies
SCM. *See* "Smart Cities Mission" program
Sedum, 23
Semiconductor materials, 114
Semi-transparent glass sheet, 62
Semi-transparent modules, 121f
Sensible heat storage, 135
Sensible storage medium, 136f
Set Point, 230
Silica aerogel, 76, 81
    blankets of. *See* Aerogel blankets
Silicone, 39
Silver coating, 63
SIMBAD Building and HVAC Toolbox, 221t
Simergy software, 229
SIM, large output files (text), 228
Simulation studies, 239–240
    computational fluid dynamics, 223–226,
        223f–225f
    energy consumption, 210–211, 210f–211f
    heat and energy balance, 213–214, 213f
    software for modeling, 215–232, 215t,
        216t–222t, 223f–225f, 231f
Single-crystal silicon solar cell, 114, 114f
SIP. *See* Structural insulated panels (SIP)
Smart Building Automation System
        (SBAS), 241
Smart cities, 126–127, 127t
"Smart Cities Mission" program, 260, 265–266,
        274, 280
Smart glasses technology, 64

Smoke, 80
Software Development Kit (SDK), 227
Software for modeling
    building energy simulation programs,
        216t–222t
    computational fluid dynamics, 223–226
    Energy Plus, 226–227
    eQUEST, 228
    FORTRAN, 229
    MATLAB, 229–232
    Open Studio®, 227
    parameters used in development, 215t
    radiance, 227–228
    Simergy, 229
    TRACE 700, 228–229
    transient system simulations
        (TRNSYS), 226
Soil–cement block masonry
        construction, 97
Solar cells, 114
    efficiency, 117–118
    glazing products, 120
    tiles, 120f
Solar chimney for ventilation, 40–41
Solar energy, absorption of, 74
Solar gain, 214
Solar heat collectors, 34
Solar heat gain coefficient (SHGC), 63, 76
Solar heating systems, 141f
Solar hot water systems, 35
Solar NZEBs, 33–34
Solar photovoltaics (PV), 203
    Click Roll Cap System, 119
    panels, 34, 38
Solar radiation, 215t
Solar reflectance (SR), 66
Solar reflective roofs, 66, 77–78
Solar shading systems, 21
Solar thermal collectors, 204
Solar Thermal Energy Lab, Centre for Green
        Energy Technology, Pondicherry
        University, 148
Solar thermal (ST) systems, 34, 203,
        246–247
Solar water heaters, 38–39
    components, 38
    parameters, 39t
Solid oxide fuel cells (SOFC), 248
Space heating systems, 139
    hot air, 140
    hot water, 139–140
SPARK (Simulation problem analysis and
        research Kernel), 222t

Specific heat capacity, aerogel blankets
    application, 163
    differential scanning calorimeter
        (DSC 8500), 161
    heat losses through walls, 162f
    NGF aerogel blanket, 161f
    potential applications, 163
    real estate cost saving, 162t
    Spaceloft® blanket, 161
    thermal mass of walls, 160
    thermal resistance of walls, 163
Stabilized mud block (SMB) filler slab
        construction, 97
Standard Test or Reporting Conditions
        (STC), 117
State Designated Agencies (SDAs), 275–276
State Energy Conservation Fund (SECF)
        Scheme, 280
State-of-the-art technologies, 37
Steady-state methods, buildings
        performance, 238
Sterling engines, 248
Structural insulated panels (SIP), 35, 242
Sulabh Awas Yojna, 280
Summer set point temperature (SST), 177
Super Efficient Equipment Program
        (SEEP), 280
Sustainability
    assessment tools, 8
    of building development concepts, 7f
    as scale, 8f
Sustainable development, 60
Sustainable Development Goals (SDG)-11, 260
Swachh Bharat, 280
Switchable glazing, 64

T

TAC Vista system, 51f
TADSIM (Tools for Architectural Design and
        Simulation), 239–240
Tenancy Lighting Assessment (TLA), 241
Thermal bridges, loss due to, 214
Thermal camera, 47, 48f
    parameters, 48t
Thermal conductance, 76, 81
Thermal conductivity (U-value), 63, 76, 83
    insulation materials, 80
Thermal emissivity, 63, 63t
Thermal energy, 63
Thermal energy storage (TES) systems,
        81, 134–135, 173–174, 173f, 204,
        250–252

campus district cooling system, Cairns
        Campus, James Cook University,
        146–147
cooling building components, 142–145
cooling building materials, 141–142
cooling in building, 140–141
domestic hot water, 138
floating pavilion, Rotterdam, The
        Netherlands, 147–148
heating in buildings, 138
latent heat storage, 135–136
sensible heat storage, 135
Solar Thermal Energy Lab, Centre for Green
        Energy Technology, Pondicherry
        University, 148
space heating, 139–140
thermochemical heat storage, 137
types, comparison, 137t
Thermal insulation, 80–82
    of roof, 61, 68–69
Thermal isolation, 37
Thermal mass, 69, 81
Thermal resistance (R-value), 62, 74
    building envelope, 61
Thermal storage systems, 175t
Thermal stratification, 66
Thermochemical heat storage, 137
Thermostat, 230
Therm Version 1.0, 239
Thin-films
    PV systems, 122
    solar cell, 114–115, 115f
Tile products, 119
Time of Use rates (TOU), 111
Traditional roofs, 79
TRANE TRACE 700 software, 228–229
Transient system simulations (TRNSYS), 35, 83,
        222t, 226, 240
Transmission loss, 214
Transparent oxide (TCO) layer, 65
Transportation, 97t
Trans wall, 62
TRNSYS, 35, 83, 222t, 226, 240
Trombe walls, 61, 74, 75f

U

United Nations Environment Programme
        (UNEP), 261
United Nations Framework Convention on
        Climate Change (UNFCCC), 237
United States Green Building Council (USGBC),
        111–112

Unventilated solar walls, 61
Urbanization pressure, 274
U Values for Opaque and Glazed
     Components, 9t

V

Vacuum glazing, 64, 77f
     commercial, 77
Vacuum insulation panels (VIP), 80, 242
Variable Air Volume (VAV) boxes, 254
Variable refrigerant flow (VRF) systems, 176t
Vegetation, 77
Ventilated or double skin walls, 62
Ventilated roofs, 66
Ventilation, 21, 205
     loss of heat, 214
Vertical ground loop heat exchanger, 41–42, 41f
VIP. *See* Vacuum insulation panels (VIP)
Viracon VUE 11–50 low-emission (Low-e)
     windows, 39

W

Wallboards, 82
Walls
     latent heat storage, 62
     lightweight concrete, 62
     passive and low energy buildings, 74–76
     passive solar, 61–62
     ventilated or double skin, 62
Warmboard radiant heat floor, 139f
Wastage of construction materials, 93, 96t
Water-conserving toilets, 7
Water-proofing membrane, 67
Water-source heat pumps, 42
Water to water heat pump (WWHP), 42, 42f, 52
Water vapor transmission, needle glass fibers
     aerogel blanket, 158–159, 158f

Web camera, 49f
     physiological sensors and, 47
Wet green roofs, 79
Whole social systems, 16f
Windows and doors (fenestrations), 63–64
Wind turbine, 34
Wood-based walls, 61, 100
World Commission on Environment and
     Development (WCED), 60
World Green Building Council (WGBC),
     60, 260
World Green Building Trends report, 14

Z

Zero-energy buildings (ZEB), 33–34, 255
Zero-energy districts, 204
Zero-energy home (ZEH), 34
Zero energy (ZØE) house design, 35
Zila Panchayat members, 278
ZØE Lab, 36, 37f
     building energy monitoring and control
          system, 50–52
     energy allocation performance, 52–54, 54f
     energy consumption and generation, 52
     flexibility, 37–38
     geothermal energy utilization, 41–43
     human thermal comfort, 47f
     indoor environment and thermal comfort
          system, 46–50
     integration of renewable energy
          technologies, 36
     multi-functions, 36–37
     rainwater harvesting and filtration system,
          44–46
     rainwater harvesting system, 45f
     solar features, 38–41
     water filtration system, 46f
     wind energy utilization, 43–44

9 780367 571863